MOLECULAR METHODS
IN ECOLOGY

# METHODS IN ECOLOGY

*Series Editors*
**J. H. LAWTON** FRS
*Imperial College at Silwood Park*
*Ascot, UK*

**G. E. LIKENS**
*Institute of Ecosystem Studies*
*Millbrook, USA*

METHODS IN ECOLOGY

# MOLECULAR METHODS IN ECOLOGY

EDITED BY

## ALLAN J. BAKER

*Curator of Ornithology*
*Royal Ontario Museum*
*Toronto*

**Blackwell**
**Science**

© 2000 by
Blackwell Science Ltd
Editorial Offices:
Osney Mead, Oxford OX2 0EL
25 John Street, London WC1N 2BS
23 Ainslie Place, Edinburgh EH3 6AJ
350 Main Street, Malden
  MA 02148 5018, USA
54 University Street, Carlton
  Victoria 3053, Australia
10, rue Casimir Delavigne
  75006 Paris, France

Other Editorial Offices:
Blackwell Wissenschafts-Verlag GmbH
Kurfürstendamm 57
10707 Berlin, Germany

Blackwell Science KK
MG Kodenmacho Building
7–10 Kodenmacho Nihombashi
Chuo-ku, Tokyo 104, Japan

The right of the Author to be
identified as the Author of this Work
has been asserted in accordance
with the Copyright, Designs and
Patents Act 1988.

First published 2000

Set by Best-set Typesetter Ltd., Hong Kong
Printed and bound by the
Alden Press, Oxford and
Northampton.

The Blackwell Science logo is a
trade mark of Blackwell Science Ltd,
registered at the United Kingdom
Trade Marks Registry

DISTRIBUTORS

  Marston Book Services Ltd
  PO Box 269
  Abingdon, Oxon OX14 4YN
  (Orders:  Tel: 01235 465500
            Fax: 01235 465555)

USA
  Blackwell Science, Inc.
  Commerce Place
  350 Main Street
  Malden, MA 02148 5018
  (Orders:  Tel: 800 759 6102
            781 388 8250
            Fax: 781 388 8255)

Canada
  Login Brothers Book Company
  324 Saulteaux Crescent
  Winnipeg, Manitoba R3J 3T2
  (Orders:  Tel: 204 837-2987)

Australia
  Blackwell Science Pty Ltd
  54 University Street
  Carlton, Victoria 3053
  (Orders:  Tel: 3 9347 0300
            Fax: 3 9347 5001)

A catalogue record for this title
is available from the British Library

ISBN 0-632-03437-8

Library of Congress
Cataloging-in-publication Data

Molecular methods in ecology / edited by Allan J. Baker.
    p.   cm. — (Methods in ecology)
    Includes bibliographical references.
    ISBN 0-632-03437-8
    1. Molecular ecology. I. Baker, Allan J. II. Series.
  QH541.15.M63 M66 2000
  577'.01'5728—dc21

For further information on
Blackwell Science, visit our website:
www.blackwell-science.com

# Contents

# Contributors

**Allan J. Baker**
Curator of Ornithology
Royal Ontario Museum
100 Queen's Park
Toronto
M5S 2C6
Canada

**Tim P. Birt**
Department of Biology
Queen's University
Kingston
Ontario
K7L 3N6

**Anthony H. Bledsoe**
Department of Biological Sciences
University of Pittsburgh
Pittsburg, PA 15260
USA

**Royston E. Carter**
Novartis Agribusiness Biotechnology Research, Inc.
PO Box 12257
3054 Cornwallis Road
Research Triangle Park
NC 27709-2257, USA

**Scott V. Edwards**
Department of Zoology
University of Washington
Box 351800
Seattle
WA, 98195, USA

**Vicki Friesen**
Associate Professor
Department of Biology
Queen's University
Kingston
Ontario
K7L 3N6
Canada


**Joe Gasper**
Department of Zoology
University of Washington
Box 351800
Seattle
WA, 98195, USA


**Richard Griffiths**
Molecular Evolution Laboratory
DEEB
Graham Kerr Building
Glasgow University
Glasgow
G12 8QQ
UK


**John Nusser**
Department of Zoology
University of Washington
Box 351800
Seattle
WA, 98195, USA


**John M. Pearce**
Alaska Biological Science Center
Biological Resources Division, US Geological Survey
1011 E. Tudor Rd
Anchorage, AK 99503
USA

**Ettore Randi**
Istituto Nazionale per la Fauna Selvatica
Via Cà Fornacetta 9
40064 Ozzano dell'Emilia (BO)
Italy

**Carol Ritland**
Department of Forest Sciences
2424 Main Mall
University of British Columbia
Vancouver
BC, V6T 1Z4
Canada

**Kermit Ritland**
Department of Forest Sciences
2424 Main Mall
University of British Columbia
Vancouver
BC, V6T 1Z4
Canada

**Kim T. Scribner**
Department of Fisheries and Wildlife
Michigan State University
East Lansing, MI 48824-1222
USA

**Frederick H. Sheldon**
Museum of Natural Science
Louisianna State University
Baton Rouge
LA 70803-3216
USA

# The Methods in Ecology Series

The explosion of new technologies has created the need for a set of concise and authoritative books to guide researchers through the wide range of methods and approaches that are available to ecologists. The aim of this series is to help graduate students and established scientists choose and employ a methodology suited to a critical look at different approaches to the solution of a problem, whether in the laboratory or in the field, and whether involving the collection or the analysis of data.

Rather than reiterate established methods, authors have been encouraged to feature new technologies, often borrowed from other disciplines, that ecologists can apply to their work. Innovative techniques, properly used, can offer particularly exciting opportunities for the advancement of ecology.

Each book guides the reader through the range of methods available, letting ecologists know what they could, and could not, hope to learn by using particular methods or approaches. The underlying principles are discussed, as well as the assumptions made in using the methodology, and the potential pitfalls that could occur–the type of information usually passed on by word of mouth or learned by experience. The books also provide a source of reference to further detailed information in the literature. There can be no substitute for working in the laboratory of a real expert on a subject, but we envisage the Methods in Ecology Series as being the 'next best thing'. We hope that, by consulting these books, ecologists will learn what technologies and techniques are available, what their main advantages and disadvantages are, when and where not to use a particular method, and how to interpret the results.

Much is now expected of the science of ecology, as humankind struggles with a growing environmental crisis. Good methodology alone never solved any problem, but bad or inappropriate methodology can only make matters worse. Ecologists now have a powerful and rapidly growing set of methods and tools with which to confront fundamental problems of a theoretical and applied nature. We see the Methods in Ecology Series as a major contribution towards making these techniques known to a much wider audience.

John H. Lawton
Gene E. Lkiens

# CHAPTER 1

# Molecular Ecology

ALLAN J. BAKER

## 1.1 Historical overview

Until the mid-1960s, when protein electrophoresis was first used to detect genetic variation in samples of individuals from different populations and species, it had not been possible for population biologists to routinely estimate multilocus parameters of direct relevance in ecology and evolution. I remember some years later the thrill of seeing for the first time how codominant markers appeared on a gel as the stain reaction precipitated a coloured dye at the sites of these genetically encoded protein products. Suddenly, the seemingly arcane theoretical quantities of population genetics leapt to life in front of our eyes and the everyday lexicon of ecologists, systematists and behavioural biologists included terms such as average heterozygosity, number of alleles per locus, percent polymorphic loci, inbreeding coefficient, genetic equilibrium, population structure, fixed allele differences, unbiased genetic distance and so on.

Molecular ecology had its genesis at this time, although the proliferation of methods and applications that spawned the journal *Molecular Ecology* followed much later. As with any new tools that open windows into the biological realm, molecular methods generated new data, discoveries and controversies that stimulated new theoretical developments such as the neutral theory of molecular evolution (Kimura 1968; King & Jukes 1969). The overwhelming influence of natural selection in biology was suddenly challenged by this theory which had the extremely beneficial effect of serving as a null hypothesis against which empirical data could be tested. The results of these tests were disappointing to those who expected an answer favouring selection over neutralism, or vice versa, and attention soon switched from the translated protein to the DNA level. This transition was possible because the innovations of molecular biology facilitated the study of variation in DNA by all biologists.

Despite criticisms of some methodological aspects of the technique of DNA–DNA hybridization, the herculean efforts of Charles Sibley and Jon Ahlquist to revise completely the phylogenetic relationships of birds based upon divergence in their DNA, provided a huge stimulus for further molecular studies in biology, especially in systematics. The generation of such large-scale phylogenies using DNA provided an 'independent' hypothesis about relationships of organisms, which could then be compared with hypotheses generated with more conventional characters from morphology, behaviour and ecology. Many ecologists and evolutionary biologists embraced the comparative method in which ecological and behavioural characters were mapped on independently derived trees (Brooks & McLennan 1991; Harvey & Pagel 1991) to study their evolutionary pathways.

Major advances in molecular approaches followed when restriction endonucleases were used to detect variation previously hidden in DNA sequences. John Avise and coworkers exploited the properties of the mitochondrial DNA molecule. With its maternal inheritance, apparent lack of recombination, large copy number, relatively small size and rapid rate of evolution, they realized that this genome could provide novel insights into the evolutionary process. To assay variation at the DNA level in this clonally inherited genome they employed restriction endonucleases to cleave it into fragments and then compared fragment profiles among individuals, populations and species. Theoretical developments inevitably followed as they demonstrated such remarkable phenomena as stochastic lineage extinction in matriarchal phylogenies, phylogeographic structuring in species and biotas, female-biased dispersal, differences between gene trees and species trees and the link between phylogenies of genes, populations and species in the hierarchy of life. Alec Jeffreys and colleagues discovered hypervariable minisatellite repeats in genomic DNA and this led to the development of DNA fingerprinting of individuals that revolutionized parentage analysis and forensics. Anonymous restriction fragment polymorphisms in the nuclear genome played an important part in the development of molecular ecology in plants and, to a lesser extent, in animals.

With the advent of the Nobel prize-winning technique of the polymerase chain reaction (PCR) by Kary Mullis, it then became possible to extend this research programme to the DNA sequence level at reasonable cost. By obviating the need for expensive and more technically demanding cloning techniques, PCR with thermostable polymerases made it possible for researchers in many more laboratories around the world to routinely amplify and sequence DNA templates from specific genes to whole organellar genomes. This enabled researchers to choose fast-evolving regions, such as the control region of the mitochondrial DNA (mtDNA) of animals, for population studies of maternal lineages and contrast them with biparentally transmitted markers such as introns in nuclear genes.

Not only did this speed our understanding of the process of molecular evolution, it also enabled much more detailed study of population structure within species and provided a range of markers for reconstructing phylogenies of species, populations and genes. By combining classical cloning and Southern hybridization techniques with PCR, a new class of markers called microsatellites was developed based on variable numbers of short repeat sequences. These markers are proving invaluable in a wide range of applications from parentage and mating systems to population genetics, conservation genetics and phylogenetic inference. Similarly, PCR-based methods were developed to sex animals molecularly based originally on randomly amplified polymorphic DNA (RAPD) and later with amplified fragment-length polymorphism (AFLP) to locate sex-specific sequences on sex chromosomes. Finally, evidence is accruing for positive selection acting on genes such as the major histocompatibility complex (MHC) in vertebrates and many other genes, thanks to sophisticated analyses made possible with DNA sequence data.

## 1.2 Choosing appropriate markers

The major objectives of this book are to present readers with the requisite molecular background to the methods used in molecular ecology in the past and the present, and to assist in the appropriate choice of markers for different problems encountered in the field. Researchers with no, or very limited, molecular training continue to employ these new methods and sometimes make poor choices when selecting markers for their studies. Rather than choosing markers that are evolving at an appropriate rate, it is common place for inexperienced workers to amplify genes based on the availability of universal primers for highly conserved genes. When studying populations of animals that invaded regions of the world since the last Ice Age 10000–14000 years ago, fast-evolving control region sequences, microsatellites, or possibly variable introns, are required. For older population or species splits, protein-coding genes and introns may be appropriate and for deep branch phylogenies a range of protein coding, tRNA and rRNA genes can be used to cover different nodes in the tree. Consult current literature and download sequences from sequence databases before choosing a class of markers to use for your problem—do not simply use someone else's primers designed for a completely different problem.

Rates of evolution in plant and animal DNAs are often very different, so a different choice of genes is usually required. Chloroplast DNA (cpDNA) is very slowly evolving and is uniparentally inherited and so it is used for deeper phylogenies, despite some complications from intragenomic recombination and strong codon bias (Page & Holmes 1998). Slowly evolving ribosomal RNA genes in nuclear DNA can also be employed at this level. However, most studies of plants at and below the level of genera will require markers like RAPDs, AFLPs, microsatellites and sequenced-tagged-sites (STS), and even allozymes. The latter three classes of markers have the additional advantage of codominance which makes them suitable for studies in intraspecific systematics and population genetics. Amplified fragment-length polymorphisms will be the markers of choice for linkage maps because they detect so many loci, but microsatellites will help to anchor these maps. Chapters 4–12 in this book provide more detailed guidelines in choosing the right markers for the right problem.

## 1.3  Organization of the book

The book is organized so that basic molecular techniques are presented in Chapters 2 (restriction analysis of DNA, library construction, Southern hybridization, probing and cloning of fragments of interest) and PCR (Chapter 3). This material is presented first so that those in need of a better grounding in molecular biology can then proceed to specific methods mentioned in later chapters. There then follow accounts of protein electrophoresis (Chapter 4), solution DNA–DNA hybridization (Chapter 5), DNA fingerprinting with minisatellites (Chapter 6) and mitochondrial DNA (Chapter 7). The last five chapters capture the newer methods and markers in the field, beginning with the MHC (Chapter 8) and proceeding through DNA markers in plants (Chapter 9) to microsatellites (Chapter 10), introns (Chapter 11) and molecular sex identification using DNA markers (Chapter 12). Appendices containing specific protocols are presented at the end of the chapter to which they pertain.

## 1.4  Future prospects

Where is the field of molecular ecology going? No doubt new and unforeseen classes of genetic markers will be developed in the years ahead, by researchers spurred on, in part, by large-scale genomics studies. The phylogenetic framework required for so much of the work we undertake in molecular ecology will continue to be driven by increasingly sophisticated analyses of DNA sequences. Already, the use of mtDNA sequences in reconstructing higher level phylogenies is proving more difficult than anyone expected but, as the processes of molecular evolution are better elucidated by the onslaught of sequence data, it is clear that more realistic models of substitution and rate variation among sites need to be incorporated into tree-building methods. The promise of slowly evolving sequences of exons from nuclear genes in providing well supported trees because of minimal homoplasy is currently being tested in many laboratories, but it remains to be seen whether problems will emerge from recombination, gene conversion and horizontal transfer of elements in nuclear genomes. Exciting new studies suggest that short interspersed nuclear elements (SINEs) might provide excellent markers for phylogenetic studies because the integration of each element into a genome is an unambiguous event which can be treated as derived character (e.g. Takahashi *et al.* 1998). The task for the future will be to isolate different elements that record the history of phylogenetic splits among related organisms.

Much progress will be made in population level studies, with a central focus on maximum likelihood approaches in coalescent theory because it exploits the information provided by DNA sequences. Not only can the data on allele frequencies be used, but also the phylogenetic component of evolution can be incorporated to interpret the history of genes, populations and species. Sequence data sets can be pruned to include sites that are internally consistent and so fit the infinite-sites model. These properties have already been used to age mutations on gene trees and to age populations. Mutations are often older than the populations in which they now reside and these methods will force biologists to obtain better estimates of mutation rates in the sequences they obtain, of effective population sizes and of migration rates. The problem of estimating gene flow among post-Pleistocene populations will probably yield to faster evolving markers and the application of techniques like nested clade analysis (Templeton *et al.* 1995), which can tease apart the confusing tangle of historical and current gene flow.

Anyone who has ever attempted to study multigene families such as the MHC can recount the difficulties they pose in characterizing specific loci. Despite these difficulties, genetic systems such as MHC will be increasingly exploited by molecular ecologists because of their functional relevance in protecting against parasites, their high polymorphism and the long persistence of alleles as a result of balancing selection. We can expect significant progress in understanding how these polymorphisms relate to fitness and how they might be involved in mate choice and the expression of secondary sexual characters.

Microsatellite markers from the nuclear and chloroplast genomes will play increasingly important and fruitful roles in studies of paternity, population structure and speciation, and conservation genetics. In response to the wealth of new genetic data generated by these markers, many new analytical and statistical methods are emerging to broaden applications of microsatellites in molecular ecology (Luikart & England 1999). Amplified fragment length polymorphism will be further exploited in the search for new classes of markers and will be the marker of choice in many studies of plants, where they already enjoy considerable success.

Plant biologists will continue to lead the way in many aspects of molecular ecology, including the evolution of mating systems, the effects of inbreeding and outcrossing and linkage mapping to study quantitative trait loci (QTLs), in part because plants are so easy to breed. The large-scale phylogenies emanating from an unprecedented collaborative effort among plant systematists to elucidate the origins and phylogeny of green plants (Deep Green) points the way for the future and will ensure that comparative studies of plants will forge ahead. Ultimately, we need to utilize genetic markers described in this book to obtain more precise estimates of quantitative genetic parameters because the latter provide crucial information on the loss of adaptive potential and the accumulation of deleterious mutations in endangered species (Lynch 1996). Studies now underway presage this synthesis—with current molecular tools available to ecologists, multigenerational pedigrees corrected for extra-pair matings can be constructed in natural populations. Additionally, molecular markers can be used routinely to sex large samples of parents and offspring in even monomorphic species, and mid-parent offspring regressions calculated for each sex to check for sexual differences in heritabilities of quantitative traits. These markers will also facilitate the detection of maternal effects if they exist. Along with these important intellectual and theoretical achievements will come practical advances in conservation and it is imperative that molecular ecologists, conservation biologists and systematists are at the forefront of research and advocacy to preserve what is left of the world's biodiversity before it is too late. As pointed out by Lynch (1996), the selective force exerted by factors such as global climatic change could potentially have a demographic cost that exceeds the replacement rate of even large populations and thus places them at risk of extinction in the long term.

## References

Brooks, D.R. & McLennan, D.A. (1991) Phylogeny, ecology and behavior. *A Research Program in Comparative Biology.* University of Chicago Press, Chicago.

Harvey, P.H. & Pagel, M.D. (1991) *The Comparative Method in Evolutionary Biology.* Oxford University Press, Oxford.

Kimura, M. (1968) Evolutionary rate at the molecular level. *Nature* **217**, 624–626.

King, J.L. & Jukes, T.H. (1969) Non-Darwinian evolution. *Science* **164**, 788–798.

Luikart, G. & England, P.R. (1999) Statistical analysis of microsatellite DNA data. *Trends in Ecology and Evolution* **14**, 253–256.

Lynch, M. (1996) A quantitative-genetic perspective on conservation issues. In: *Conservation Genetics: Case Histories from Nature* (eds J.C. Avise & J.L. Hamrick), pp. 471–501. Chapman & Hall, New York.

Page, R.D.M. & Holmes, E.C. (1998) *Molecular Evolution: a Phylogenetic Approach.* Blackwell Science Ltd, Oxford.

Takahashi, K., Terai, Y., Nishida, M. & Okada, N. (1998) A novel family of short interspersed repetitive elements (SINEs) from cichlids: the patterns of insertion of SINEs at orthologous loci support the proposed monophyly of the four major groups of cichlid fishes in Lake Tanganyika. *Molecular Phylogenetics and Evolution* **15**, 391–407.

Templeton, A.R., Routman, E. & Phillips, C. (1995) Separating population structure from population history: a cladistic analysis of the geographical distribution of mitochondrial DNA haplotypes in the tiger salamander, *Ambystoma tigrinum. Genetics* **140**, 767–782.

# CHAPTER 2

# General Molecular Biology

ROYSTON E. CARTER

## 2.1 Introduction

The analysis and measurement of variation, whether it be between individuals, populations or even species, is fundamental to many ecological studies. Molecular techniques that measure DNA variation are very powerful, and are increasingly employed in many of these studies as demonstrated by the rapid exploitation of DNA techniques, such as genetic fingerprinting and polymerase chain reaction (PCR) amplification, into traditional field studies (see Table 2.1). New techniques are continually being developed, and thus properly prepared and stored samples may be useful for analysis by these new methods. The primary aim of this chapter is to describe the molecular basis of these techniques, so that the detailed procedures in later chapters can be more fully appreciated and understood. In particular, emphasis will be given to sampling regimens, specimen storage, DNA extraction and steps involved in DNA-hybridization-based techniques. Many of these steps are also relevant to PCR-based techniques, but special considerations are described in detail in Chapters 3, 9 and 10. Also, in section 2.2.2, a background to basic cloning techniques will be presented. This will describe the basic steps, the key enzymes and the vectors used. However, it is beyond the scope of this chapter to provide detailed protocols.

## 2.2 Sample sources for DNA

The success or failure of molecular analyses often depends upon collecting the most suitable samples and storing them in such a way as to minimize damage. If suitable samples are collected and properly preserved, nucleic acids are very versatile and can be used in several different analyses. Almost all cells contain a copy of the genome, and most molecular techniques require only relatively small amounts of DNA, typically less than 5 µg. Consequently, very small biological samples are necessary, permitting nondestructive sampling of large numbers of individuals.

This is not to say, however, that all cell or tissue types are equally suitable. Some cells produce secondary metabolites that can interfere with the enzymes used to manipulate DNA; for example, high concentrations of melanin inhibit

**Table 2.1** List of the basic steps that are common to a wide range of molecular analyses used for ecological studies

1 Sample collection
2 Sample storage
3 Extraction of DNA
4 Enzymic modification of DNA (usually restriction enzyme digestion)
5 Separation of DNA fragments by gel electrophoresis
6 Immobilization of DNA by blotting
7 Preparation of labelled probes
8 Hybridization of the labelled probe to the immobilized DNA and stringency washing
9 Interpretation of results

*Other important steps involved in some analyses*
1 PCR amplification
2 DNA sequencing
3 Molecular cloning

PCR, polymerase chain reaction.

*Taq* polymerase and can interfere with PCR amplification, and various carbo-hydrates and polyphenols produced by some plants and animals inhibit many other DNA modification enzymes. Some tissues, such as liver, are meta-bolically very active, and after death release potent degradative enzymes which can rapidly damage DNA unless the sample is handled carefully (i.e. frozen). Some materials such as bone or woody plant tissues are very hard and therefore difficult to work with. In these instances, extra steps may be necessary to extract DNA such as grinding the sample in liquid nitrogen, and sometimes the specimen must be disrupted in order to preserve it prior to DNA extraction.

In those instances where the organisms being studied are not to be sacri-ficed, blood specimens are usually the easiest and most convenient samples. For nonmammalian species that have nucleated red cells, as little as 1 or 2 microlitres are sufficient (although usually more is taken), and can be col-lected by simple venepuncture (Sooter 1954; McClure & Cedeno 1955). For mammalian species, in which only the white cells are nucleated, it may be necessary to collect 1–2 mL to ensure that sufficient DNA can be recovered. In some instances, however, other tissues may be preferred; for example, small skin biopsies, often in the form of ear plugs, are taken from many small mammals, especially from species that are routinely tagged in such a manner. Also muscle and liver biopsies may be taken surgically by suitably experienced and authorized researchers. Although not an immediately obvious source, 'ancient DNA' (for review see Thomas & Paabo 1993), which may be tens, hundreds or even thousands of years old, has been isolated from bone and teeth and used for PCR amplification and sequencing (Hagelberg & Clegg 1991) in evolutionary studies. For most plant species leaves (or needles) are the most obvious source, but sometimes flowers, fruit or root tissues may be more suitable or convenient. Young plant tissues are usually preferable because they contain more cells per weight, and typically contain less polyphenols and polysaccharides, but, in some instances, the 'best' specimen will not be available and compromises must be made.

It should be noted that various local, national and international laws and regulations govern the collection and use of biological samples; for example, the humane collecting of blood samples and/or sacrifice of animals and collection and transport of samples from endangered species or from regions of high biodiversity, such as the Amazon basin. Also, regulations exist that govern the transport of any biological specimens which may harbour pathogens (to man or agricultural species). For reviews of these issues see Cann *et al.* (1993) and Sytsma *et al.* (1993) and check with appropriate governmental agencies.

### 2.2.1 Preservation of samples

When properly handled, DNA is very robust, but if samples are mistreated it is prone to damage by: nucleases, chemical degradation, extremes of pH, mechanical shearing, excessive heat and even strong light. Traditional methods developed in museum collections for the preservation of biological material from plant, invertebrate and vertebrate sources are usually based upon substances such as formalin or Carnoy's solution. Although these are good at retaining the structural integrity of the samples, they are not ideal for the preservation of samples from which nucleic acids are to be extracted. Large genomic DNA is usually not present, but DNA adequate for PCR-based techniques is often obtained.

For ecological studies, which do not rely upon existing museum collections, freshly collected samples are ideal, but even these must usually be stored and preserved prior to the extraction of nucleic acids. For these samples many preservative buffers have been described, in practice some of these do not always perform well. Often the ratio of preservative to sample must be strictly maintained for optimal performance, and this may be difficult to accomplish under arduous field conditions. The two most common principles that are exploited by these regimes are chelation and dehydration. Magnesium ions are essential requirements for the activity of nucleases so buffers that contain agents such as ethylenediamine tetraacetic acid (EDTA) are able to chelate the free magnesium ions and therefore prevent the nucleases from degrading the cellular nucleic acids. It is very common for such buffers also to include detergents, such as sodium dodecyl sulphate (SDS), that lyse the cells to ensure that the nucleases are released and the magnesium is cleared before detrimental damage can occur.

Good quality DNA can usually be prepared from samples protected in this way, but the extraction procedures may need to be modified as the released nucleic acids can make the suspension viscous and difficult to handle without causing damage by shearing forces acting on the DNA. Based on experience, it is often difficult to remove the contaminating proteins from such highly viscous suspensions, and such samples may need to be purified by additional solvent extractions to obtain sufficiently clean DNA for further analysis.

Alternatively, nucleases can be rendered inactive by dehydration of the sample. Several methods popular with field biologists have been described. The most common is to place the specimen in several volumes of absolute ethanol. Other alcohols have been tried but are less efficient: both the yield and quality of the DNA rapidly diminish after prolonged storage. The ethanol system works particularly well for blood: a small drop is suspended into 1 mL of absolute ethanol in a screw-capped tube. Similar systems have also been used satisfactorily to store whole or minimally disrupted insects (R. Post, personal communication), and even vertebrates (although these were first roughly chopped, Borowsky 1991). Such material has yielded adequate DNA after storage for several years at room temperature, although some loss of quality has been recorded (see Seutin *et al.* 1990). This may be attributable to contaminants found in some alcohol (Ito 1992), and it is therefore advisable for batches of alcohol to be tested first, and the most suitable kept for sample preservation. As soon as convenient, the material should be extracted or stored at −80 °C.

A buffer consisting of a saturated sodium chloride solution supplemented with 10% dimethyl sulphoxide (DMSO) has been described for storing relatively large tissue samples. This material has been used for DNA fingerprinting and mitochondrial DNA (mtDNA) analysis, even after prolonged storage (Amos & Hoelzel 1990). However, blood placed in this buffer rapidly clots and becomes difficult to handle, complicating the subsequent DNA extraction.

Chelation and dehydration are mostly used for preserving blood or other biological fluids. Solid tissue must be disrupted to allow complete lysis to occur, eliminating pockets of cells that may gradually release their nucleases and damage the DNA before the divalent ions can be chelated. Protocols for handling dried blood stains have been developed by forensic scientists, and some field biologists have successfully adopted air-drying blood onto blotting paper or glass microscope slides as a method of preserving biological material. Similar procedures are suitable for plant materials. Alternatively, leaves and the like can be dried after harvesting by air drying, use of a herbarium, lyophilization or with chemical desiccants such as anhydrous $CaSO_4$ (Drierite J.T. Baker, Inc., New Jersey) or silica gel (see Sytema *et al.* 1993 for a review), and stored at room temperature (Harris 1993; Laulier *et al.* 1995).

A final way to preserve biological samples for later analysis is to keep them alive. I have successfully extracted high quality DNA from avian blood that has been stored in Alseviers solution for many weeks before the start of the extraction process. The remainder was then frozen and has subsequently been used successfully for the isolation of more DNA. Many other samples may be suitable for similar short-term live storage in appropriate isotonic media. Similarly, plant material may be collected as live cuttings.

Photochemical degradation of DNA can be a problem, so it is advisable for all samples to be stored away from strong light, and, if possible, in a cool place. Repeated cycles of freezing and thawing rapidly diminish both yield and quality of recovered DNA (Ross *et al.* 1990). It is good practice to subdivide samples so that it is not necessary to thaw them repeatedly. These simple precautions can have a dramatic effect on the success of any later analysis.

It is advisable to determine an appropriate preservation systems in the laboratory with specimens similar to those anticipated in the field before the field season, and to check the quality of DNA that can be extracted.

### 2.2.2 DNA extraction

The extraction of nucleic acids in good yield with only minimal damage is a very important part of most molecular analyses, especially for more demanding procedures such as DNA fingerprinting and PCR of large DNA templates. Usually this is relatively straightforward, but it can be the cause of problems, particularly if the samples have been mistreated. Essentially, the aim of the various extraction protocols is to lyse the cells, and then to gently remove the proteins and other cellular components so that the DNA may be recovered in a relatively intact and pure form.

The two most widely used procedures are based on removing the contaminants by precipitation with either organic solvents or high ionic strength salt solutions to leave a relatively pure solution of DNA. This is then precipitated from solution by dehydration with ethanol (Wallace 1987a, 1987b), and re-dissolved into buffer at a suitable concentration for the job at hand.

Methods for the extraction of nonhuman DNA have been adapted for plants (Murray & Thompson 1980; Doyle & Doyle 1990; Rogers *et al.* 1996; Lee & Nicholson 1997) and fungi (Raeder & Broda 1985). Some techniques offer rapid throughput (e.g. Bowtell 1987; Reymond 1987), while others eliminate hazardous phenol extractions (Longmire *et al.* 1987; Miller *et al.* 1988; Grimberg *et al.* 1989; Thompson *et al.* 1990). Some use specialized equipment to ease the process, e.g. gel barrier tubes (Moreno *et al.* 1989; Thomas *et al.* 1989). Several commercial companies (e.g. Qiagen, Molecular Biosystems Inc. and Nucleon) have now developed efficient and reasonably cost-effective kits and reagents for the extraction of DNA from different sources; for example, blood and cell culture kits (Qiagen) and plants PhytoPure (Nucleon). These have increasingly found widespread use in ecological studies. These procedures provide consistently high quality and high yields of DNA.

It must be noted that, because the genome size and complexity of different organisms vary considerably, different amounts of DNA are necessary for some organisms to accomplish similar analyses. Also, that different yields of DNA are produced from seemingly similar starting materials. Preliminary experiments using readily available laboratory specimens should be conducted prior to collecting field material, to ensure the materials collected are appropriate.

## 2.3  Enzymes used in molecular biology

The tools of molecular biology for manipulating and modifying DNA are enzymes. One of the most important classes of enzymes are *restriction endonucleases* (see Table 2.2 for other enzymes and principal uses). These are used to cut DNA at very specific points, either as a preparative step in gene cloning, or, most importantly, for ecological research for analytical purposes, usually as an early step in Southern blot hybridization techniques. For example, restriction enzymes are used to resolve specific minisatellite alleles during DNA fingerprint analysis and to identify specific restriction fragment-length polymorphism (RFLP) (Fig. 2.1) or amplified fragment-length polymorphisms (AFLPs, see Chapter 9).

### 2.3.1  Properties of restriction enzymes

Restriction enzymes cleave double-stranded (ds) DNA in a very specific manner, by recognizing specific nucleotide motifs and cutting either within (type II) or beside (type I or III) the sequence. Neither type I nor type III enzymes are widely used for molecular analysis. Most restriction enzymes used have recognition sites that are short (4–6 bp) which show dyad symmetry (i.e. recognition sequences are palindromic). However, some recognition sequences are longer, e.g. *Sfi*I (12 bp) or *Not*I (8 bp), or the recognition site can be degenerate, e.g. *Hinf*I (GANTC where N can be any nucleotide). The position of cleavage within the recognition site also varies with the enzyme. It may be at the centre of symmetry (e.g. *Hae*III and *Sma*I) and so leave blunt ends, or cleavage may occur at equivalent points either side of the axis of symmetry in each strand (e.g. *Hind*III and *Pst*I). These enzymes thus produce DNA fragments with protruding single-stranded termini, called 'cohesive' or 'sticky' ends' because they have the facility for joining up again.

Generally the recognition site for a given restriction enzyme is unique to that particular enzyme. However, a few DNA sequences are recognized by different restriction enzymes (called isoschizomers) isolated from different

**Table 2.2** Common enzymes used in molecular biology and their uses

| Enzyme class | Examples | Common uses |
|---|---|---|
| Restriction enzymes | *Eco*RI, *Bam*HI, *Pst*I, *Hae*III, *Sau*3A | 1 DNA digestion for cloning<br>2 Diagnostics and validation of clone structure<br>3 Digestion of genomic DNA for RFLP analysis and other Southern blot techniques<br>4 Characterization of AFLP alleles |
| Thermostable<br>*Taq* polymerase<br>*Pfu* polymerase | | 1 Analytical PCR of specific DNA fragments of microsatellites, RAPDs, or AFLPs<br>2 Preparative amplification of DNA for cloning substrates<br>3 Blunting DNA fragments by end-filling |
| Non-thermostable polymerases | DNA polymerase I<br>Klenow fragment<br><br>Sequenase<br>Reverse transcriptases | 1 Used for preparation of DNA probes<br>2 End-filling restriction sites for blunt end cloning<br>3 Used for manual and automated sequencing<br>4 Synthesis of first and second strand for cDNA preparation |
| Ligases | T4 DNA ligase | Joining of DNA fragments either at blunt or sticky ends for cloning applications |
| RNA polymerases | T7, T3, SP6 | *In-vitro* transcription for probe synthesis |
| Phosphatases | CIAP, SAP | Dephosphorylation of DNA fragments to prevent ligation, a common strategy for preventing vector self-ligation, especially when cloning into a single cut site and for cloning blunt-ended DNA fragments |
| Kinases | T4 polynucleotide kinase | 1 End-labelling DNA, RNA or oligonucleotides<br>2 Addition of 5′ phosphates prior to ligation |
| Proteases | Proteinase K, Pronase E<br><br><br>Lysozyme | 1 Degradation of cellular proteins during nucleic acid extraction<br>2 Removal of added enzymes, e.g. removal of restriction enzyme during cloning experiments<br>3 Bacterial lysis |
| Other nucleases | Dnase I<br>Rnase A | 1 Removal of DNA, e.g. cDNA synthesis<br>2 Removal of RNA, e.g. background reduction when using Riboprobe, or removal of cellular RNA during DNA extraction |

RFLP, restriction fragment-length polymorphism; PCR, polymerase chain reaction; AFLP, amplified fragment-length polymorphism; CIAP, calf intestinal alkaline phosphatase; SAP, shrimp alkaline phosphatase.

organisms, e.g. *Sau*3A (from *Staphylococcus aureus*) and *Mbo*I (*Moraxella bovis*). Both cleave the same sequence (GATC), but vary in their ability to cut methylated DNA; *Sau*3A is able to cut the 'Dam' methylated sequence G$^m$ATC but *Mbo*I cannot. In addition to isoschizomers, the recognition site of one

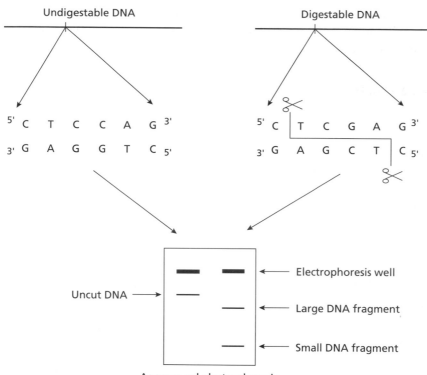

**Fig. 2.1** Diagram demonstrating how a single base change within a restriction site can result in a restriction fragment-length polymorphism (RFLP). A change from C to G within a region of DNA carried by two different individuals creates a novel restriction site which can be detected as an RFLP.

restriction enzyme may contain the recognition sequence of another, e.g. the *Bam*H1 site (GGATCC) contains the *Sau*3A/*Mbo*I site (*GGAT*CC). This is an important consideration when manipulating and cloning DNA.

Restriction enzymes generally have very great specificity for their own recognition sequences, but under inappropriate conditions this specificity may break down. In the presence of >10% glycerol (added as a cryopreservant to commercial enzyme preparations), some restriction enzymes can cut at alternative sites. This phenomenon is called 'star activity'. Although cutting at these secondary sites is relatively inefficient, after prolonged incubation the effects may become significant. Star activity will affect the efficiency of cloning strategies, and can be a cause of spurious bands or high background in genomic Southern blots, but can easily be avoided. Usually, 1–2 μL of enzyme at 10 U μL$^{-1}$ is used for digestion, but by increasing the total volume of the reaction to, say, 50 μL, more enzyme can be added while keeping the concentration of added glycerol below 10%. Alternatively, higher concentration of enzyme ($40\,U\mu L^{-1}$ instead of normal $10\,U\mu L^{-1}$) and/or longer incubation times can be used.

### 2.3.2 Cloning

Many of the techniques described throughout this book are dependent upon gene cloning and manipulation. Cloning is essentially the introduction of a foreign piece of DNA into an autonomously replicating DNA molecule, called a vector. Several different types of vector are available such as plasmids or bacteriophages (see below). The vectors containing the foreign DNA are then introduced into a suitable host so that it may be propagated. This step is called either transformation, transfection or infection, depending upon what sort of vector and introduction method is used (see section 2.10).

### 2.3.3 Vectors and libraries

Recombinant DNA molecules are used for a number of purposes such as to isolate sequences for analysis or for the preparation of probes, etc. A number of different types of vector are available that are ideal for preparing DNA for different purposes; for example, creating libraries or for preparing substrates for probe manufacture. Libraries are collections of recombinant clones that are usually produced to isolate specific DNA fragments for further analysis or for use as probes. There are two general types of library. Genomic libraries are derived from large DNA fragments (possibly whole genomes), usually as a set of random, overlapping restriction fragments. Complementary DNA (cDNA) libraries represent the messenger RNA (mRNA) molecules expressed by the cells. The mRNAs of the cells are copied into a single-stranded DNA molecule by an enzymatic process called reverse translation, and then the RNA is removed and a complementary strand of DNA is synthesized using the first strand as a template. In both cases the new recombinant DNAs are cloned into appropriate vectors and maintained in bacterial hosts for screening and recovery (Fig. 2.2). For example, a specific gene fragment wanted as a probe from a species under investigation may be isolated by screening a suitable library by hybridization with a previously isolated probe from another species. This homologue can then be isolated, purified and grown independently of the rest of the library. Most of the probes used in ecological studies were originally isolated in this way from some form of genomic library.

The main classes of vectors used for these purposes include plasmids, bacteriophage such as M13 and lambda insertion and replacement vectors, P1, cosmids and yeast artificial chromosomes (YACs). The choice of vector is largely dictated by the expected sequence copy number, complexity and insert size.

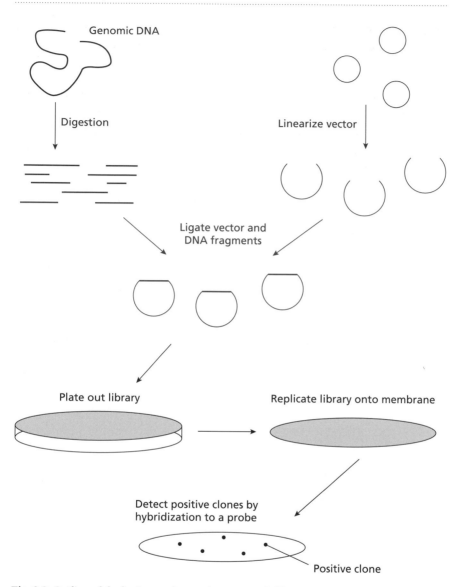

**Fig. 2.2** Outline of the basic steps in creating a genomic library. Genomic DNA is digested with a restriction enzyme and the fragments are then ligated into a linearized vector. The recombinant molecules are introduced into suitable bacteria and then propagated on nutrient agar. Colonies or plaques are transferred to hybridization membranes and these are hybridized to a suitable probe. Clones carrying the appropriate sequences are then detected by autoradiography.

Current plasmid cloning vectors have been derived from naturally occurring plasmids, the most commonly used vectors (e.g. pUC series, Messing 1983; pMB, Bishop & Davies 1980; and pEMBL, see Dente *et al.* 1985 for details). Plasmids are able to harbour DNA sequences of virtually any size, however, if small inserts are cloned then very large numbers of clones are needed to represent entire genomes, and the library becomes unwieldy and difficult (and costly) to screen. Large inserts, however, are selectively disadvantageous to the bacterial host, and become lost or underrepresented from libraries. Plasmid cloning vectors are therefore used mainly for subcloning sequences that have been isolated from libraries for further manipulation.

A family of cloning vectors have been created from single-stranded bacteriophages, mainly M13 (but also Fd, Ff and f1) (Messing 1983; Carlson & Messing 1984). M13 and other single-stranded DNA (ssDNA) vectors are used primarily for subcloning experiments, particularly when sequencing studies are involved. This is because ssDNA is the traditional substrate for dideoxy-sequencing (Sanger *et al.* 1977). Single-stranded bacteriophages are also used for preparing ssDNA hybridization probes (Hu & Messing 1982; Messing & Vieira 1982; and see Jeffreys *et al.* 1985 for an example). Phagemids (Zagursky & Baumeister 1987) are a relatively new family of multifunctional vectors (e.g. pGem series (Promega) and the pTZ series (Pharmacia)), which combine properties of plamids such as antibiotic resistance, blue/white colour screening, inducible expression systems and high copy number, with those of the phages: M13 origin of replication (for ssDNA production) and often RNA polymerase transcription promoters (e.g. T7, T3 and SP6 (for Riboprobe preparation)). These vectors are now preferred over the conventional ss-phage systems for most subcloning experiments.

Other classes of vectors based on bacteriophage lambda (λ) and newer derivatives called cosmids are used primarily for isolating genomic DNA or cDNA for the preparation of large insert libraries.

Insertion vectors such as λgt11 are based on wild-type λ (λ$^{wt}$). Their use relies upon the fact that λ can efficiently package DNA molecules larger than the wild-type genome into infective phage particles. Typically an additional 9 kbp (approx.) can be inserted into one of several unique cloning sites and propagated during normal replication. These vectors are used primarily for the preparation of cDNA libraries (see Huynh *et al.* 1985 for details). Replacement vectors such as λGem11 (Promega), λL47 (Loenen & Brammar 1980) and EMBL3 (Frischauf *et al.* 1983), are able to carry larger fragments of foreign DNA (typically between 9 and 23 kbp). This is because only about 60% of the λ genome is essential for lytic propagation, and the remainder can be removed and replaced with foreign DNA. Lambda replacement vectors are more versatile, and are widely used for genomic library construction (see Kaiser & Murray 1985; and Frischauf 1987 for reviews and methods). Cosmid vectors, such as pWE15, are vectors which combine properties of plasmids and bacteriophage λ. A short, specific DNA sequence of λ called the *cos* site (and about 200 bp of DNA either side) is all that is required for *in vitro* packaging of DNA (Collins & Hohn 1978) provided that they are separated by 38–52 kbp of DNA. This sequence has been inserted into conventional 4–5 kbp plasmids and permits efficient cloning of DNA fragments up to 35 kbp. After packaging and infection, cosmids behave as large plasmids (screened as colonies not plaques, etc.), and they are maintained and handled as such. There are now several different cosmid cloning systems available, including charomids (Saito & Stark 1986). Cloning in cosmids is, however, intrinsically more difficult than in λ (for a review of cosmid cloning see Little 1987).

Even larger pieces of DNA can be cloned in more specialized vectors. Up to 100 kbp (but typically in the range 75–95 kbp) can be cloned with high efficiency using P1 vectors (Sternberg 1990, 1991), and up to 1 Mbp (but more typically 100–600 kbp) in YACs (Cooke 1987; see Burke 1990 for reviews). Yeast artificial chromosomes in particular are currently of great use for the isolation of large fragments of individual chromosomes as part of the human genome mapping programme and other similar genome projects.

Plasmid and phagemid clones are the most commonly encountered in ecological studies. However, sequences isolated from phage and cosmid libraries may be used in some instances, but these will usually need to be subcloned into more appropriate vectors to be useful. For details of the screening, handling and manipulation of DNA sequences cloned in libraries see Sambrook et al. (1989).

## 2.3.4 Restriction digestion and analysis of genomic DNA

The frequency of a given restriction enzyme recognition site in a DNA sequence depends upon its length, as short sequences will be common and longer sequences will occur less often. A tetranucleotide sequence will occur on average every $4^4$ base pairs (256 bp), while a hexanucleotide sequence will occur every $4^6$ (4096) bp. However, DNA is not a random string of nucleotides, and therefore in some species a given sequence will occur more frequently than expected, while in others it will be rare. Also, the density of restriction sites varies in different parts of the genome. Consequently the choice of restriction enzyme can have a great bearing on the success of any analysis.

When genomic DNAs are digested with restriction enzymes they are cleaved into thousands or even millions of different fragments at the sites recognized by those enzymes. Let us consider a simple case in which two otherwise identical DNAs differ from each other only in that one possesses a restriction site and is therefore cleaved into two fragments, while the other does not and remains uncleaved (Fig. 2.1). This is an example of a restriction fragment-length polymorphism (RFLP). The different DNA fragments can be resolved by gel electrophoresis, as the DNA fragments will migrate at different rates. If there is insufficient DNA for direct detection, the DNAs are hybridized to a radioactive probe, which allows them to be detected by autoradiography.

Restriction fragment-length polymorphism data produced in this way have been commonly used in ecological studies.

## 2.4  Separating restriction fragments by electrophoresis

After DNA has been digested by restriction enzymes, the complex mixture of DNA fragments can be analysed. The usual first step in analysis is to separate the fragments by electrophoresis on an agarose gel. In free solution, DNA fragments move anodally (positive) in an electric field, at a constant speed independent of size. However, a gel matrix will retard the movement of the fragments, and the magnitude of the retardation will depend to a great extent on their size, being great for long molecules and small for short molecules. This 'molecular sieving' serves to separate DNA fragments according to their length, unlike proteins which are separated according to charge and conformation. In practice, the resolution of DNA fragments through agarose gels depends upon molecular size of the DNA fragments, their conformation, the agarose concentration and the applied current. Linear DNA fragments are believed to migrate end-on through pores in the gel, and the rate of migration is proportional to $Log_{10}$ of the molecular weight. The rate of migration of a given DNA fragment is dependent upon the size of the pores in the gel matrix, which, in turn, is a function of the agarose concentration of the gel. The linear expression which describes this relationship is:

$$Log\,\mu = Log\,\mu_0 - K_r \tau$$

(Mobility of DNA = free electric mobility − retardation coefficient × gel concentration).

The retardation coefficient is related to properties of the gel and the size and shape of the DNA molecules. Most manufacturers of agarose provide information relating to the optimal agarose concentration for resolution of particular DNA sizes, but the ideal conditions are best determined empirically. Most of the DNAs analysed by agarose gel electrophoresis are linear fragments generated by restriction digestion of either genomic (itself linear), or plasmid or mitochondrial (both covalently closed circular DNAs). However, covalently closed (I) and nicked circular (II) molecules behave differently compared with linear (III) DNA. The magnitude of the difference depends primarily upon the agarose concentration, but also on the field strength, the buffer composition and the degree of superhelicity of the DNA. When the voltage is low, the rate of migration of a DNA fragment is proportional to the voltage. However, as the voltage is increased, the mobility of large molecules rises differentially and so the resolution diminishes. Optimal separation occurs when the voltage gradient does not exceed 5 V cm$^{-1}$ (for review see Sambrook *et al.* 1989). Agarose gel electrophoresis is frequently used to estimate the size of DNA fragments. Size standards of known molecular weight are run in adjacent lanes to DNAs to be sized. Because DNA migrates in an approximately $\log_{10}$ linear manner, a simple plot of $\log_{10}$ of the molecular mass of the standards against their mobility can be constructed, and the molecular mass of unknowns can then be estimated by interpolation. Other more sophisticated algorithms are also available (e.g. Sealy & Southern 1990) if very precise estimates are required. Small DNA fragments, including PCR amplified microsatellites, are usually separated using polyacrylamide gels (see Chapter 9), and very large DNA fragments, such as chromosome-sized fragments of DNA, are separated by methods such as pulsed-field gel electrophoresis (PFGE).

### 2.4.1  Gel properties

The typical agarose gel system consists of a tank of elecrophoresis buffer with electrodes at each end, into which is positioned a horizontal agarose bed with slots towards one end into which the DNA samples are loaded. Two general sizes of gel are recognized, known as minigels and maxigels. Minigels are small and can be run fast but offer poor resolution. They are typically used for assays, checking digests, isolating specific DNA fragments or in situations where maximum resolution is not paramount. Maxigel systems are larger and are able to offer much improved resolution over the minigels. They are used for RFLP and fingerprinting studies where DNA is to be blotted and analysed in hybridization studies.

Several types and grades of agarose are available which differ in their mechanical strength, gelling temperature and the presence of impurities. Ideally the gel should provide good separation of DNA fragments and be mechanically strong in order to withstand handling. It should be chemically pure (agarose preparations contain variable amounts of sulphated polysaccharide chains, which electrostatically retard DNA migration and can also prove inhibitory to many modification enzymes used on the DNA after it has been gel purified). A 'general purpose' agarose such as Seakem LE (FMC Biopsoducts, Maine) is usually chosen. This proves suitable for separating DNA fragments over the range of 50 000 bp to 300 bp. However, more specialized agaroses may be used if the DNA fragments are very large or very small. Low-temperature-gelling agaroses, in which hydroxyethyl groups have been introduced, are particularly useful for isolating specific DNA fragments, especially if further enzymatic modification to the DNA is envisioned, such as labelling it for use as a probe or ligation into a vector for cloning.

### 2.4.2  Electrophoresis buffers

Two main electrophoresis buffers are used for agarose gel electrophoresis, known as Tris-acetate (TAE) and Tris-borate (TBE), a third Tris-phosphate

(TPE) buffer is little used. For DNA analysis, buffers are conveniently prepared as concentrated stock solutions and then diluted to about 50 M for use. There is little difference between the buffers for most routine separations, but for mainly historical reasons TAE is more widely used, even though its buffering capacity is lower, and it tends to become exhausted during prolonged electrophoresis. For some applications buffer recirculation (or periodic replacement) is often recommended, however, at low or moderate field strength ($1–2\,V\,cm^{-1}$) it is not usually necessary, particularly if large volumes of buffer are used. Because analytical minigels are often run at higher field strength (typically $10\,V\,cm^{-1}$), TBE is sometimes preferred.

### 2.4.3  Running an agarose gel

DNA samples are mixed with a concentrated loading dye, usually a mixture of a high molecular weight polysaccharide to make the sample dense (e.g. sucrose or ficol), and one or two coloured dyes (bromophenol blue and xylene cyanoll FF), which allow the migration to be monitored. Very often EDTA is added to stop enzymatic reactions and to protect the DNA from nuclease damage. The mixture is then heated to 65 °C for 10 min to denature any sticky ends that may have become associated since the restriction digest, quenched on ice, and carefully loaded into the slots of the gel immersed in electrophoresis buffer. Electrophoresis is commenced at a voltage of around 40–100 V, depending on the gel length and required resolution.

## 2.5  Detecting DNA

Silver staining is often used for the detection of restriction fragments during the analysis of purified mtDNA resolved through polyacrylamide gels (10–30 pg; Tegelstrom 1986). For agarose gels ethidium bromide (detection limits 1–5 ng) is the most common stain for detecting DNA. It is an organic molecule containing a planar group that is able to intercalate between the stacked bases of the DNA. When excited by ultraviolet (UV) radiation, the DNA/ethidium bromide complex fluoresces in the orange-red region of the spectrum (max. at 590 nm). It can be used to detect double- or single-stranded DNA or RNA, although the sensitivity for single-stranded nucleic acid is much lower than for double-stranded.

Either the ethidium bromide ($0.5\,\mu g\,mL^{-1}$) is incorporated into the gel and running buffer prior to the electrophoresis, or, alternatively, the ethidium bromide is diffused into the gel after electrophoresis is complete, simply by soaking the gel. The former method works well and saves time. Electrophoresis can be stopped and the progress monitored, but care should be taken because binding of the dye can cause the migration of linear DNA to be reduced by up to 15%, and circular DNA molecules can run aberrantly depending upon their concentration (Waye & Fourney 1990). Consequently, if accurate size estimates are required, it is normal to stain the gel after electrophoresis is complete. Post-electrophoretic staining can be performed by adding ethidium bromide ($0.5\,g\,mL^{-1}$) to a small volume of electrophoresis buffer and immersing the gel for 20–60 min, depending upon the concentration and thickness of the gel. For the detection of very small amounts of DNA (<10 ng), background fluorescence can be reduced by soaking the gel in 1 M $MgSO_4$. Gels are best illuminated from beneath using a UV transilluminator (Sealy & Southern 1990). Lamps emitting at 254, 302 or 366 nm are available. The 254 nm lamp offers the highest sensitivity but can rapidly damage the DNA by photo-nicking, this can affect the suitability of the DNA for further processing. For most applications, such as photography, 366 nm lamps are preferred. Permanent records can be made conveniently by photographing the gel, but an orange (e.g. Kodak 23A) and a UV opaque filter must be attached to the camera (Sealy & Southern 1990).

For some analyses it is possible to incorporate radioactive nucleotides into the DNA, e.g. for mtDNA RFLP analysis by end-fill labelling followed by autoradiography (sensitivity about 2–5 ng), or by end-labelling PCR primers before amplification. Alternatively, specific sequences can be identified by hybridization to a suitable probe, and then detected either by autoradiography or by a specific colorimetric enzyme reaction depending upon the nature of the probe (<1–10 pg).

## 2.6 Blotting

To detect specific sequences from within complex mixtures of restriction fragments, the resolved DNA is denatured into single strands and hybridized to a specific labelled probe. Under appropriate conditions, this will bind to the target sequence and detection of the probe will reveal the presence, position and organization of the target. Fragments of DNA retained in a gel can be directly hybridized to a radiolabelled probe. However, it is more usual to transfer the DNA fragments from the fragile gel to a more solid support, such as nylon membranes. This is accomplished by some form of blotting process, which transfers the DNA fragments from the gel to the membrane while faithfully retaining their relative positions.

The original, and still the most widely used blotting procedure, was developed by Southern (1975), and is both efficient and straightforward (see Fig. 2.3, and Protocol 6 on page 39). Double stranded DNA (dsDNA) fragments in a gel are denatured *in situ* by alkali, then the gel is neutralized in high salt buffer to maintain the single-strandedness of the DNA. A membrane is closely juxtaposed onto the gel and the DNA fragments are eluted and deposited onto the surface of the membrane carried in a stream of high salt buffer pulled through

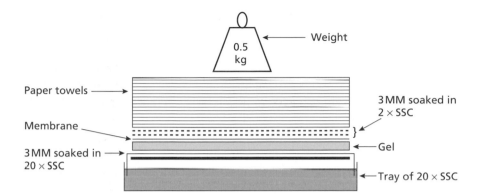

**Fig. 2.3** Schematic representation of the set-up for Southern blotting.

the gel by capillary action. A detailed protocol for Southern blotting suitable for most analyses is provided in Protocol 6 (page 39).

An alternative blotting procedure, known as alkaline transfer, is used by some. It varies from the standard protocol of Southern blotting in that the gel is not neutralized, and transfer is accomplished in an alkaline solution. In the original technique (Reed & Mann 1985), 0.4 M sodium hydroxide was used as a transfer solution. However, loss of hybridization efficiency was reported by several groups who developed improved regimens (Rigaud *et al.* 1987; Broad *et al.* 1988; Scheenberger *et al.* 1988; Chorazy & Edlind 1990). In addition to these capillary blotting procedures, mechanical systems have been developed and are sold commercially, such as vacuum blotting (Hybaid Ltd., Middlesex, UK), pressure blotting (Stratagene, La Jolla, California) and electroblotting (Amersham Pharmacia Biotech, Little Chalford, UK).

Various synthetic polymers are now used to prepare blotting membranes including nylon and polyvinyl difluoride (PVDF). These offer much greater mechanical strength and are usually more chemically resistant and have higher DNA binding capacity. In addition, some polymers are chemically modified to give them a net positive charge, which results in a very high affinity for negatively charged DNA. It is claimed that the DNA is fixed to the membrane during the blotting process and results in superior resolution. Personal experience suggests that fixing the DNA is still an important step if high resolution hybridization signals are required. Not all manufacturers perform rigorous quality control on their membranes and great variation is often encountered, not only between batches but also within a single roll, particularly for the positively charged membranes. Another problem that may be encountered with positively charged membranes is that they might contribute to high, nonspecific background hybridization unless precautions are taken.

## 2.7 Fixing

Once the DNA has been transferred to the membrane it must be immobilized so that it will not wash off during the subsequent hybridization steps. The procedure most used is still baking *in vacuo*; however, several different UV cross-linking and alkaline fixation protocols have been described.

## 2.8 Hybridization

Molecular hybridization offers a very sensitive (<1 pg of target sequence), rapid (<24 h) and convenient (simple and relatively inexpensive) opportunity to study nucleic acid structure and function. These techniques were developed to study gene structure and function but are increasingly being applied in other studies, such as for the diagnosis of heritable diseases, detection of viral and bacterial pathogens, and the analysis of genetic variation in ecology and evolution.

### 2.8.1 Probes

Hybridization probes are used to detect specific sequences in immobilized DNA. They can take several forms, but essentially they are single-stranded nucleic acids, either DNA, RNA or short chemically synthesized DNA chains (oligonucleotides). Most probes in current use are labelled by incorporating radioactive nucleotides, usually $^{32}$P, but increasingly nonradioactive moieties are being used. Radioactive nucleotides can be incorporated into the nucleotide chain, or alternatively attached to the 5′ or 3′ terminus (see Table 2.1 for examples). The principles of the major labelling procedures will be presented below. However, it is recommended that commercial labelling kits be used for probe manufacture.

Recombinant DNA techniques make it possible to obtain any sequence as a dsDNA, ssDNA, RNA transcript or synthetic oligonucleotide. So, the choice of labelling technique will be dictated, in the most part, by the type of investigation being undertaken. Probes can be prepared with very high, moderate or low amounts of radioactivity within them (specific activity; cpm µg$^{-1}$). The specific activity will affect the sensitivity of the probe and hence the sort of target that can be detected (e.g. a single copy gene, an abundant multicopy repetitive sequence or a high copy mtDNA).

Single-stranded probes, either RNA or ssDNA, can offer significant advantages over double-stranded probes for some applications. The absence of a competing second strand in the hybridization can result in improved kinetics. The length of the probe can be predetermined by cutting the 3′ end of the probe template with a suitable restriction enzyme, and this can be utilized to prevent cross-hybridization to the flanking sequences of the target. RNA probes, also known as riboprobes, offer improved hybrid stability, and hence the opportunity to use more stringent hybridization conditions. For riboprobes, there is an additional advantage; an efficient method exists for reducing a nonspecifically bound probe from the membrane by specifically degrading it with Rnase (this treatment will not damage the RNA–DNA hybrids).

### 2.8.2  Double-stranded probes

Double-stranded DNA, such as a cloned gene in a plasmid or an isolated PCR fragment, can be labelled to high specific activity of around $10^8$ or $10^9$ cpm μg$^{-1}$ with deoxynucleotide 5'-[$^{32}$P] triphosphates by either nick translation or random priming. Nick-translated probes can be labelled to about $10^8$ cpm μg$^{-1}$ but they tend to be a little inconsistent, and so are not used much nowadays. Random priming (Feinberg & Vogelstein 1984), uses as the substrate double-stranded DNA, which is heat denatured into single strands. Then random sequence oligonucleotides (usually hexanucleotides or octanucleotides) are allowed to anneal to the single-stranded DNA and the primers are extended using Klenow polymerase. Random priming is very reliable, and probes typically are labelled to specific activity of $10^9$ cpm μg$^{-1}$ or more.

Double-stranded DNA probes can also be generated by PCR simply by incorporating radiolabelled nucleotides into the PCR reaction. Double-stranded DNA probes must be heat-denatured at 100 °C and quenched on ice to separate the single strands immediately prior to their use. The ability of the complementary strands to reanneal and hence be unavailable to hybridize to the target can be a disadvantage. However, under suitable conditions partial reannealing can allow the formation of hyperpolymers, and so result in an amplification of the signal.

### 2.8.3  Single-stranded probes

To prepare ssDNA probes, it is generally necessary to have the required probe sequence as a single-stranded template. Originally, this would involve cloning the desired probe sequence into one of a series of recombinant bacteriophage M13 vectors that have been designed for DNA sequencing studies, but now phagemids are more common substrates. In the usual system a specific short oligonucleotide primer (usually the M13 sequencing primer) will be annealed to its complementry strand upstream of the cloned probe sequence. Extension from the primer catalysed by Klenow polymerase results in a partially double-stranded molecule. This is then digested with selected restriction enzymes to liberate the double-stranded fragment of interest. This is then denatured and used as a hybridization probe. This method allows the preparation of very 'hot' probes that are of a specified length, see Jeffreys *et al.* 1985 for example. An alternative method of making single-stranded DNA probes from M13 or derivatives has been described (Hu & Messing 1982), but is not routinely used.

Polymerase chain reaction can also be used to generate single-stranded DNA probes by adjusting the ratio of the two specific primers. If one primer is present at a limiting concentration, the amplification will become biased towards the strand with the excess primer and only this will be amplified (and labelled) to a significant level.

Alternatively, ssRNA probes (riboprobes) can be prepared from a DNA template by *in vitro* transcription from an upstream transcription promoter using DNA-dependent RNA polymerases from bacteriophages (either T3, T7 or SP6). These are very specific for their own promoter sequences, so, if cloned DNA fragments are inserted into phagemid vectors which contain these promoters, *in vitro* transcription can be directed to prepare large quantities of RNA copies of the cloned fragment. In the presence of radiolabelled ribonucleotides, very high specific activity probes can be prepared. As with ssDNA probes, the size of the probe can be defined most conveniently by linearizing the substrate with a restriction enzyme before transcription. If extreme sensitivity is required, the DNA template can be removed by DNase I treatment to prevent it from competing for the probe, after which the RNA is cleaned and used in the hybridization.

### 2.8.4  Synthetic oligonucleotide probes

Increasingly, probes consisting of labelled synthetic oligonucleotides are being used for DNA fingerprinting and other studies (see Chapter 6). Usually, they are end-labelled with [32P] ATP catalysed by T4 polynucleotide kinase (Van der Sande *et al.* 1973; for protocol Ausebel *et al.* 1997). They may also be concatenated and labelled in a specific priming reaction (May & Wetton 1991).

### 2.8.5  Labels

For most hybridization studies the probe is radioactively labelled, usually with 32P nucleotide that is incorporated into the molecule but occasionally attached to one or other end. Radiolabelled probes usually offer the best sensitivity, and are easy and convenient to produce in a well-equipped molecular biology laboratory. However, many small ecology laboratories may not possess the facilities to undertake radioactive labelling. For these workers, non-radioactive alternatives do exist. Usually biotin or digoxygenin moieties are incorporated into the probe; these are recognized in the hybrids by antibodies specific to the moiety. Enzymes conjugated to these antibodies, such as horseradish peroxidase (HRP) or alkaline phosphatase (AP) are able to catalyse specific reactions, resulting in the production of insoluble products which precipitate onto the membrane. Some of these are coloured and can be seen directly, others are luminescent (chemiluminescent) and are detected by autoradiography.

### 2.8.6 Hybrid stability

Nucleic acid hybridization is a reversible process. Nucleic acid duplexes can be dissociated by heating or by chaotropic agents, then allowed to reanneal. Manipulation of the parameters that affect the stability of a hybrid can be exploited to allow perfect and imperfect hybrids to be discriminated. The melting temperature ($T_m$) is the temperature at which half of the hybrids have melted into single strands. It is affected by the concentration of monovalent ions, the base composition (%G+C), the length of the shortest chain and the concentration of helix destabilizing agents, such as formamide. Equations that describe the influence of these parameters on stability of perfectly matched DNA–DNA, RNA–DNA and RNA–RNA hybrids, as well as for oligonucleotide–DNA hybrids (14–20 nucleotides), are described and explained in Wahl *et al.* 1987. Hybrids of RNA–RNA are the most stable, followed progressively by RNA–DNA, DNA–DNA and the least stable are oligonucleotide–DNA hybrids. If mismatched nucleotides are present, they will reduce the stability of the hybrid, and the magnitude of this reduction will depend upon the number and position of the mismatching bases. Such instability is particularly important if short oligonucleotide probes are used: $T_m$ will fall by approximately 5 °C for every 1% mismatch. For hybrids around 150 bp or greater, the fall in $T_m$ will be about 1 °C.

### 2.8.8 Hybridization kinetics

The rate at which hybridization occurs effectively follows pseudo-first order rate kinetics, because under normal experimental conditions the probe will be vastly in excess over the target. The time taken for half the filter-bound DNA to anneal to the probe is:

$$t_{1/2} = \ln 2 / KC$$

Where $K$ is the first-order rate constant (L×mol nucleotides s$^{-1}$), and $C$ is the probe concentration (mol nucleotides×L$^{-1}$). $K$ is a function of the probe length ($l$), molecular complexity (number of nucleotides in a nonrepeating sequence) and the nucleation rate constant $K_N'$ itself affected by temperature, viscosity and pH.

$$K = K'nN \, l \, 0.5 \text{ N}^{-1}$$

Under standard conditions defined as $T_m$ −25°C, 0.18 M cation concentration and a single-stranded probe length of 500 bases, the rate constant can be approximated to:

$$K = 10^6 \ N^{-1}$$

Usually for hybridization studies, the probe is in excess, so the probe concentration and its complexity are more important. In practice, ionic strength has little effect providing that it is above 0.4 M. The effects of pH are also small (<1.3-fold in the range of pH 5.00–9.15). Providing the probe is greater than 150 nucleotides in length, the maximum rate occurs in 1 M NaCl at 25°C below the $T_m$. The implications are that low complexity probes at high concentration yield the lowest hybridization times.

## 2.8.8 Hybridization protocols

Essentially, hybridization is the binding of the probe to its target, and it is accomplished simply by washing the membrane on which the genomic DNA is immobilized in a buffer solution containing the labelled probe. Salt concentration and temperature are controlled to regulate the 'stringency' of the hybridization, that is, the degree of homology required between the probe and target for stable hybrids to form. The specific activity and concentration of a probe will affect the quality of the signal achieved and so hybridizations are performed in minimal volumes. Traditionally, membranes were sealed into plastic bags and small volumes of prehybridization solution added. More recently, mechanical rotisserie-type hybridization ovens have become popular. A single membrane is placed into a glass tube and a small volume of hybridization solution is then added. It is then incubated in the oven, and the movement of the rotisserie constantly bathes the membrane. These ovens have capacity for several tubes and so allow many membranes to be hybridized simultaneously with different probes, and greatly reduce the risks of contamination during the hybridization and subsequent washing procedures. However, when large numbers of membranes are to be hybridized to the same probe it is more convenient to hybridize them together. In such circumstances, hybridization chambers are used. These may be specially constructed devices or, as is common, a plastic sandwich or cake box may be used.

Particularly when concentrated probes are used, and when several membranes are to be hybridized together, it is desirable to agitate the hybridization solution constantly. Hybridization ovens do this automatically, but for hybridizations in bags or chambers it is advisable to perform the incubation in a shaking waterbath.

Prehybridization steps which serve to reduce nonspecific hybridization, and hence reduce background, are usually necessary, but the length of time required will depend on the nature of the probe, the membrane, the strength of signal and the blocking agents. Blocking agents are substances or cocktails of substances that are able to bind to charged sites on the membrane surface and so prevent the nonspecific binding of the labelled probe. Those most commonly used are Denhardt's solution (Denhardt 1966), Blotto (Johnson *et al.* 1984) and 7% SDS (Church & Gilbert 1984). Usually prehybridization is carried out for several hours at the stringency chosen for the hybridization; the prehybridization fluid becomes the hybridization solution simply by adding the probe to it. The hybridization is then carried out for 12–16h, conveniently overnight. Sometimes kinetic enhancers such as dextran sulphate (Wahl *et al.* 1979) or polyethylene glycol (Renz & Kurz 1984) are included to promote hyperpolymer formation. After this, the unhybridized probe is poured away and stringency washing is begun. During this stage the probe is dissociated from hybrids according to the stringency chosen.

In most experimental designs, low stringency is used during the hybridization step (e.g. 4×SSC (standard sodium choride/sodium citrate buffer, recipe for 20 × SSC on page 45)) to promote the formation of strong hybrids, and then they are washed at a higher stringency (normally 1×SSC or less) to remove weakly hybridized probe. For some studies, such as DNA fingerprinting, it is better to use the same stringency for hybridization and subsequent washes so as to minimize the variation between experiments where poorly washed membranes possess artefactual bands that have not been adequately removed during the washing stages.

### 2.8.9  Deprobing and rehybridization

Often it is desirable to hybridize more than one probe to a set of DNAs, and so it is necessary to probe individual membranes repeatedly by removing already hybridized probe from the membrane without removing the immobilized target DNA. Positively charged nylon membranes have a strong affinity for nucleic acids. So it is necessary to wash the membrane in two changes of 0.4 M NaOH solution for approximately 1h, and then to neutralize the membranes in two changes of 0.2 M Tris pH 7.5, 0.1×SSC, 0.1% SDS for 1h. The membrane can then be rehybridized to a new probe as before.

Membranes that are to be stripped and rehybridized should not be allowed to dry out, as this makes complete removal of the old probe much more difficult. It is usual, therefore, to wrap damp (not wet) membranes in Saranwrap for autoradiography to prevent them from drying.

### 2.8.10  Detection and measurement of hybrids

Hybrids between radiolabelled probes and target sequences attached to the membrane can be analysed by counting the amount of radiation they emit by liquid scintillation. But in the majority of hybridization studies the presence of radioactive hybrids is detected by exposing the hybridized membrane to a photographic emulsion and subsequent photographic development (autoradiography). The hybridized membrane is juxtaposed to a sheet of X-ray film, usually in a special cassette. Exposure required may be from hours to several weeks, depending upon the amount of radiation emitted. If very low levels of radiation are to be detected, then the signal can be amplified by using 'intensifying screens' (Swanstrom & Shank 1978). When using intensifying screens, it is necessary to perform the exposure below −70 °C to prolong the period of fluorescence. Increases of 10–14-fold are possible when used appropriately, but with a small reduction in the resolution (Bonner 1987). High-technology methods such as storage phosphors (Johnston *et al.* 1990) and solid state detectors probably offer some improvements in speed, resolution and sensitivity, and also offer improved data manipulation, but they are often not economically viable alternatives for small laboratories because of the high initial capital outlay involved.

## 2.9 DNA sequencing

Many forms of DNA variation of interest to the field ecologist may not be easily detected using RFLP type analyses, and it is appropriate to determine the actual DNA sequence of a limited stretch of DNA. Two methods of sequencing DNA are used by molecular biologists, the first involves specific chemical cleavage of DNA chains (Maxam & Gilbert 1977), the other involves the synthesis of a DNA template that is specifically terminated at individual residues (Sanger *et al.* 1977). Only the Sanger dideoxynucleotide system is commonly used. Essentially a small fragment of DNA of interest is denatured to form two single strands. A short oligonucleotide (primer) complementary to one strand (+strand) is allowed to anneal, and this is then used to prime synthesis of the −strand, using the +strand as a template. Four separate enzyme-catalysed synthesis reactions are performed in which specific nucleotide analogues (dideoxynucleotides) are included, and if these become incorporated into the growing chain, they terminate the chain preventing further elongation. This results in four sets of nested DNA fragments, each ending in a specific dideoxynucleotide. The four reactions are separated on a high resolution denaturing polyacrylamide gel. Radioactive nucleotides ($^{35}$S, $^{33}$P or $^{32}$P), which are incorporated during the reaction, allow the sequence to be read from an autoradiograph of the gel (Fig. 2.4). This manual sequencing system is relatively slow and labour intensive and is used only for small-scale projects. Recent advances in technology permit the incorporation of nucleotides labelled with fluorescent analogues into the nascent DNA strand. After the labelled fragments have been resolved through a denaturing polyacrylamide gel, the fluorescence is read by a laser scanning device. This

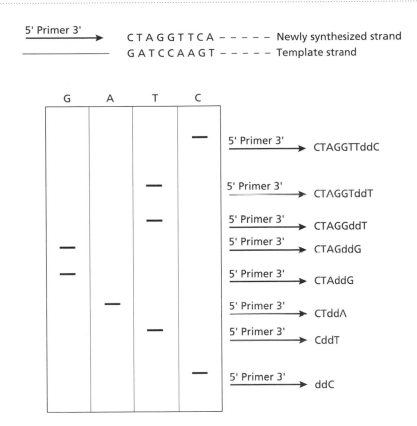

**Fig. 2.4** Simplified representation of dideoxy sequencing.

technology is the basis for various automated fluorescent sequencing devices. These are expensive, but provide high throughput, and individual molecular biology laboratories are increasingly investing in the technology; alternatively most universities or departments have core laboratories which perform DNA sequencing for modest fees.

## 2.10 Other recombinant DNA technologies

Bacteria are propagated as hosts for recombinant DNA molecules, allowing large quantities of the specific DNA to be 'manufactured' which can then be used for preparation of hybridization probes, or the creation of other novel recombinant molecules. Special properties of the bacteria (or the vectors they harbour) allow strong selection for the particular recombinants (e.g. antibiotic resistance), and simplified screening protocols (e.g. blue/white colour screening). Yet others reduce the opportunity for undesirable genetic recombination to occur.

To express some of these specific functions it is often necessary to grow or maintain the bacteria harbouring the recombinant DNA under rigorously controlled selection conditions. Several different regimes for the viable maintenance of the strains are employed, each offering advantages and disadvantages in terms of length of viable storage and accessibility. Details of these methods can be found in any good molecular biology techniques manual (e.g. see Sambrook *et al.* 1989 or Ausebel *et al.* 1997).

There are three general methods for the transfer of recombinant DNA into bacteria. Transformation is the transfer of naked plasmid DNA; transfection is the transfer of naked bacteriophage DNA. Both methods require pretreatment of the bacteria to make them 'competent' for DNA transfer. Many different protocols exist, but most are based on treating the cells with simple salt solutions such as $CaCl_2$ or $RbCl_2$ (Mandel & Higa 1970; Cohen *et al.* 1972; Chung & Miller 1988; Martin & Burke 1989; see also Hanahan 1985 for a review). Many strains of *Escherichia coli*, ready competent, can be purchased from various vendors. Alternatively, simple and very efficient electroporation methods have become popular (Jacobs *et al.* 1990), but require specialized electroporator equipment (e.g. Biorad, Hercules, California). Bacteriophage λ and cosmid DNA is large (~ 50 kb) and does not transform efficiently. In order to introduce these DNA into suitable host bacteria, *in vitro* packaging and infection are used. In this system, protein extracts containing all of the components required for packaging λ DNA are prepared from specific mutant *E. coli* strains and are mixed with DNA. The DNA becomes encapsulated into infective particles and is used to infect suitable hosts. Again, commercial packaging extracts are available from several vendors (e.g. Gigapack: Stratagene, La Jolla, California; Packagene: Promega, Madison, Wisconsin).

### 2.10.1 Plasmid extraction methods

Recombinant plasmids contained in suitable host bacteria are the usual substrates for the preparation of hybridization probes, and are used as cloning and subcloning vectors. Plasmid DNA needs to be purified from bacterial genomic DNA, other cell components and the growth media. The preparation of large quantities of plasmid is often necessary, and for this reason a wide variety of protocols have been developed that allow plasmid DNA of a quality suitable for most of these applications to be prepared. Methods for preparing plasmid DNA are available in most good molecular biology technique manuals, or alternatively, many affordable commercial plasmid DNA extraction kits are available, which produce DNA from various scale cultures.

### 2.11 Safety

A short note on laboratory safety should be included. A molecular biology laboratory can pose hazards to human health and the environment; for example, many chemicals used routinely in molecular biology techniques are toxic or otherwise hazardous: unpolymerized acrylamide is a potent neurotoxin; some commonly used chemicals are known or suspected mutagens or carcinogens, even at small doses (ethidium bromide used as a DNA-specific labelling dye, for example); and chloroform used for some methods of DNA extraction. Common acids and bases are used for preparing various buffers and can cause severe chemical burns, as can phenol. The improper use of equipment such as centrifuges, UV light sources and electrophoresis equipment can also prove hazardous. Various radioisotopes are often used as molecular tracers in the preparation of hybridization probes, and improper use can cause significant individual and environmental contamination. And of specific concern for field ecologists is the possible introduction of pathogens from biological field specimens into the environment. These issues are the subject of numerous local, national and international laws. However, it is the individual responsibility of those undertaking the work to recognize and understand the hazards and take appropriate action to minimize risk. This will include being properly trained in experimental procedures, using protective clothing (laboratory coat, gloves and eyeware) or other devices such as radiation shielding, and, where necessary, gaining permission from the appropriate authorities to undertake the experiments.

## Protocol 1: Methods for sample storage

### 1a  Frozen whole blood

Whole blood, taken by venepuncture, is transferred to a sterile Eppendorf tube and frozen as soon as possible, preferably at $-80\,^\circ$C, but for most instances $-20\,^\circ$C will suffice. Some loss of DNA integrity occurs if samples are repeatedly thawed and re-frozen (Ross $et\,al.$ 1990).

### 1b  Blood storage in extraction buffer

If possible, appropriate-sized aliquots of fresh whole blood (routinely $15\,\mu$L of avian blood, or $1$–$2\,$mL of mammalian blood) are resuspended in $500\,\mu$L ($5\,$mL if mammalian) of $1\times$SET extraction buffer (sodium chloride/EDTA/Tris, see page 45) then frozen at $-80\,^\circ$C. Individual tubes can then be removed and processed without thawing the stock sample.

### 1c  Blood storage in lysis buffer

Longmire $et\,al.$ (1988) describe a lysis buffer into which whole blood is resuspended for long-term storage at ambient temperature. It contains SDS, which lyses cells, and EDTA, which chelates $Ca^{2+}$ ions and therefore prevents nuclease activity. Laboratory experiments with this buffer, however, show that if excess blood is resuspended, the suspension becomes very viscous and consequently difficult to manipulate and clean efficiently.

## 1d  Blood in ethanol

Absolute ethanol is a powerful dehydrating agent. If small volumes of fresh whole blood are suspended, individual erythrocytes remain intact and separate. Nucleases and other degradative enzymes remain inactive and allow high molecular weight DNA to be extracted with relative ease. Usually one drop (approx 100 μL) of avian blood is suspended into 1 mL of ethanol.

## Protocol 2:  Extraction of high molecular weight DNA from various sources

### 2a  DNA extraction from whole blood

The following protocol has been developed specifically for the extraction of genomic DNA from avian blood samples, but can easily be modified for other samples by adjusting the volumes of starting material.

1    Prepare fresh isotonic lysis stock solution. For each sample use:
    650 μL of 1×SET (20×stock; 3 M NaCl, 1 M Tris, 20 mM EDTANa$_2$, pH 8.0).
    10 μL of SDS solution (25% w/v sodium dodecyl sulphate).
    20 μL of Proteinase K solution (10 mg ml$^{-1}$).

2    Dispense 650 μL into individual 1.5 mL microfuge tubes, then add either 25 μL of fresh or freshly thawed whole blood. Immediately mix it thoroughly but gently into the lysis buffer by inversion to prevent clumping. Incubate overnight at 55 °C.

For mammalian species substitute an equivalent number of isolated white cells or finely ground solid tissues for whole blood.

The commencement of lysis is usually immediately evident as an increase in viscoscity as DNA is released. At 37 °C proteolysis is more rapid; however, at 55 °C the endogenous nucleases are relatively inactive and thus less prone to cause damage to the DNA.

3    Remove the tubes from the water bath and centrifuge them briefly (e.g. 5 s at 14000×g) to collect all the contents to the bottom of the tube.

4    Add 500 μL of phenol (buffered to pH 7.5–8.0 by equilibrating with 1 mol Tris) to each sample. Then mix by gentle inversion for approximatcly 20 min. **Do not vortex**, as this will shear the DNA.

5    Centrifuge the samples at high speed (14000×$g$) for 3 min to resolve the organic/aqueous emulsion.

**6**  Carefully pipette the aqueous phase (upper layer) to a fresh tube. Take care not to disturb the 'white' interface. Discard the organic phase.

Use wide-bore pipette tips, prepared by cutting 2–3 mm off a normal blue tip, to manipulate the viscous DNA solution, in order to minimize shear damage to the DNA.

**7**  Repeat organic solvent extractions until no precipitate appears at the interface, first with phenol then with phenol/chloroform (phenol:chloroform:isoamyl alcohol, 24:23:1 v/v), and finally remove residual traces of phenol by a single extraction with chloroform (chloroform:isoamyl alcohol, 23:1 v/v).

**8**  Recover 500 μL of the aqueous phase to a fresh tube and add 2× volumes (1 mL) of absolute ethanol. Mix by repeated inversion (the DNA can usually be seen as a floating white blob). Incubate at −80 °C for 10–15 min, or −20 °C for 30–60 min.

**9**  Centrifuge the sample at high speed (14000×g) for 10 min to pellet the DNA.

**10**  Discard the supernatant, and wash the pellet with 500 μL of 70% (v/v) ethanol and re-pellet as in step 9.

**11**  Re-centrifuge the pellet as in step 9. Drain away the supernatant and evaporate all traces of ethanol, either by placing the uncapped tubes at 37 °C for 30–60 min, under a heat lamp for 10 min or by drying in a speed-vac or other vacuum device.

**12**  Dissolve the DNA into 150 μL of TE (10 mM Tris, 1 mM EDTA pH 8.0) buffer overnight at 55 °C. Typical DNA concentrations are in the range of 0.5–2.0 μg μL$^{-1}$.

**13**  Store the solution at 4 °C.

### 2b  DNA extraction from other material

DNA can readily be extracted from a variety of other sources, such as solid tissue obtained from corpses, and in these cases the tissue is ground to a fine powder in liquid nitrogen using mortar and pestle, and then resuspended in SET buffer. Such samples are then processed as for blood.

### 2c  DNA extraction from ethanol-preserved material

The strong dehydrating properties of ethanol that are exploited in this preservation system result in a blood sample that is difficult to pipette; consequently it is washed in several volumes of 1×SET two or three times and recovered by a short centrifugal spin prior to commencing the extraction proper, by resuspending the sample in 1×SET and continuing the normal extraction procedure.

## 2d  DNA extraction from lysis buffer

Field samples of blood suspended in lysis buffer are frequently very viscous. An approximately 100 μL aliquot is harvested and resuspended in 500 μL of 1×SET buffer. Proteinase K is added, the mixture is incubated overnight and then processed as above. However, this material often requires several additional organic extractions to obtain clean DNA, so that the yield may be low.

## Protocol 3:  Restriction of genomic DNA

Usually for population genetics and ecological studies, several samples need to be compared and this example shows how the DNA from 10 different individuals would be digested for the analysis of RFLPs or DNA fingerprinting.

1    Label 10 microfuge tubes and place them in a suitable rack.

2    Using a suitable micropipette (e.g. Gilson or Finnpipette) dispense into the bottom of each tube 10 μL of appropriate DNA solution. If the DNA solution is very viscous it may be necessary to cut the end from the pipette tips to facilitate the transfer.

3    Prepare a stock solution of restriction reagents adequate to digest each sample, e.g. for 10 samples prepare sufficient for 11. Dispense the following 'ingredients' into a fresh tube held at room temperature. (Do not prepare the stock on ice, spermidine will cause the DNA to precipitate. The enzyme should be added last.)

4    Mix the contents of the stock tube by centrifuging briefly.

5    Dispense 10 μL of the reaction stock onto the inside edge of each tube. Mix the stock and DNA by centrifuging briefly.

6    Incubate the tubes at an appropriate temperature for the enzyme used, usually 1–5 h, or even overnight.

For RFLP analysis where the presence or absence of a restriction site is to be determined, the above regime will usually be sufficient. However, techniques such as DNA fingerprinting have an absolute requirement for completely cut DNA. The large number of bands routinely analysed need to be as distinct as possible to ensure correct interpretation (see Chapter 6). Consequently, as a precaution against enzyme exhaustion (Crouse & Amorese 1986), the above regime is extended for DNA fingerprint analyses, as follows.

7    Usually after 6–8 h restriction, an additional 1 μL of restriction enzyme is added to the reaction, and the reaction is allowed to continue overnight. Before terminating the reaction, an aliquot (1–2 μL) is examined on an analytical minigel.

**Table 2.3** Restriction of genomic DNA

| Reagent | Volume per sample | Total volume |
| --- | --- | --- |
| ×10 reaction buffer | 2 μL | 22 μL |
| Spermidine HCl (40 mmol) | 2 μL | 22 μL |
| Sterile distilled water | 5 μL | 55 μL |
| Restriction enzyme | 1 μL | 11 μL |

## Protocol 4:  Detecting DNA fragments

### 4a  Staining gels with ethidium bromide

Ethidium bromide may be included in the gel and running buffer at $1 \, mg \, mL^{-1}$, or by soaking the gel for 10 min in $2 \, \mu g \, mL^{-1}$ ethidium bromide solution, then washing in distilled water. For very small amounts of DNA, stain overnight in the dark at 4 °C to reinforce the fluorescence.

### 4b  End-filling of restriction sites

One or more $\alpha^{32}P$ nucleotides can be used, and the same protocol is used if 3' or 5' extensions are to be labelled.

Prepare 1 μg of digested DNA in 20 μL, add 2 μCi (148 kBq) of one or more of the nucleotides, and 1 μL of a 2-mM solution of each of the other nucleotides as needed. Add 1 U of Klenow polymerase and make up to 25 μL. Labelling is efficient in medium- or high-salt restriction buffer; if low-salt buffer is used for the restriction, add NaCl to 50 mM. Incubate at room temperature for 30 min, then terminate by adding 1 μl of 0.5 M EDTA. Unincorporated nucleotides can be removed by phenol, phenol/chloroform extraction. A third of the mixture can be electrophoresed through an agarose or polyacrylamide gel, which is then dried and autoradiographed.

## Protocol 5:  Preparation and running of agarose gels

### 5a  Agarose minigel

To prepare 100 mL of a 0.8% agarose minigel suitable for analysing the restriction fragments of a plasmid digestion or for checking the digestion of genomic DNA.

1  Into a suitable receptacle (e.g. 200 mL Duran flask) measure 0.8 g of agarose (e.g. Seakem LE) and add 100 mL of 1×TBE buffer. Allow the agarose to hydrate at room temperature for 5–10 min then boil in a microwave

(medium setting in a 650W machine). Ensure that the agarose has dissolved completely, then cool to 55 °C by immersion in a waterbath. If required, 5 µL (10 mg mL⁻¹) ethidium bromide can be added to the warm agarose solution.

**2**  Seal the edges of the gel base with tape and suspend the comb approximately 1mm above the floor of the tray. Pour the agarose to a depth of 3–4mm. Allow to cool and set, and then remove the tape and comb. Fill the electrophoresis tank with 1×TBE buffer (add ethidium bromide to 0.5 µg mL⁻¹) if required. Immerse the gel in the buffer, ensure that the sample wells are filled, and that no bubbles of air are trapped.

**3**  The DNA solution is mixed with electrophoresis tracking dye (see below), incubated at 65 °C for 10 min and then quickly quenched on ice. This ensures that fragments with compatible sticky ends which may have associated are denatured into independent fragments. The DNA solution is then pipetted into the wells of the gel, together with suitable molecular weight markers (e.g. λ restriction fragments). The apparatus is attached to the power source and electrophoresis commences at 40–100 V. Progress is monitored by the motion of the tracking dyes. If ethidium bromide was incorporated into the gel, progress can also be monitored by periodic observation of the gel under UV.

### 5b  Agarose maxigel

Essentially the procedure is the same as for preparing a minigel but more care must be taken to minimize variation between gels and thereby ensure consistency.

Agarose is added to 1×TAE to give a gel of known volume and concentration and is then mixed thoroughly and allowed to hydrate. The bottle is weighed so that water lost by evaporation can be replaced. The agarose is then melted in a microwave as before. When it has cooled below boiling, distilled water is added to replace that lost as vapour. The gel is cooled to 55 °C and poured into the gel mould. For DNA fingerprinting, it is not usual to include ethidium bromide to avoid variation in the electrophoretic mobility, but most RFLP studies do, thereby allowing the migration of the marker fragments to be monitored.

### 5c  Running agarose gels

DNA samples (together with suitable molecular weight markers, such as λ restriction fragments) are mixed with a loading dye, usually a mixture of a high molecular weight polysaccharide to make the sample dense, one or two coloured dyes which allow migration to be checked, and very often EDTA to protect the DNA from nuclease damage (see Sambrook *et al.* 1989 for various recipes). The mixture is then heated to 65 °C for 10 min to denature any sticky ends that may have become associated since the restriction. The samples are quenched on ice and carefully loaded into the slots of the gel, which has been immersed in electrophoresis buffer in the gel tank. The lid is placed onto the rig, and electrophoresis is commenced at a voltage of around 40–100 V depending on the application. Progress is usually monitored by the motion of the tracking dyes. If ethidium bromide was incorporated into the gel, progress can also be monitored by periodic observation of the gel under UV.

## Protocol 6: Southern blotting

1   Remove the gel from the electrophoresis tank and stain with ethidium bromide if necessary, then view on a UV transilluminator.

2   Invert the gel (the DNA is close to the bottom of the gel) and place it in a suitable tray. A Perspex sheet the same size as the gel is a useful gadget if large or flimsy gels are to be handled (see Fig. 2.3).

3   Cover the gel with 0.2 M HCl solution and agitate gently to ensure it can move freely in the buffer. Many people maintain constant agitation using a rocking platform, but this is not strictly necessary providing the tray is agitated occasionally. Depurination is continued for approximately 15 min, depending on the gel concentration and thickness, and can be monitored by the pH change in the gel as indicated by the blue to yellow transition of bromophenol blue indicator in the tracking dye. It should be noted that acid depurination is temperature-dependent, and so at elevated temperatures the rate of depurination will be much greater and may result in a reduced hybridization signal (Bettecken & Kress 1990).

4   Pour away the depurination solution and replace it with several volumes of the denaturing solution, which for conventional Southern blotting is 1.5 M NaCl, 0.5 M NaOH. Soak the gel for 30–45 min depending upon the gel concentration and thickness; again occasional agitation is advisable. The bromophenol dye will revert from yellow to blue, but it is usual to continue the treatment beyond the colour transition to ensure that complete denaturation has occurred.

5   Remove the denaturation buffer and replace with several volumes of the neutralization solution (1 M Tris (pH 8.0), 1.5 M NaCl). A period of 45–60 min is required to ensure that the gel is properly neutralized. This stage is most important if nitrocellulose membranes are to be used, as NaOH retained in the gel from the denaturation step will damage the nitrocellulose, reducing DNA binding and hence the hybridization signal. Such damage does not occur for nylon membranes, in fact, alkaline transfer protocols use NaOH as a capillary transfer buffer. Because differences in DNA fingerprint patterns have been

noticed between alkali and Southern-blotted replicates, this is a precaution advised in order to obtain reproducible results.

**6**   Set up the blot as shown in Fig. 2.3. A shallow tray is partially filled with 20×SSC (20×SSC: 3 M NaCl, 0.3 M trisodium citrate, pH 7.0). A piece of Whatman 3MM chromatography paper is wrapped around a sheet of glass or Perspex and is supported over the reservoir of 20×SSC so that the ends of the 3MM are in the buffer and can act as wicks to maintain a flow of the transfer buffer. The Whatman paper is wetted in 20×SSC and any air bubbles between the paper and the glass platform are squeezed out by gently rolling a pipette over the surface. The gel is then placed onto the surface of the paper with the original bottom of the gel (DNA side) uppermost, and any trapped air bubbles are removed. A membrane is carefully placed onto the surface of the gel and again any air bubbles are removed. If a nitrocellulose membrane is used, it should have been prewetted in 2×SSC but most nylon membranes are hydrophilic so prewetting is not necessary. The membrane can be marked or labelled with a soft graphite or chinagraph pencil for orientation and subsequent identification. Around the gel is placed an impermeable barrier of Saranwrap to prevent short circuiting of the transfer buffer flow. Onto the membrane are then placed two sheets of 3MM paper, slightly larger than the gel, which have been prewetted in 2×SSC. Trapped air is removed and a stack of absorbent paper towels are placed onto the surface. Finally a flat platform is added to the top and a weight, approximately 0.5 kg, is added. Transfer buffer is drawn from the reservoir by capillary action and passes through the gel, eluting the single-stranded DNA as it it goes, which it then deposits onto the adjacent surface of the membrane. This process takes 2–16 h depending on the agarose concentration of the gel, the thickness of the gel and the size of the DNA fragments being transferred. It is usual, however, to allow blotting to continue overnight.

**7**   After blotting, the DNA has been deposited onto the membrane surface. It is, however, not strongly attached and so it is necessary to fix the DNA to prevent it from washing off. Several methods of fixing have been described, the most common being to bake *in vacuo*, but UV or alkaline crosslinking are alternatives. The blotting stack is dismantled and the membrane is separated from the gel and rinsed briefly in 2×SSC to remove adhering agarose. The membrane is allowed to air dry. It is then placed between protective sheets of Whatman 3MM and baked at 80 °C under a partial vacuum for up to 2 h, although 30–60 min is adequate totally to immobilize the DNA.

Always wear gloves or use forceps when handling the hybridization membrane, grease from fingerprints will prevent proper wetting which will then prevent probe hybridization.

## Protocol 7:  Storage and recovery of bacterial cultures

### 7a  Storage of colonies on a nutrient agar plate

Single colonies of recombinant bacteria that have been recently transformed are best picked while still small (<2 mm) and streaked onto fresh plates so as to maintain strong selection. If sealed and inverted, they remain viable for 1–2 weeks at 4 °C.

### 7b  Preparation of a stab culture

A single, well-isolated colony is picked with an inoculating loop and then stabbed 2–10 mm into liquid broth (LB) agar in a 5 mL screwcap vial. The cap is tightened. It can then be stored in the dark for several months.

### 7c  Recovery from stab culture

1   Remove the cap from the stab culture, and streak some bacteria onto a fresh selective media plate (e.g. LB ampicillin) to allow single, well-distributed colonies to grow. Incubate the plate overnight, inverted, at 37 °C.
2   Pick a single, well-isolated colony and re-streak it onto a fresh plate (e.g. LB ampicillin). Incubate the plate overnight, inverted, at 37 °C. These plates remain viable for several weeks at 4 °C.

### 7d  Preparation of an 'overnight' culture

Inoculate 8 mL of liquid broth (e.g. LB ampicillin) with a single colony from a fresh plate. Incubate at 37 °C with gentle agitation overnight (e.g. 10 r.p.m.).

### 7e  Preparation of a glycerol stock

Place 2.5 mL of freshly prepared overnight culture into a sterile 5 mL screw-cap vial. Add 1.5 mL of sterile 80% glycerol solution. Mix thoroughly by vor-texing briefly. Glycerol stocks can be stored frozen at –20 or –80 °C for several years.

### 7f  Recovery from a glycerol stock

It is not necessary to thaw frozen glycerol stocks in order to recover the bacterial strain. The glycerol stock is removed from the freezer and rapidly uncapped, the frozen surface of the stock is gently scraped with a sterile inoculating loop and then used to streak onto a fresh LB plate containing a suitable antibiotic if selection is required. The plate is then incubated, inverted, overnight at 37 °C.

## Protocol 8:  Introduction of plasmids into bacteria

Two methods for preparing bacteria competent for transformation with plasmid constructs are commonly used.

### 8a  CaCl$_2$ method

1   Inoculate 100 mL of LB with 0.5 mL of a fresh overnight culture.

2   Grow cells to an approximate density of $5 \times 10^7$ cells per ml ($A_{550}$ is 0.15–0.20) (typically 1–2 h, but note that absorbance will depend on the bacterial strain used).

3   Cool the culture on ice then pellet the cells in a Corex tube by centrifugation at 4 °C for 5 min at 4000 g in a Sorval GSA rotor.

4   Decant the supernatant and resuspend the cells in 10 mL of ice-cold 0.1 mol CaCl$_2$. Re-pellet the cells as previously.

5   Decant the supernatant. Resuspend the cells in 4 mL of ice cold 0.1 mol CaCl$_2$. The cells may be stored frozen at –70 °C. Storage for 24–48 h at 4 °C increases transformation efficiency 4- to 6-fold (Dagert & Ehrlich 1979).

### 8b  RuCl$_2$/CaCl$_2$ method

1   Inoculate 20 mL of LB with 0.5 mL of a fresh overnight culture.

2   Grow cells to an approximate density of $10^7$ cells per ml ($A_{550}$ is 0.15–0.20) (typically 1–2 h, but note that absorbance will depend on the bacterial strain used). The $A_{550}$ corresponding to $5 \times 10^7$ for rec+ is approximately 0.5, and for rec+ is approximately 0.2.

3   Pellet the cells in a Corex tube by centrifugation at 4 °C for 5 min at 5 K in a Sorval SS34.

4   Decant the supernatant and resuspend in 1 mL of solution A (10 mM MOPS (3-[N-morpholino] propanesulphonic acid), pH 7.0, 10 mmol RuCl$_2$), then take the volume to 10 mL. Re-pellet cells as previously.

5   Decant the supernatant. Resuspend the cells in 1 mL of solution B (10 mM MOPS, pH 6.5, 50 mM CaCl$_2$, 10 mM RuCl$_2$).

6   Incubate on ice for 30 min. Pellet cells.

7   Decant supernatant and drain the tube. Resuspend the cells in 1 mL of solution B.

### 8c  Transformation

The transformation procedure of Kushner (1978) can be used to introduce the foreign DNA into the bacteria.

**1**   Add 3 µL of fresh dimethyl sulphoxide (DMSO) to 200 µL of competent cells (10- to 20-fold increase in transformation efficiency).

**2**   Add 10–20 µL of ligation mixture containing 0.15–0.20 µg of DNA. Then incubate on ice for 30 min.

**3**   **(optional step)** Heat shock the mixture at 42 °C for 1–2 min.

**4**   Add 4 mL of media and incubate for 1 h at 37 °C (to allow the expression of antibiotic resistance).

**5**   Pellet cells and resuspend in 100–200 µL of broth ready to be spread onto a plate.

## Protocol 9:  Extraction of plasmid DNA from bacterial cultures

### 9a  Colony lysis

This procedure is designed to allow the very rapid screening of a large number of recombinants, e.g. from a 'shotgun cloning' experiment. The DNA produced by this protocol is usually electrophoresed intact and is used for insert size comparison.

### 9b  Miniprep

**1**   From a freshly prepared overnight culture take 1.5 mL and place it in a 1.5 mL Eppendorf tube. Pellet the cells at high speed for 2 min. Remove the supernatant, and the cells from a second 1.5 mL aliquot may be pelleted into the same tube.

**2**   The cells are then resuspended into 100 µL of 'Miniprep' buffer and incubated for 5 min at room temperature.

**3**   To the suspension is added 300 µL of freshly prepared lysis mix, the mixture is then incubated for a further 5 min at room temperature.

**4**   150 µL of precooled 5 mol potassium acetate solution is added, and the contents of the tube mixed gently by inversion then incubated for an additional 5 min on ice.

**5**   Protein, chromosomal DNA and SDS will precipitate; they are then pelleted by high speed centrifugation for 10 min.

**6**   The supernatant is carefully recovered and 200 µL of Tris/EDTA buffer (TE) is added. The aqueous solution of plasmid DNA is then extracted by phenol, phenol/chloroform and chloroform extractions as for genomic DNA.

**7**   To the final aqueous phase is added 2× its volume of cold (−20 °C) ethanol and it is mixed briefly by inversion and then incubated at −20 °C for 30–60 min.

**8**   The DNA is recovered by high speed centrifugation for 10 min. The ethanol is discarded and the pellet dried either *in vacuo* or at 37 °C.

**9**   The DNA may then be resuspended in 20 µL of TE at 55 °C for 10–30 min.

### 9c  Maximiniprep

**1**   Spin down 25 mL aliquots of culture in 30 mL centrifuge tubes at 6 K for 6 min in a Sorval SS34 rotor. Discard the supernatant, add a further 25 mL and repeat the spin.

**2**   Discard the supernatant then resuspend the pellet in 1 mL of 'Miniprep' buffer. Incubate at room temperature for 10 min.

**3**   Digest cell walls by adding approximately 20 mg of lysozyme and incubating at 4 °C for a further 30 min.

**4**   Complete cell lysis by adding 2 mL of freshly prepared 0.2 M NaOH, 1% SDS and mix gently, continue incubation. RNA can be removed by digestion at this point by adding 20 µL of Rnase A.

**5**   Precipitate the genomic DNA and cell debris by adding 1.5 mL of pre-cooled 5 M potassium acetate. Mix gently and continue incubation for 5 min.

**6**   Pellet the cellular debris by centrifuging at 10 000 g for 10 min.

**7**   Transfer the supernatant, and clean with phenol, phenol/chloroform and chloroform.

**8**   Precipitate the plasmid DNA by adding 2× volumes of cold ethanol. Mix gently and leave for several hours at −70 °C, then pellet it at 10 000 g for 10 min.

**9**   Remove the supernatant and wash the pellet with 70% ethanol, discard the ethanol.

**10**   Dry the pellet and resuspend in TE at 55 °C.

### Protocol 10:  Double restriction digestion of plasmid DNA

Often multiple reactions can be performed in a single buffer; sometimes, however, the buffer of one enzyme is unsuitable for another. If this is so, it may be possible to cut the DNA sequentially, firstly in the more dilute buffer, then to modify the buffer so that the second enzyme will work. Alternatively, after the first digestion the DNA must be recovered and the first buffer components removed before subsequent digestion with other enzymes.

Given below is an example of cutting a recombinant plasmid with two restriction enzymes having different buffer requirements to isolate a DNA fragment, which may, for example, be used for labelling by random priming.

*N.B. Always wear gloves when handling DNA as potent nucleases are present in sweat.*

**1** Using a suitable micropipette (e.g. Gilson or Finnpipette) dispense into an Eppendorf or other suitable microfuge tube 10 µL of plasmid DNA solution containing 1–10 µg of DNA.

**2** Add the following reagents, ensuring no cross contamination takes place:

2 µL of ×10 reaction buffer

2 µL of Spermidine HCl (40 mmol)

1 µL of Rnase A (60 U)

5 µL of sterile distilled water

1 µL of *Hind*III

Rnase A is added to remove contaminating RNA; if the RNA has been removed previously during plasmid preparation then this step can be omitted.

**3** Mix the contents briefly by centrifugation, then incubate for 1–2 h at the appropriate temperature for the enzyme.

**4** An aliquot of 1–2 µL can be removed and analysed on an agarose minigel to check that the digestion is complete, but this is usually not necessary.

**5** Add 1 µL of 0.5 M NaCl solution and 1 µL of *Eco*RI enzyme. Mix briefly and then continue incubation.

**6** The DNA fragments (vector and insert) are separated electrophoretically on an agarose gel, and the insert fragment is purified (for details of isolation methods see Sambrook *et al.* 1989). This material is then ready for further manipulation, e.g. radiolabelling for use as a probe, or ligation into a vector for propagation.

## Recipes

**20× SET**

(3 M NaCl, 1 M Tris, 20 mM EDTA Na$_2$, pH 8.0)

Dissolve 17.53 g of NaCl and of Tris into 80 mL of distilled water, add 4 mL EDTA Na$_2$ solution and pH to 8.0. Make up to 100 mL.

*Proteinase K (10 mg mL$^{-1}$).* Add 100 mg of proteinase K to 10 mL of distilled water. Aliquot 150 µL into Eppendorf tubes. Freeze at −20 °C. Use fresh aliquots each time.

*Ethidium bromide (10 mg mL$^{-1}$).* Add 1 g of ethidium bromide to 100 mL of distilled water. Stir on a magnetic stirrer for several hours to ensure that it has dissolved completely. Transfer to a dark bottle and store in the dark at 4 °C.

*20× SSC.* Dissolve 175.3 g of sodium chloride and 88.2 g of tri-sodium citrate into approximately 800 mL of water. Adjust the pH to 7.0 (actually not absolutely necessary), make volume up to 1 L.

*50× TAE.* Dissolve 242 g of Tris base into 700 mL of distilled water. Add 100 mL of 0.5 mol EDTA (pH 8.0) solution and 57.1 mL of glacial acetic acid. Make up to 1 L. Dilute to 1× when needed.

*5× TBE.* Dissolve 54 g of Tris base and 27.5 g of boric acid into 700 mL of distilled water. Add 20 mL of 0.5 mol EDTA (pH 8.0). Make up to 1 L. Dilute to 1× when needed.

## Alseviers solution

Dissolve 0.55 g of citric acid, 4.2 g of sodium chloride and 20.5 g of dextrose into 800 mL of distilled water. Make up to 1 L. Sterilize by autoclaving at 69 kPa for 10 min.

For other buffer recipes see, for example, Sambrook *et al.* 1989 or Ausebel *et al.* 1997.

## References

Amos, W. & Hoelzel, A.R. (1990) DNA fingerprinting of cetacean biopsy samples for individual identification. *Report of the International Whaling Commission, Special Issue* **12**, 79–85.

Ausebel, F.M., Brent, R., Kingston, R.E *et al.* (eds) (1997) *Current Protocols in Molecular Biology.* John Wiley and Sons Inc., Chichester.

Bishop, J.O. & Davies, J.A. (1980) Plasmid cloning vectors that can be nicked at a unique site. *Molecular and General Genetics* **179**, 573–580.

Bonner, W.M. (1987) Autoradiograms: 35-S and 32-P. *Methods in Enzymology* **152**, 55–61.

Borowsky, R. (1991) Field preservation of fish tissues for DNA fingerprint analysis. *Fingerprint News* **3**, 8–9.

Bowtell, D.D.L. (1987) Rapid isolation of eukaryotic DNA. *Analytical Biochemistry* **162**, 463–465.

Broad, T.E., Forrest, J.W. & Pugh, P.A. (1988) Effect of NaOH on hybridization efficiency of DNA. *Trends in Genetics* **4**, 146.

Burke, D.T. (1990) YAC cloning options and problems. *Genetic Analysis Techniques and Applications* **7**, 94–99.

Cann, R.L., Feldman, R.A., Freed, L.A., Lum, J.K. & Reeb, C.A. (1993) Collection and storage of vertebrate samples. *Methods in Enzymology* **224**, 39–51.

Carlson, J. & Messing, J. (1984) Efficiency in cloning and sequencing using the single-stranded bacteriophage M13. *Journal of Biotechnology* **1**, 253–264.

Chorazy, P.A. & Edlind, T.D. (1990) Artifactual bands associated with alkaline transfer. *Nucleic Acids Research* **18**, 3101.

Chung, C.T. & Miller, R.H. (1988) A rapid and convenient method for the preparation and storage of competent bacterial cells. *Nucleic Acids Research* **16**, 3580.

Church, G.M. & Gilbert, W. (1984) Genomic sequencing. *Proceedings of the National Academy of Sciences USA* **81**, 1991–1995.

Cohen, S.N., Chang, A.C.Y. & Hsu, L. (1972) Nonchromosomal antibiotic resistance in bacteria: Genetic transformation of *Escherichia coli* by R-factor DNA. *Proceedings of the National Academy of Science USA* **69**, 2110–2114.

Collins, J. & Hohn, B. (1978) Cosmids: a type of plasmid gene-cloning vector that is package-able *in vitro* in bacteriophage heads. *Proceedings of the National Academy of Sciences USA* **75**, 4242–4246.

Cooke, H. (1987) Cloning in yeast: an appropriate scale for mammalian genomes. *Trends in Genetics* **3**, 173–174.

Denhardt, D.T. (1966) A membrane-filter technique for the detection of complementary DNA. *Biochemical and Biophysics Research Communications* **23**, 641–646.

Dente, L., Sollazzo, M., Baldari, C., Cesareni, G. & Cortese, R. (1985) The pEMBL family of single-stranded vectors. In: *DNA Cloning* (ed. D.M. Glover), pp. 101–107. IRL Press, Oxford.

Doyle, J.J. & Doyle, J.L. (1990) Isolation of plant DNA from fresh tissue. *Focus* **12**, 13–14.

Feinberg, A.P. & Vogelstein, B. (1984) A technique for radiolabelling restriction endonucle-ase fragments to high specific activity. *Analytical Biochemistry* **137**, 266–267.

Frischauf, A.-M. (1987) Construction and characterization of a genomic library in l. *Methods in Enzymology* **152**, 190–199.

Frischauf, A.-M., Lehrach, H., Poustka, A. & Murray, N. (1983) Lambda replacement vectors carrying polylinker sequences. *Journal of Molecular Biology* **170**, 827–842.

Grimberg, J., Nawoschik, S., Belluscio, L., McKee, R., Turck, A. & Eisenberg, A. (1989) A simple and efficient non-organic procedure for the isolation of genomic DNA from blood. *Nucleic Acids Research* **17**, 8390.

Hagelberg, E. & Clegg, J.B. (1991) Isolation and characterization of DNA from archaeological bone. *Proceedings of the Royal Society, London B* **244**, 45–50.

Hanahan, D. (1985) Techniques for transformation of *E. coli*. In: *DNA Cloning*, Vol. 1 (ed. D.M. Glover), pp. 109–135. IRL Press, Oxford.

Harris, S.A. (1993) DNA analysis of tropical plant species: an assessment of different drying methods. *Plant Systematics and Evolution* **188**, 57–64.

Hu, N.-T. & Messing, J. (1982) The making of strand-specific M13 probes. *Gene* **17**, 271–277.

Huynh, T.V., Young, R.A. & Davis, R.W. (1985) Construction and screening cDNA libraries in λgt10 and λgt11. In: *DNA Cloning* (ed. D.M. Glover), pp. 49–78. IRL Press, Oxford.

Ito, K. (1992) Nearly complete loss of nucleic acids by commercially available highly purified ethanol. *Biotechniques* **12**, 69–70.

Jacobs, M., Wnendt, S. & Stahl, U. (1990) High-efficiency electro-transformation of *Escherichia coli* with DNA from ligation mixtures. *Nucleic Acids Research* **18**, 1653.

Jeffreys, A.J., Wilson, V. & Thein, S.L. (1985) Hypervariable 'minisatellite' regions in human DNA. *Nature* **314**, 67–73.

Johnson, D.A., Gautsch, J.W., Sportsman, J.R. & Elder, J.H. (1984) Improved technique utilizing nonfat dry milk for analysis of proteins and nucleic acids transferred to nitrocel-lulose. *Gene Analysis Techniques* **1**, 3–8.

Johnston, R.F., Picket, S.C. & Barker, D.L. (1990) Autoradiography using storage phosphor technology. *Gene Analysis Techniques* **11**, 355–360.

Kaiser, K. & Murray, N.E. (1985) The use of phage lambda replacement vectors in the construction of representative genomic libraries. In: *DNA Cloning* (ed. D.M. Glover), pp. 1–47. IRL Press, Oxford.

Laulier, M., Pradier, E., Bigot, Y. & Periquet, G. (1995) An easy method for preserving nucleic acids in field samples for later molecular and genetic studies without refrigerat-ing. *Journal of Evolutionary Biology* **8**, 657–663.

Lee, M.L. & Nicholson, P. (1997) Isolation of genomic DNA from plant tissue. *Nature Biotech-nology* **15**, 805–806.

Little, P.F.R. (1987) Choice and use of cosmid vectors. In: *DNA Cloning* (ed. D.M. Glover), pp. 19–42. IRL Press, Oxford.

Loenen, W.A.M. & Brammar, W.J. (1980) A bacteriophage lambda vector for cloning large DNA fragments made with several restriction enzymes. *Gene* **20**, 249–259.

Longmire, J.L., Albright, K.L., Lewis, A.K., Meinke, L.J. & Hildebrand, C.E. (1987) A rapid and simple method for the isolation of high molecular weight cellular and chromosome-specific DNA without the use of organic solvents. *Nucleic Acids Research* **15**, 859.

McClure, H.E. & Cedeno, R. (1955) Techniques for taking blood samples from living birds. *Journal of Wildlife Management* **19**, 477–478.

Mandel, M. & Higa, A. (1970) Calcium-dependent bacteriophage DNA infection. *Journal of Molecular Biology* **53**, 159–162.

Martin, R. & Burke, J.F. (1989) Multiple *E. coli* transformations on a single bacterial plate. *Nucleic Acids Research* **17**, 8386.

Maxam, A.M. & Gilbert, W. (1977) A new method for sequencing DNA. *Proceedings of the National Academy of Sciences USA* **74**, 560–564.

May, C.A. & Wetton, J.H. (1991) DNA fingerprinting by specific priming of concatenated oligonucleotides. *Nucleic Acids Research* **19**, 4557.

Messing, J. (1983) New M13 vectors for cloning. *Methods in Enzymology* **101**, 20–78.

Messing, J. & Vieira, J. (1982) A new pair of M13 vectors for selecting either strand of double-digest restriction fragments. *Gene* **19**, 269–276.

Miller, S.A., Dykes, D.D. & Polesky, H.F. (1988) A simple salting out procedure for extracting DNA from human nucleated cells. *Nucleic Acids Research* **16**, 1215.

Moreno, R.F., Booth, F.R., Hung, I.H. & Tilzer, L. (1989) Efficient DNA isolation within a single gel barrier tube. *Nucleic Acids Research* **17**, 8393.

Murray, M.G. & Thompson, W.F. (1980) Rapid isolation of high molecular weight plant DNA. *Nucleic Acids Research* **8**, 4321–4325.

Raeder, U. & Broda, P. (1985) Rapid preparation of DNA from filamentous fungi. *Letters in Applied Microbiology* **1**, 17–20.

Reed, K.C. & Mann, D.A. (1985) Rapid transfer of DNA from agarose gels to nylon membranes. *Nucleic Acids Research* **13**, 7207–7221.

Renz, M. & Kurz, C. (1984) A colorimetric method for DNA hybridization. *Nucleic Acids Research* **13**, 3435–3444.

Reymond, C.D. (1987) A rapid method for the preparation of multiple samples of eucaryotic DNA. *Nucleic Acids Research* **15**, 8118.

Rigaud, T., Grange, T. & Pictet, R. (1987) The use of NaOH as transfer solution of DNA onto nylon membranes decreases the hybridization efficiency. *Nucleic Acids Research* **15**, 857.

Rogers, H.J., Burns, N.A. & Parkes, H.C. (1996) Comparison of small scale methods for the rapid extraction of plant DNA suitable for PCR analysis. *Plant Molecular Biology Reporter* **14**, 170–183.

Ross, K.S., Haites, N.E. & Kelly, K.F. (1990) Repeated freezing and thawing of peripheral blood and DNA in suspension: effects on DNA yield and integrity. *Journal of Medical Genetics* **27**, 569–570.

Saito, I. & Stark, G.S. (1986) Charomids: Cosmid vectors for efficient cloning and mapping of large or small restriction fragments. *Proceedings of the National Academy of Sciences USA* **83**, 8664–8668.

Sambrook, J., Fritsch, E.F. & Maniatis, T. (1989) *Molecular Cloning: a Laboratory Manual*, 2nd edn. Cold Spring Harbor Laboratory Press, Cold Spring Harbor.

Sanger, F., Nicklen, S. & Coulson, A.R. (1977) DNA sequencing with chain terminating inhibitors. *Proceedings of the National Academy of Sciences USA* **74**, 5463–5467.

Scheenberger, R.G., Gorman, S.W. & Cullis, C.A. (1988) Effect of NaOH on hybridization efficiency of southern transferred DNA. *Trends in Genetics* **4**, 328.

Sealy, P.G. & Southern, E.M. (1990) Gel electrophoresis of DNA. In: *Gel Electrophoresis of Nucleic Acids: a Practical Approach* (eds D. Rickwood & B.D. Hames). pp. 51–98. IPL Press.

Seutin, G., White, B.N. & Boag, P.T. (1990) Preservation of avian blood and tissues for DNA analyses. *Canadian Journal of Zoology* **69**, 82–90.

Sooter, C.A. (1954) A technique for bleeding nestling birds by cardiac puncture for viral studies. *Journal of Wildlife Management* **18**, 409–410.

Southern, E.M. (1975) Detection of specific sequences among DNA fragments separated by gel electrophoresis. *Journal of Molecular Biology* **98**, 503–517.

Sternberg, N.L. (1990) Alternatives to YACs. *Genetic Analysis Techniques and Applications* **7**, 126–137.

Sternberg, N.L. (1991) Cloning high molecular weight DNA fragments by the bacteriophage P1 system. *Trends in Genetics* **8**, 11–16.

Swanstrom, R. & Shank, P.R. (1978) X-ray intensifying screens greatly enhance the detection by autoradiography of the radioactive isotopes $^{32}$P and $^{125}$I. *Analytical Biochemistry* **86**, 184.

Sytsma, K.J., Givnish, T.J., Smith, J.F. & Hahn, W.J. (1993) Collection and storage of land plant samples for macromolecular comparison. *Methods in Enzymology* **224**, 23–37.

Tegelstrom, H. (1986) Mitochondrial DNA in natural populations. An improved routine for the screening of genetic variation based on sensitive silver staining. *Electrophoresis* **7**, 226–230.

Thomas, S.M., Moreno, R.F. & Tilzer, L.L. (1989) DNA extraction with organic solvents in gel barrier tubes. *Nucleic Acids Research* **17**, 5411.

Thomas, W.K. & Paabo, S. (1993) DNA sequences from old tissue remains. *Methods in Enzymology* **224**, 406–419.

Thompson, J.D., Cuddy, K.K., Haines, D.S. & Gillespie, D. (1990) Extraction of cellular DNA from crude cell lysate with glass. *Nucleic Acids Research* **18**, 1074.

Van der Sande, J.H., Kleppe, K. & Khorana, H.G. (1973) Reversal of bacteriophage T4 induced polynucleotide kinase action. *Biochemistry* **12**, 5050–5055.

Wahl, G.M., Berger, S.L. & Kimmel, A.R. (1987) Molecular hybridization of immobilized nucleic acids: Theoretical concepts and practical considerations. *Methods in Enzymology* **152**, 399–407.

Wahl, G.M., Stern, M. & Stark, G.R. (1979) Efficient transfer of large DNA fragments from agarose gels to diazobenzyloxymethyl-paper and rapid hybridization by using dextran sulfate. *Proceedings of the National Academy of Sciences USA* **76**, 3683–3687.

Wallace, D.M. (1987a) Large- and small-scale phenol extractions. *Methods in Enzymology* **152**, 33–41.

Wallace, D.M. (1987b) Precipitation of nucleic acids. *Methods in Enzymology* **152**, 33–41.

Waye, J.S. & Fourney, R.M. (1990) Agarose gel electrophoresis of linear genomic DNA in the presence of ethidium bromide; band shifting and implications for forensic identity testing. *Applied and Theoretical Electrophoresis* **1**, 193–196.

Zagursky, R. & Baumeister, K. (1987) Construction and use of pBR322 plasmids that yield single stranded DNA for sequencing. *Methods in Enzymology* **155**, 139–155.

# CHAPTER 3

# Polymerase Chain Reaction

TIM P. BIRT AND ALLAN J. BAKER

## 3.1 Introduction

Research in the fields of population and evolutionary ecology (and many others) has been revolutionized by the introduction of the polymerase chain reaction (PCR). Using this technology, the researcher can amplify, *in vitro*, specific DNA fragments in virtually unlimited quantity (Saiki *et al.* 1985). Variation among individual organisms in the resulting PCR products can then be readily characterized using any of a number of techniques. Hence, the great power of PCR is that it presents to ecologists investigating a wide variety of phenomena, a window into the realm of biological variation at its most fundamental level, the level of DNA base sequence. Until the introduction of PCR, this window was much more restricted because of the inherent limitations of methods previously available for analysing DNA sequence variation within and among populations (e.g. classical cloning, sequencing and restriction analysis). During the past decade PCR has become firmly established among the basic molecular tools of biological research.

Conceptually, PCR is elegant in its simplicity. In practice, the researcher soon discovers that PCR is anything but simple because of the many interacting variables that influence the outcome. In this chapter we discuss some technical aspects of PCR to provide information useful to researchers new to the field. We also consider some applications relevant to the broad field of ecology.

## 3.2 Technical considerations

The PCR involves enzymatic synthesis of many copies of a particular DNA (or sometimes RNA) sequence. The DNA region to be amplified is determined by the base sequences of a pair of oligonucleotide primers which are complementary to binding sites situated on either side of the target sequence. These primer binding sites reside on opposite strands of the template DNA so that bound primers have their 3′ hydroxyl ends orientated towards each other. DNA polymerase-mediated extension of each annealed primer therefore proceeds in the direction of the other primer. Each cycle begins by heating the reaction mixture to a temperature high enough to cause the template DNA strands to dissociate (usually 94 or 95°C). The temperature is then lowered sufficiently to permit annealing of the primers to the appropriate sequences of the template strands (typically 45–60°C). Finally, the temperature is elevated to achieve optimal polymerase activity (70–72°C) and extend the annealed primers as directed by the template strands. As the temperature cycle is repeated, each new strand can act as template and the amount of product increases geometrically (Fig. 3.1).

### 3.2.1  Laboratory requirements

A thermal cycler is the central piece of equipment and the researcher has many makes and price tags from which to choose. High quality machines are capable of rapid and precise heating and cooling with little temperature variation across the heating block. Some machines have interchangeable blocks that allow some versatility with respect to the format of reaction vessels that can be accommodated. A very useful option available on most machines is a heated bonnet that prevents condensation on the inside of the tube lids. Without this feature a mineral oil overlay will usually be required for routine applications. More importantly, excessive evaporation/condensation will alter the conditions of the reaction mixture, particularly for reactions with small volumes.

Additional equipment basic to most other DNA laboratories includes: items for preparing and pipetting solutions (balance, pH meter, stirrers, pipetters, glassware, etc.); electrophoresis chambers (for agarose and polyacrylamide gels); power supplies; microcentrifuge; ultraviolet (UV) transilluminator; camera; a source of deionized water; and access to an autoclave.

### 3.2.2  DNA polymerases

The use of DNA polymerases isolated from thermophilic bacteria has simplified and improved PCR because these enzymes are not rapidly denatured by high temperature and, therefore, retain their activities even after repeated exposure to temperatures sufficient to achieve DNA denaturation. The most widely used enzyme today, *Taq* polymerase, was isolated from the thermophilic bacterium *Thermus aquaticus*. This enzyme is now available commercially in engineered and native forms. In addition to *Taq* DNA polymerase, enzymes derived from other thermophilic bacteria and having additional activities such as 3' to 5' exonuclease activity and reverse transcriptase activity are available. Long PCR involves an optimized mix of *Taq* and more processive enzymes like *Pfu*I which amplify templates up to about 20 kb.

### 3.2.3  Primer design

Primer sequence is one of the most critical determinants of success or failure of

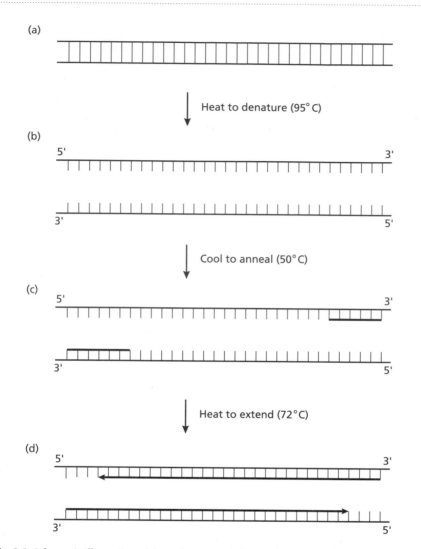

**Fig. 3.1** Schematic illustration of the polymerase chain reaction. (a) Double-stranded DNA containing regions complementary in base sequence to specific oligonucleotide primers is heated to a temperature sufficient to cause dissociation of the two strands. (b) Cooling of the resulting single-stranded DNA then permits annealing of the primers to their respective binding sites on alternate strands. (c) Temperature is then elevated to the activity optimum of *Taq* polymerase, which then extends the annealed primers. (d) Repeated temperature cycling results in a geometric increase in amplified product.

PCR. Generally, primers are 20–24 base pairs (bp) in length, although shorter or longer primers are sometimes used. The most obvious requirement for primers to work well is that their base sequences match closely those of the template binding sites. For some applications, workers can use primers designed for other species, some of which work in a variety of taxa, but more often primers will need to be designed to meet specific needs. This is a relatively simple matter if a reference sequence is available for the species of interest, but in many cases no such sequence will be available. In this situation sequences from other species (sometimes distantly related) will need to be aligned as references. By searching for highly conserved areas in the gene of interest (among several organisms if possible) one can usually design primers that work satisfactorily, as some degree of mismatch is tolerable so long as there is a good match at the 3' end where the polymerase binds. If the primers are directed to part of a protein-coding gene, it is common practice to have the last base at the 3' end pair to the second base position in a codon as these positions are generally less variable than first and third positions. The least desirable situation is for the last base at the 3' end to fall on a third position of a codon, particularly a four-fold degenerate codon because of the higher probability of sequence variation at this position.

Secondary structure in PCR primers can cause problems so researchers should take care to avoid sequence complementarity within primers. Similarly, base sequences should be selected so that primer pairs do not complement each other. In addition, primers containing repeats should be avoided. Innis & Gelfand (1990) recommend that primer pairs have a 50–60% G + C content and similar melting temperatures (temperature at which the primer and template dissociate, assuming perfect match). The latter can be estimated by counting 2°C for each A and T and 4°C for each C and G. Because melting temperature increases with G + C content, primers with high G + C content can be annealed at higher temperature with greater resulting specificity than primers with low G + C content. A second benefit of high melting temperatures in primers is that temperature cycling is more rapid, resulting in shorter reaction times.

Primers cannot be guaranteed to work even when the common rules-of-thumb for design are followed. Base sequences at template binding sites will vary unpredictably in some species or in some individuals within a species, resulting in an apparent null allele. The presence of null alleles can complicate a population survey by causing scoring errors to be made. An individual heterozygous for a null allele will appear homozygous for the other allele, or, in relatively rare instances, null homozygotes will not amplify at all. Null alleles are quite common in microsatellite loci (e.g. Paetkau & Strobeck 1997). Finally, in some instances primers fail to work for no obvious reason.

### 3.2.4  Sample preparation

The sensitivity of PCR permits amplification of specific DNA targets, even if they are present at very low concentration in a complex mixture of DNA sequences. For this reason samples need not be highly purified to achieve success, although those with greater concentrations of target generally give less trouble. Amplifiable templates have been recovered from a wide range of unlikely sources such as a 7000-year-old human brain preserved in a Florida peat bog (Pääbo *et al.* 1988).

Many protocols have been described for preparation of nucleic acids and the nature of the sample will dictate which is most appropriate. Because highly purified DNA is often not required, a common approach for PCR applications involving fresh tissues is to prepare relatively crude DNA samples using proteinase K digestion in the presence of detergent. The following procedure has been used successfully for different tissue types from a variety of organisms and is rapid enough to allow many samples to be processed per day. Soft tissue (approximately 25 mg, finely chopped) or whole blood (10–20 µL) is incubated for 6 h (or overnight) at 55°C in a volume of 750 µL of lysis buffer containing 100 mmol Tris (pH 8.0), 10 mmol ethylene diamine tetraacetic acid (EDTA), 100 mmol NaCl, 0.1% sodium dodecyl sulphate (SDS) and 5 µg proteinase K. The resulting solution is then extracted twice with Tris-saturated phenol, once with phenol/chloroform (1 : 1) and once with chloroform. Some protocols then recommend precipitation of the DNA with ethanol, but we usually find this step unnecessary. In many cases amplification will be more successful if the resulting DNA preparation is diluted 10- or 100-fold with water to dilute inhibitory substances such as haemoglobin and cytochromes. One microlitre of this solution usually provides adequate template for amplification of mitochondrial or nuclear loci. Kawasaki (1990) described a simplified method suitable for whole blood (and presumably other soft tissues) that does not involve phenol extraction. After digestion, samples are simply incubated at 95°C for 10 min to inactivate the proteinase K and are then ready to use. DNA suitable for PCR can also be prepared from tissues fixed in ethanol (Smith *et al.* 1987) or in formalin, although DNA from the latter is usually very difficult to amplify. Many tissue collections therefore contain samples that are potentially useful for long-term studies involving species that cannot be collected or otherwise disturbed at present.

Although specific target sequences can be amplified from complex DNA mixtures, fewer problems will be encountered if the concentration of target sequence can be increased relative to nontarget sequences. This is not difficult if the region of interest resides in the mitochondrial DNA genome. Tissue homogenates enriched for mitochondria can be prepared using differential centrifugation, or highly purified mitochondrial DNA (mtDNA) can be isolated on caesium chloride density gradients (Carr & Griffith 1987). Polymerase chain reaction from such samples will require fewer cycles and problems stemming from nonspecific priming to nuclear sequences will be reduced. These methods, however, require fresh tissue samples and are not appropriate with degraded samples.

In parallel with the rise of PCR since the mid-1980s is an increased interest in analysis of DNA recovered from various ancient sources (e.g. mummified human and other remains preserved by various means). DNA from these sources provides a historical perspective to a variety of investigations (Thomas *et al.* 1990; Cooper *et al.* 1992). However, material is invariably present in small quantities and is degraded to some extent, so extraction procedures must take this into account. The reader is referred to several chapters in Herrmann & Hummel (1994) for specific methods for recovering DNA from such sources.

### 3.2.5 Sample PCR protocol

The following protocol is typical of those used for routine double-stranded amplification from good quality templates. Amplification of more difficult templates may be more successful under different conditions (see Table 3.1).

Reaction volumes are commonly 10–25 µL, although smaller or larger volumes are sometimes used. Reaction constituents are combined into a cocktail and aliquoted among an appropriate number of tubes. A template is then added and finally a drop of mineral oil to prevent evaporation if the thermal cycler is not equipped with a heated bonnet. A typical temperature profile consists of a 30 s denaturation step at 94°C followed by a 30 s annealing step at 55°C, and a 30 s extension step at 72°C. In most instances 30 cycles will result in ample product, although many workers use 35 or 40 cycles. Product will appear as a bright band of appropriate size after electrophoresis through agarose (usually 2%) containing 1 µg mL$^{-1}$ ethidium bromide (Fig. 3.2).

### 3.2.6 PCR optimization

Because optimal reaction conditions vary for different applications of PCR, the researcher will usually need to identify these empirically. Varying reaction parameters such as $Mg^{2+}$ and deoxyribonucleoside triphosphate (dNTP) concentrations or annealing temperature can dramatically influence the yield of PCR product. Proper optimization of conditions can therefore make the difference between success and failure.

**Table 3.1** Conditions for amplification

| Reaction constituent | Concentration |
| --- | --- |
| Tris, pH 8.8 | 67 mmol |
| M$_g$Cl$_2$ | 2 mmol |
| Primers | 1 µmol (each) |
| dNTPs | 100 µmol (each) |
| *Taq* polymerase | 0.5–1.0 units |

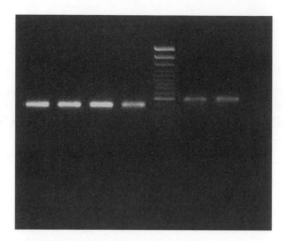

**Fig. 3.2** Double-stranded polymerase chain reaction (PCR) products visualized by ethidium bromide staining following electrophoresis through a 2% agarose gel. The samples in the four left lanes are a 361 bp fragment of the mitochondrial DNA control region (3′ end) amplified from common murres (*Uria aalge*). The two lanes on the right contain fragments of an intron (approximately 500 bp) from δ-crystalline amplified from pigeon guillemots (*Cepphus columba*). Lane 5 contains a 100 bp size standard ladder; the smallest band is 100 bp long.

Stock solutions of dNTPs are usually purchased at a concentration of 100 mmol (pH 7.0). These are aliquoted to convenient volume and stored at −20°C. Concentrations of dNTPs used in reaction mixtures vary; Innis & Gelfand (1990) recommend a range of 20–200 µmol each. Higher concentrations can be used but lower values provide greater specificity and lower base misincorporation. One locus that amplifies more efficiently with a relatively high dNTP concentration (400 µmol each) is part of a class II major histocompatibility complex (MHC) gene in marbled murrelets (*Brachyramphus marmoratus*) and other alcids (V.L. Friesen, unpublished data). This is four times the concentration we use for most other loci.

Concentrations of inorganic ions, especially $Mg^{2+}$, are also important to the outcome of PCR experiments. In addition to binding with template DNA, primers and dNTPs, $Mg^{2+}$ is required for DNA polymerase activity. Workers need to ensure that $Mg^{2+}$ is present in reaction mixtures at sufficient concentrations to ensure there are enough unbound ions to meet enzyme requirements. As the concentrations of dNTPs and template can vary, there is no single optimum $Mg^{2+}$ concentration, but we have found that a concentration of 2 mmol generally gives satisfactory results. Concentrations above or below this may be desirable in certain applications. The concentration of $Mg^{2+}$ is especially important for amplification of microsatellite loci; in particular, the number and intensity of stutter and background bands produced is strongly influenced by this factor (Fig. 3.3).

0.5mM        1.5mM        2.0mM        2.5mM        3.5mM

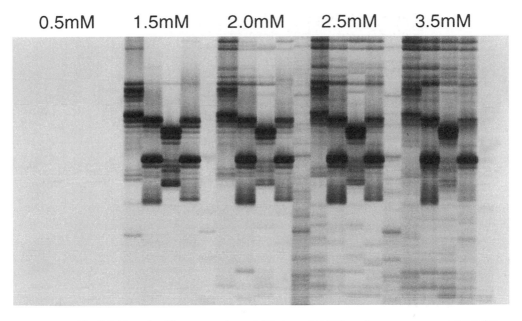

**Fig. 3.3** Dinucleotide-repeat microsatellite locus 14b29 from four common murres under differing concentrations of magnesium (shown across top of figure). Higher concentrations resulted in enhancement of background bands, while no amplified product is evident at 0.5 mmol. The optimal concentration appears to be 1.5 mmol.

Some suppliers of thermostable DNA polymerases recommend addition of KCl and/or a detergent to reaction mixtures. At a concentration of 50 mmol, KCl will enhance the stability of the primer–template complex, but higher concentrations will reduce enzyme activity. Detergent will improve enzyme stability but causes foaming in the reaction mixture which makes pipetting difficult.

A critical factor in any reaction is annealing temperature. The goal is to find a temperature high enough to prevent primers from annealing to nontarget sequences, yet low enough to ensure that primers are able to anneal at the appropriate sites. In any given situation this parameter will be determined by several variables, including the degree of primer–template match, base composition of the primer and the concentrations of other constituents of the reaction mixture. As indicated above, one can readily estimate melting temperature based upon base composition, but in practice this must be determined empirically for each situation, especially when the experiment involves a new species with unknown template-binding-site sequences. In general, an annealing temperature of 50–60°C is suitable. Occasionally we have annealed as low as 37°C, using highly purified mtDNA as template without experiencing nonspecific priming. The probability of nonspecific priming at such a low annealing temperature would be much higher if a complex DNA mixture had been used as a template source. We have also used annealing temperatures as high as 72°C, using long primers with perfect sequence matches (effectively a two-step temperature cycle).

Primer extension is usually carried out at 72°C. At this temperature nucleotides are incorporated at a rate of at least $35 \, s^{-1}$. Extension times therefore need not be longer than 30 s for most applications, although many workers increase extension times during the later cycles, and particularly after the final cycle, to ensure that templates are fully amplified.

Effective denaturation of duplex sequences is very important for efficient amplification. The melting temperature of DNA is influenced by salt concentration as well as base composition (high G + C content increases melting temperature). As with annealing temperature, a balance is sought that will ensure complete strand dissociation, without imposing undue thermal stress on the DNA polymerase and dNTPs. *Taq* polymerase has remarkable thermostability, but activity does diminish with prolonged exposure to high temperature. In most applications a denaturing temperature of 94°C for a duration of 30 s is sufficient.

Occasionally, loci are encountered that are difficult to amplify routinely using standard conditions. Success can sometimes be improved in these cases by adding an adjunct to the reaction mixture. Several materials have been suggested including formamide, ammonium sulphate, glycerol and dimethyl sulphoxide (DMSO). Some of these materials are thought to reduce secondary structures in the templates or improve DNA denaturation and are useful for G + C rich templates. Dimethyl sulphoxide has been found useful for amplifying a region of a class II MHC locus in alcids (V.L. Friesen, unpublished data). Adjuncts such as bovine serum albumin (BSA) and gelatine are often recommended as adjuncts to PCR mixtures. These proteins may improve enzyme stability.

For nuclear genes, such as the 3.1 kb exon in the recombination activating gene (*RAG-1*), for example, much better amplification is achieved using a hot-start touchdown protocol, in which the initial 10 min melting step is followed by five cycles annealing at 61°C, five cycles at 59°C, five cycles at 57°C, and the final 25 cycles at 55°C or 57°C. The initial steps improve the specificity of the PCR and thus give cleaner final products (see Groth & Barrowclough 1999 for a version of this protocol).

### 3.2.7 Contamination

The remarkable sensitivity of PCR allows the investigator to address questions that, until recently, could not be answered. This high degree of sensitivity demands that great care be exercised by laboratory workers to avoid contamination of samples and reactions. Contamination with unwanted DNA is most serious when amplifying from a few template molecules, as a small amount of contaminating template can easily take over a reaction. Contamination generally arises from inadvertent introduction of unwanted DNA (often from human sources) to reagent stocks or reaction tubes, or more often from carry-over of product from previous amplifications. Kwok & Higuchi (1989) and Kwok (1990) have outlined a number of precautions that, if followed, will prevent most problems associated with contamination.

Laboratory supplies, such as microcentrifuge tubes and pipette tips, must be clean and sterile. In addition, implements used for sample preparation (tissue grinders, blood sampling supplies, scalpel blades, etc.) must be free from DNA. Solutions used to prepare PCR reaction mixtures, including buffers, salts and deionized water, are autoclaved before use (note that dNTPs, primers and *Taq* polymerase are not autoclaved). Laboratory workers must wear disposable gloves which should be changed periodically, especially after moving between work areas. Dividing reagent stocks such as dNTPs and primers into aliquots will reduce the number of times any tube will be sampled and hence reduce the opportunity for DNA carry-over. Tubes containing solutions should be centrifuged briefly before opening, and caps should be removed carefully to control spills and aerosols. Care must be exercised when pipetting to prevent contamination of the pipetting device. Ideally, separate sets of pipettes should be used for sample preparation and formulation of reaction mixtures, and these should be cleaned regularly.

The use of appropriate controls is important to ensure that amplified products are actually the desired target sequences. Negative controls (containing all reaction components except template DNA) that give rise to reaction product indicate the presence of contaminating DNA. If the contaminant is present in some reagent in very low concentration, several negative control tubes may be necessary for detection. When dispensing the reaction mixture into reaction tubes, the negative controls should be pipetted last. Positive controls, containing template DNA known to amplify with the primers in question, are useful when evaluating the performance of the primers with new templates and of the system in general.

### 3.2.8 Alternative PCR methods

In some instances, PCR methods are required that differ somewhat from the typical approach outlined above; for example, methods have been devised to amplify complementary DNAs (cDNAs) in situations in which limited sequence information prevents design of specific primer pairs. The RACE (rapid amplification of cDNA ends) protocol, used to amplify cDNA fragments situated between a single internal region of known sequence and the unknown 3' and/or 5' ends, is useful when the target gene is expressed in low abundance (Frohman *et al.* 1988). This method relies on tailing the cDNA with a homopolymer bound to a specific 'anchor' sequence. The anchor sequence then acts as the second primer binding site for amplification and also contains restriction sites that facilitate subsequent cloning of the amplified product. This method was used by Kasahara *et al.* (1992) to amplify a portion of the 3' end of the *Mhc* class II *A* gene in the nurse shark.

Ligation-anchored PCR is a variation of the RACE method that is used to amplify cDNAs with unknown 5' sequences. T4 RNA ligase is used to link an anchor oligonucleotide to the 3' end of the first-strand cDNA. The anchor and a region of known sequence towards the 5' end of the cDNA then function as PCR priming sites (Troutt *et al.* 1992). Edwards *et al.* (1995) used this approach to explore the evolutionary dynamics of avian *Mhc* class II β chain genes (see Chapter 8).

Inverse PCR (Ochman *et al.* 1990) is a method of amplifying unknown DNA sequences flanking a region whose sequence is known (the core region). The DNA to be amplified is first cut with a restriction enzyme(s) with recognition sites outside the core region and the fragments are then circularized using T4 DNA ligase. PCR can then proceed using the circularized fragments as templates according to standard protocols, except that orientation of the primers must be reversed.

## 3.3 Detecting variation

Having amplified a DNA target, the researcher must then use some method to characterize sequence variation in the PCR products. The simplest approach is to visualize fragments using agarose gel electrophoresis followed by staining with an intercalating dye, usually ethidium bromide. This method is appropriate when the variation results from fragment size heterogeneity or if scoring is based on presence/absence of specific bands. Most often this approach is taken when anonymous nuclear loci such as randomly amplified polymorphic DNA (RAPD) are the genetic markers of interest. More commonly, this approach is not adequate and a more sensitive detection method is indicated.

In most applications involving mtDNA or chloroplast DNA, variation can be characterized by direct sequencing of amplified fragments, most often using the same primers employed for amplification as sequencing primers. The effectively haploid nature of these genomes usually (but not always) ensures that sequencing is not complicated by the presence of more than a single PCR product. One approach to sequencing is to perform a second amplification under conditions that generate single-stranded product (Gyllensten & Erlich 1988). Single-stranded DNA amplification is accomplished by reducing the concentration of one primer, typically to 1/100, that is used in the double-stranded amplification. Double-stranded product from the first amplification is used as a template in the second amplification. A simple method for preparing template for the second PCR is to excise a portion of the band of double-stranded product from the agarose gel following electrophoresis and then dissolve it in water (100 μL). A small amount of this diluted sample (1–2 μL) can then serve as template for the single-stranded amplification (Kocher *et al.* 1989). Reaction constituents remaining after amplification must then be removed and the single-stranded product concentrated before sequencing. Several commercial products are available for this procedure but we have found that single-stranded product can be purified and recovered inexpensively by simply precipitating in isopropanol (Maniatis *et al.* 1982). Sufficient product can be amplified in a 100 μL reaction for two to three sequencing reactions. The limiting primer in the second reaction then serves as the sequencing primer.

A favoured and cost-effective alternative to single-stranded sequencing of products from a second PCR is to amplify a larger amount of double-stranded product in the first round and then employ double-stranded sequencing. We have found that a 25 μL amplification of mtDNA target usually generates sufficient product for two to three sequencing reactions. Following amplification the entire reaction mixture is electrophoresed in low-melting agarose (usually 2%) and the PCR product band is excised from the gel and purified. Purification is easily accomplished by binding the amplified product to silica beads under high salt concentration, followed by a series of washes, and finally eluting the fragments under low salt conditions. An alternative method for releasing the fragments from the agarose is to digest the latter enzymatically. Commercial kits are available for both methods. A cheaper method is to place each gel slice in a separate filter tip inside a microtube tube and spin for 10 min at 8000 r.p.m. in a microcentrifuge.

An additional benefit of PCR has been the development of a new approach to chain termination sequencing (Sanger *et al.* 1977). Cycle sequencing is actually a modification of the PCR designed to generate high quality sequencing ladders (Brow 1990). The best method we have used for manual cycle sequencing employs labelled dideoxynucleoside triphosphates ([$\alpha$-$^{33}$P]ddNTPs). The advantage of this method is the elimination of spurious bands on autoradiograms, caused by premature termination of chains, because only correctly terminated chains are labelled. Other sequencing methods that employ radioactive labels incorporate the latter throughout the chains, hence the signal produced by improperly terminated chains is visible on the resulting autoradiograms. Automated sequencing technology is becoming more accessible to molecular ecologists, and on our LICOR 4200 (LI-COR, Inc., Nebraska Canada) bidirectional sequencer we have eliminated artefacts associated with M13-tailed primers and have obtained very high quality readings of up to

1485 bp by increasing the extension temperature in the PCR-sequencing protocol to 62–65°C and shortening the duration of this step to 15–30 s.

The great strength of PCR is the ability to characterize variation at the DNA
sequence level, but this does not mean that the researcher must sequence
products of every reaction; several shortcuts are available to circumvent this
effort and expense. Many workers use restriction endonucleases to assay variation in PCR products. This method is less expensive and does not require
radioactive labels but is less sensitive than direct sequencing, hence some
variation will not be detected.

Very sensitive electrophoretic methods have been adopted by a number
of population ecologists to reduce the amount of time, effort and expense
devoted to sequencing PCR products. One such method employs modified
polyacrylamide gels to separate single-stranded DNA fragments based on conformational polymorphism (SSCP analysis, Lessa & Applebaum 1993). Products of PCR are labelled either by direct incorporation of labelled dNTPs or by
using end-labelled primers. Labelled PCR products are combined with formamide and NaOH and heated to ensure that strands are denatured. Samples
are then cooled rapidly on ice and loaded onto gels. This electrophoretic
method is capable of detecting almost all variation in PCR products because
differences in nucleotide sequence result in different secondary structures
adopted by the DNA strands. Analysis using SSCP is suited for both haploid or
diploid systems, and thus genetic surveys can include loci from both nuclear
and cytoplasmic genomes (see Chapter 11 for examples).

Additional electrophoretic methods currently available are useful for
reducing the amount of DNA sequencing required in population genetic
analyses. These include heteroduplex analysis, denaturing gradient gel electrophoresis (DGGE) and temperature gradient gel electrophoresis (TGGE)
(see Lessa & Applebaum 1993 for an informative review). Using these
methods (including SSCP analysis) the researcher can score individuals
according to genotype for any number of loci. Following the survey, representatives of each allelic variant can be selected for sequencing if desired. This
dramatically reduces the amount of effort that must be devoted to DNA
sequencing and enables population surveys to be conducted rapidly (e.g.
Friesen *et al.* 1997).

## 3.4  Future prospects

There is no end in sight for the use of PCR methods in ecological research.
These methods provide readily available genetic markers suitable for a variety
of applications. Virtually any field of ecological research based on molecular
genetic information is fertile ground for application of PCR. Large-scale studies
are increasingly utilizing newer electrophoretic methods, such as SSCP and

DGGE analysis, for resolving allelic variation in large population samples because of their reduced cost relative to DNA sequencing. The latter methods also readily permit inclusion of many diploid loci (in which variation is not primarily a result of length heterogeneity) in population surveys of genetic variation. We expect these studies to become more frequent in future, in part because the ever greater availability of sequence information simplifies primer development. However, we also foresee much increased use of automated sequencers and eventually DNA chips because optimized protocols involving the PCR are reducing costs below manual methods, which were previously cheaper. The great advantage provided by sequence data is that they provide a treasure trove of information about past and present forces that have shaped the evolution of organisms, which can be deciphered by increasingly sophisticated mathematical approaches such as coalescent analysis. Whatever methods for detecting variation in PCR products gain favour in the future, there is no doubt that the PCR procedure itself will continue to hold a position of central importance among molecular methods in ecology.

## References

Brow, M.A.D. (1990) Sequencing with *Taq* DNA polymerase. In: *PCR Protocols, a Guide to Methods and Applications* (eds M.A. Innis, D.H. Gelfand, J.J. Sninsky & T.J. White), pp. 189–196. Academic Press, San Diego.

Carr, S.M. & Griffith, O.M. (1987) Rapid isolation of animal mitochondrial DNA in a small fixed-angle rotor at ultrahigh speed. *Biochemical Genetics* **25**, 385–390.

Cooper, A., Mourer-Chauviré, C., Chambers, G.K., von Haeseler, A., Wilson, A.C. & Pääbo, S. (1992) Independent origins of New Zealand moas and kiwis. *Proceedings of the National Academy of Sciences USA* **89**, 8741–8744.

Edwards, S.V., Wakeland, E.K. & Potts, W.K. (1995) Contrasting histories of avian and mammalian *Mhc* genes revealed by class II B sequences from songbirds. *Proceedings of the National Academy of Sciences USA* **92**, 12200–12204.

Friesen, V.L., Congdon, B.C., Walsh, H.E. & Birt, T.P. (1997) Intron variation in marbled murrelets detected using analysis of single stranded conformational polymorphism. *Molecular Ecology* **6**, 1047–1058.

Frohman, M.A., Dush, M.K. & Martin, G.R. (1988) Rapid production of full-length cDNAs from rare transcripts. amplification using a single gene-specific oligonucleotide primer. *Proceedings of the National Academy of Sciences USA* **85**, 8998–9002.

Groth, J.G. & Barrowclough, G.F. (1999) Basal divergences in birds and the phylogenetic utility of the nuclear RAG-1 gene. *Molecular Phylogenetics and Evolution* **12**, 115–123.

Gyllensten, U.B. & Erlich, H.A. (1988) Generation of single-stranded DNA by the polymerase chain reaction and its application to direct sequencing of the *HLA-DQA* locus. *Proceedings of the National Academy of Sciences USA* **85**, 7652–7656.

Herrmann, B. & Hummel, S. (1994) *Ancient DNA*. Springer-Verlag, New York.

Innis, M.A. & Gelfand, D.H. (1990) Optimization of PCRs. In: *PCR Protocols, a Guide to Methods and Applications* (eds M.A. Innis, D.H. Gelfand, J.J. Sninsky & T.J. White), pp. 3–12. Academic Press, San Diego.

Kasahara, M., Vazquez, M., Sato, K., McKinney, E.C. & Flajnik, M.F. (1992) Evolution of the major histocompatibility complex: Isolation of class II *A* cDNA clones from the cartilaginous fish. *Proceedings of the National Academy of Sciences USA* **89**, 6688–6692.

Kawasaki, E.S. (1990) Sample preparation from blood, cells, and other fluids. In: *PCR Protocols, a Guide to Methods and Applications* (eds M.A. Innis, D.H. Gelfand, J.J. Sninsky & T.J. White), pp. 146–152. Academic Press, San Diego.

Kocher, T.D., Thomas, W.K., Meyer, A., *et al.* (1989) Dynamics of mitochondrial DNA evolution in animals: Amplification and sequencing with conserved primers. *Proceedings of the National Academy of Sciences USA* **86**, 6196–6200.

Kwok, S. (1990) Procedures to minimize PCR-product carry-over. In: *PCR Protocols, a Guide to Methods and Applications* (eds M.A. Innis, D.H. Gelfand, J.J. Sninsky & T.J. White), pp. 142–145. Academic Press, San Diego.

Kwok, S. & Higuchi, R. (1989) Avoiding false positives with PCR. *Nature* **339**, 237–238.

Lessa, E.P. & Applebaum, G. (1993) Screening techniques for detecting allelic variation in DNA sequences. *Molecular Ecology* **2**, 121–129.

Maniatis, T., Fritch, E.F. & Sambrook, J. (1982) *Molecular Cloning, a Laboratory Manual.* Cold Spring Harbor Laboratory, New York.

Ochman, H., Medhora, M.M., Garza, D. & Hartl, D.L. (1990) Amplification of flanking sequences by inverse PCR. In: *PCR Protocols, a Guide to Methods and Applications* (eds M.A. Innis, D.H. Gelfand, J.J. Sninsky & T.J. White), pp. 219–227. Academic Press, San Diego.

Pääbo, S., Gifford, J.A. & Wilson, A.C. (1988) Mitochondrial DNA sequences from a 7000-year old brain. *Nucleic Acids Research* **16**, 9775–9787.

Paetkau, D. & Strobeck, C. (1997) The molecular basis and evolutionary history of a microsatellite null allele in bears. *Molecular Ecology* **4**, 519–520.

Saiki, R.K., Scharf, S., Faloona, F., *et al.* (1985) Enzymatic amplification of β-globin genomic sequences and restriction site analysis for diagnosis of sickle cell anemia. *Science* **230**, 1350–1354.

Sanger, F., Nicklen, S. & Coulson, A.R. (1977) DNA sequencing with chain-terminating inhibitors. *Proceedings of the National Academy of Sciences USA* **74**, 5463–5467.

Smith, L.J., Braylan, R.C., Nutkis, J.E., Edmundson, K.B., Downing, J.R. & Wakeland, E.K. (1987) Extraction of cellular DNA from human cells and tissues fixed in ethanol. *Analytical Biochemistry* **160**, 135–138.

Thomas, W.K., Pääbo, S., Villablanca, F.X. & Wilson, A.C. (1990) Spatial and temporal continuity of kangaroo rat populations shown by sequencing mitochondrial DNA from museum specimens. *Journal of Molecular Evolution* **31**, 101–112.

Troutt, A.B., McHeyzer-Williams, M.G., Pulendran, B. & Nossal, G.J.V. (1992) Ligation-anchored PCR: a simple amplification technique with single-sided specificity. *Proceedings of the National Academy of Sciences USA* **89**, 9823–9825.

CHAPTER 4

# Protein Electrophoresis

ALLAN J. BAKER

## 4.1 Introduction

The separation of enzymatic and nonenzymatic proteins by their charge and molecular weight as they migrate through an electrical field is referred to as protein electrophoresis. This technique dates from the 1950s, when starch gel electrophoresis (Smithies 1955) was coupled with the direct visualization of separated proteins on these gels via specific histochemical stains (Hunter & Markert 1957). The utility of protein electrophoresis in detecting genetical variation in natural populations was first demonstrated when Hubby & Lewontin (1966) and Lewontin & Hubby (1966) assayed variation in *Drosophila pseudoobscura*, and Harris (1966) performed the same assays for humans. These papers galvanized biologists into adopting molecular genetic approaches to their studies because it was now possible routinely to study genetic variation directly.

## 4.2 Basic principles of protein detection

Proteins that can be separated in a gel medium on the basis of their net charge, size and shape can be classified into a general category of isozymes and a more specific set called allozymes. Isozymes are different molecular forms of an enzyme that are encoded by different loci, and allozymes are molecular forms of an enzyme that are encoded by different alleles at a specific gene locus. A distinction can also be made between enzymatic proteins and general proteins that are nonenzymatic.

   Proteins are synthesized by the transcription and translation of the genetic code in DNA. Individual codons encode amino acids which are linked together by covalent peptide bonds to form polypeptides. The 20 amino acids each have unique side chains differing in size, shape and sometimes their charge. Some are basic and positively charged, others are acidic and negatively charged, and the remainder are uncharged. The net charge on a protein determines its movement through a gel matrix during electrophoresis. However, the net charge also varies with pH because the carboxyl groups become negatively charged at low pH, and the amino groups become positively charged at high pH. The isoelectric point of a protein is the point at which the positive and negative charges are balanced, and thus it does not migrate in an electric field.

With an appropriate selection of pore size in the gel matrix, the migration of proteins can also be affected by their size and shape. This is because they can have secondary structure as a result of folding or by the formation of α-helices that result from hydrogen bonding between the amino acids. Additional folding can also alter the shape of the protein and give its tertiary structure. Electrophoretic separation of many proteins can be further influenced by their quaternary structure formed by cohesive forces holding more than one polypeptide chain together, such as hydrophobic interactions, hydrogen bonds, disulphide bridges, ionic bonds or van der Waal forces (Murphy *et al.* 1996). Under suitable conditions the rate of movement increases with the net charge and strength of the electrical field, and declines with the size of the molecule. Positively charged proteins (cations) migrate towards the cathode, and negatively charged proteins (anions) move towards the anode.

In scoring bands on gels that have been visualized by staining reactions, it is important to know the quaternary structure of each enzyme because it determines the so-called zymogram or pattern of bands expected for homozygotes and heterozygotes, thus making it possible to score genotypes directly from stained gels. The simplest banding pattern occurs when an enzyme is monomeric, being composed of one polypeptide chain. In a diploid organism, homozygotes have the same allele at a locus, and thus only one band appears on the gel. Heterozygotes have two bands, one encoded by each allele. Most enzymes are multimers, however, and yield multibanded heterozygotes which are formed by random association of gene products in predictable proportions (Fig. 4.1). Heterozygotes for dimeric enzymes have three bands with staining intensities of $1:2:1$, corresponding to random association of homomers A and B to give 1 AA: 2 AB: 1 BB. Similarly, trimers composed of three polypeptides yield heterozygotes with expected band staining intensities of 1 (AAA): 3 (AAB, ABA, BAA): 3 (BBA, BAB, ABB): 1 BBB, and so on for more complex heteromers.

Complex banding patterns can arise with multilocus enzyme systems because heteromers may be formed between gene products from different loci. If different alleles are segregating at even one of these loci, then a large number of isozymes can be formed; for example, the number of bands expected for the two *Ldh* loci expressed in the liver of vertebrates is 15 if one locus is polymorphic for two alleles (see Fig. 4.1). Not all of these will be seen on gels unless the loci are equally expressed in the tissue under study, and it may be necessary to examine several different tissue types to understand fully the genetic control of isozymes. Nonenzymatic proteins, of which the structure is unknown, require careful interpretation of banding patterns. Inheritance studies are usually required before definitive genetic interpretations of zymograms can be made.

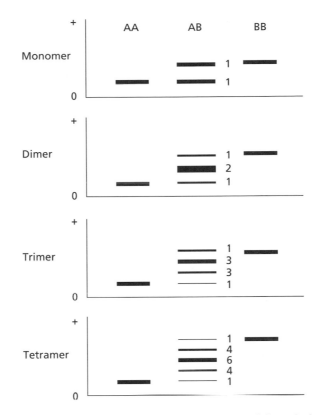

**Fig. 4.1** Diagram of isozyme banding patterns expected at a single locus for homozygotes and heterozygotes for proteins according to their subunit composition. Staining intensities expected under random association of subunits are indicated numerically alongside each band. Redrawn from Murphy *et al.* (1996).

To ensure homology of enzymes across different studies, a standard nomenclature for enzymes has been adopted, based on a three letter code (e.g. *Adh* for alcohol dehydrogenase) that is often followed by an EC or Enzyme Commission number (1.1.1.1 for *Adh*) (International Union of Biochemistry 1984). A standard convention is to designate the most anodal locus in a multilocus enzyme system as '1', and the successively more cathodal ones with increasing integers (e.g. *Mdh*-1 and *Mdh*-2).

## 4.3 Staining reactions

Specific histochemical staining is used to visualize the location of particular proteins after they have been separated on a gel matrix. The basic principle is to utilize the ability of proteins such as enzymes to act on a substrate along with a dye that precipitates at the site of the enzyme-catalysed reaction (Harris & Hopkinson 1976; Barrowclough & Corbin 1978; Cole & Parkin 1979).

Proteins can also be visualized with fluorescence of methylumbelliferone under 340 nm ultraviolet (UV) light, or by zones of nonfluorescence of nicotinamide adenine dinucleotide (NAD) in a background of fluorescent nicotinamide adenine dinucleotide hydroxide (NADH). Specific chemical formulae for staining reactions are given in Harris & Hopkinson (1976) and Richardson *et al.* (1986). A good compilation of staining recipes is available in Murphy *et al.* (1996).

## 4.4 Methods of electrophoresis

### 4.4.1 Horizontal starch gel electrophoresis

Because it is the least expensive and has the highest throughput, starch gel electrophoresis is by far the most favoured method. Gels are made by heating a mixture of hydrolysed starch and a solution containing an ionic buffer in an Ehrlenmeyer flask. Once the heated starch solution becomes noticeably viscous and just begins to boil on the bottom of the flask, vacuum pressure from a simple water tap aspirator is applied to remove any gas from the gel. After all gas bubbles are removed, the gel solution is poured into a mould of appropriate size and allowed to cool and thus solidify.

Tissue homogenates for each organism are loaded individually onto paper wicks (about 6×3 mm), which are inserted into wells made in the gel with a metal comb, or alternatively in an origin made by slicing the gel across its width. The loaded gel is covered with Saranwrap to prevent dehydration, and it is then placed on the bridge of an electrophoresis apparatus between buffer chambers. An ice pack is placed on top of the gel to attenuate heating during the run that could otherwise denature the heat-sensitive proteins. An electrical current is then applied to the gel, mediated by appropriately chosen buffer solutions in the cathodal and anodal chambers. Running conditions for commonly assayed proteins in my laboratory are listed in Table 4.1.

When the run is completed, the gel is removed from the apparatus and sliced horizontally with a commercially available slicer to produce three to five slices. Each slice is then stained separately for one to three loci. By running six gels per day, one investigator can comfortably assay genetic variation at 30–35 loci in a population sample of 30 organisms. At the end of a survey of different populations or species, alleles at each locus have to be calibrated by running all variants alongside one another on the same gel.

A common practice in most laboratories is to screen samples for genetic variation using single-pass electrophoresis, in which one set of conditions (buffers, pH, temperature, time) are employed. Some studies have experimented with multipass electrophoresis by repeatedly re-running samples under different electrophoretic conditions, sometimes in combination with

**Table 4.1** Conditions employed for starch gel and cellulose electrophoresis

| Buffer | Starch V cm$^{-1}$ | Duration | Cellulose acetate V cm$^{-1}$ | Duration |
|---|---|---|---|---|
| Amine-citrate (morpholine) | 7.8 | Overnight 16 h | 13.3 | 30 min |
| Amine-citrate (propanol) | 7.8 | Overnight 16 h | 13.3 | 30 min |
| Lithium-borate | 16.7 | Day 5 h | | |
| Sodium-phosphate | | | 30.0 | 20 min |
| Tris-borate-EDTA-MgCl$_2$ | 16.7 | Overnight 16 h | | |
| Tris-borate-EDTA-lithium | 13.9 | Overnight 16 h | | |
| Tris-citrate II | 6.1 | Overnight 16 h | | |
| Tris-glycine | | | 30.0 | 20 min |
| Tris-maleate | | | 30.0 | 20 min |
| Tris-phosphate | 6.9 | Overnight 16 h | | |

EDTA, ethylenediaminetetraacetic acid.

heat denaturation treatment. The rationale for doing this is that a potentially large number of protein polymorphisms remain undetected by standard conditions (e.g. Singh *et al.* 1976; Bonhomme & Selander 1978; Keith 1983). The generalization that emerges from more stringent assays of protein variation is that more alleles are discovered at loci known to be polymorphic previously, rather than the detection of new variation at monomorphic loci (Ramshaw *et al.* 1979).

### 4.4.2 Cellulose acetate electrophoresis

In this method (see Hebert & Beaton 1989) electrophoresis is performed on commercially available cellulose acetate plates. Before electrophoresis, the plates are soaked in buffer solution, and samples are applied by stamping them on with a multisample applicator. Because cellulose acetate has a large pore size, it does not have any noticeable sieving action on the proteins migrating through it, and they are therefore separated on charge alone. Advantages of this technique are:

1  run times are very short, ranging from about 15–45 min;
2  very small samples can be used because it is very sensitive; and
3  it has higher resolution than starch (Easteal & Boussey 1987) for many (but not all) isozymes.

Disadvantages are its higher expense and lower throughput relative to starch gel electrophoresis.

### 4.4.3 Polyacrylamide gel electrophoresis

Electrophoresis can also be carried out in polyacrylamide gels that separate proteins on both size and charge (Chrambach & Rodbard 1971). Gels are prepared by the polymerization of monomers of acrylamide and bisacrylamide catalysed by ammonium persulphate and tetramethyl ethylenediamine (TEMED). By varying the concentration of acrylamide and bisacrylamide, the pore size in the gel can be varied, and thus its sieving action on isozymes can be controlled. The major advantage of this method is its high resolving power, but this has to be weighed against the expense and toxicity of the gels, and the need to run multiple gels to match the number of genotypes and loci that can be scored from one starch gel.

### 4.4.4 Agarose gel electrophoresis

Agarose gel electrophoresis is now routinely used to separate low molecular weight DNA, and it can also be used to separate proteins (Harris & Hopkinson 1976). Pore size is controlled by the agarose concentration in the gel, with higher amounts decreasing pore size and increasing the sieving effect. Agarose of high purity is required to prevent the occasional adsorption of proteins in the gel, and this dramatically increases the expense of the method. Additionally, the migration of cationic isozymes can be accelerated and anionic ones retarded or reversed by the phenomenon of electroendosmosis, whereby charged groups in the gel cause a reversed buffer flow (Murphy *et al.* 1996).

## 4.5  Examples of data

Some examples of typical gels are provided in Fig. 4.2 to illustrate a range of results. Population surveys of a single species give gels that are simple to interpret and score, whereas comparisons of different species usually generate gels with many alleles that have to be calibrated very carefully against one another on the same gel. It is very important to check for sex-linked inheritance at each locus; for example, in birds females are the heterogametic sex (ZW) whereas males are homogametic (ZZ). Thus only males will have heterozygotes at sex-linked loci such as aconitase (*Acon-1*) (Baverstock *et al.* 1982). Failure to recognize sex linkage can bias estimates of allele frequencies and average heterozygosity, leading to erroneous calculations of population genetic parameters.

Once banding patterns are interpreted correctly for each locus and the genoytypes are scored, genetic analysis is usually carried out using computer packages such as BIOSYS-1 (Swofford & Selander 1981), GENEPOP (Raymond & Rousset 1995) and DISPAN (Ota 1993), that collectively provide a sophisticated suite of analytical tools.

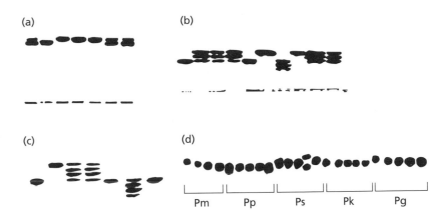

**Fig. 4.2** Scanned images showing examples of gel banding patterns obtained with starch gel electrophoresis. (a) Eastern meadowlark (*Sturnella magna*) showing *Est2* monomer. (b) Red-bellied dace (*Phoxinus eos*) showing *Gpi* dimer in the bottom locus. (c) House finch (*Carpodacus mexicanus*) showing *Sdh* tetramer. (d) Species of white pine (Pm, *Pinus monticola*; Pp, *P. peuce*; Ps, *P. strobus*; Pk, *P. koraiensis*; Pg, *P. griffithii*) showing variation in *Sdh* dimer (from Chagala 1991). (a) and (b) are examples from cellulose acetate gels, and (c) and (d) are from starch gels.

## 4.6 Analysis and interpretation

### 4.6.1 Assumptions in interpreting zymograms

There are three major assumptions in interpreting banding patterns on gels. First, the changes in mobility of enzymes or general proteins on gels are the result of corresponding changes in the DNA sequence that codes for them, and are thus controlled genetically. Second, it is assumed that enzyme expression is codominant, meaning that products of both alleles at a locus in a diploid organism are expressed. Third, allele frequencies are assumed to be in Hardy–Weinberg equilibrium. If the observed numbers deviate significantly from Hardy–Weinberg expectations, then either the banding patterns are not genetically controlled or the assumptions of the principle are not met. Nonrandom mating, selection for particular genotypes, and strong gene flow from adjacent populations are the probable causes of such deviations; for example, when working with asexual organisms, genotype frequencies are often not in Hardy–Weinberg equilibrium (Hebert & Beaton 1989) because of nonrandom mating of genetically different clones.

### 4.6.2 Genetic variability within populations

Results of multilocus assays of genetic variability within populations can be expressed with a number of measures such as average heterozygosity, the number of alleles per locus, and the percentage of loci that are polymorphic. A locus is considered to be polymorphic when the frequency of the most common allele is less than or equal to either 0.95 or 0.99. If large samples are being analysed, the latter criterion is preferred because it allows for the occurrence of very rare alleles in populations. Average heterozygosity can be estimated by a direct count of all the heterozygotes at all loci divided by the total number of genotypes scored. This estimate is also subject to sampling error because much of its variance is generated by the interlocus variation in heterozygosity. A theoretically preferable estimate using Hardy–Weinberg proportions is given by:

$$H = 1 - \sum x_i^2$$

where $x_i$ is the frequency of the $i$th allele at a locus (Nei 1975), and an unbiased estimate is provided by Nei & Roychoudhury (1974).

### 4.6.3 Among-population differentiation

A usual first step in determining whether populations are differentiated genetically is to carry out heterogeneity tests on allele frequencies at each locus in turn. Common tests are the heterogeneity chi-square test as featured in BIOSYS-1 or the $G$-test (Sokal & Rohlf 1981).

Multilocus measures of genetic differentiation include the most frequently used genetic distances of Nei (1978) and Rogers (1972). There are advantages to each of these indices, and many studies use them in concert. The advantage to Nei's genetic distance is that it is conceptually appealing because it is based on the probability of identity of alleles in different populations. The normalized probability that these alleles are identical is referred to as Nei's genetic identity ($I$), where:

$$I = J_{xy} / (J_X J_Y)^{1/2}.$$

The probability that alleles will be identical ($J_{xy}$) in populations $X$ and $Y$ is simply $\Sigma x_i y_i$, where $x_i$ and $y_i$ are allele frequencies at the $i$th locus in the two populations. $J_X$ and $J_Y$ are the probabilities that alleles drawn from the two populations are identical ($\Sigma x_i^2$ and $\Sigma y_i^2$, respectively). Nei's standard genetic distance is defined as:

$$D = -\log_e I.$$

As pointed out by Nei (1987), this formulation can be rewritten as:

$$D = D_{XY} - (D_X + D_Y)/2,$$

where $D_{XY}=-\log_e J_{XY}$, $D_X=-\log_e J_X$, and $D_Y=-\log_e J_Y$. Thus the among-population distance is corrected for the average within-population distance. When sample sizes are small, the estimate is biased. An unbiased estimate can be computed by replacing $\Sigma x_i^2$ and $\Sigma y_i^2$ with $(2n\Sigma x_i^2-1)/(2n_X-1)$ and $(2nY\Sigma y_i^2-1)/(2n_Y-1)$, respectively (Nei 1978).

The disadvantage of Nei's genetic distance is that it does not effectively resolve branch tips in a tree of populations or very closely related taxa, and it is nonmetric and cannot be used in metric techniques like principal coordinates analysis. For the above reasons, Rogers' (1972) genetic distance ($D_R$) is sometimes preferred, and is given by:

$$D_R = \left[1/2\sum(x_i - y_i)^2\right]^{1/2}$$

where $x_i$ and $y_i$ are allele frequencies as above. The disadvantage of this metric is that it is not proportional to evolutionary time or to the number of gene substitutions. For comprehensive treatments of other distance measures and their properties, the reader should consult Nei (1987) and Swofford et al. (1996).

### 4.6.4 Genetic population structure

The genetic population structure of a species is a measure of the genetic cohesiveness of conspecific populations (Rockwell & Barrowclough 1987); it reflects how effective gene flow is in 'homogenizing' genotypic arrays within a species. Commonly, species are made up of local subpopulations connected by only low levels of gene flow (migration). Within each of these subpopulations, mating is usually approximately random. This approximation will only hold for a species if there is extensive gene flow and the whole population is essentially panmictic.

Population structure is usually measured using Wright's F-statistics, $F_{IS}$, $F_{IT}$ and $F_{ST}$ (Wright 1965, 1978). The three statistics are related as follows:

$$1 - F_{IT} = (1 - F_{IS})(1 - F_{ST}).$$

$F_{ST}$ is an estimate of the subdivision of the population into subpopulations or demes, and is computed as the among-population component of genetic variance standardized by the limiting variance for complete fixation (Wright 1978). In BIOSYS-1, the estimate of $F_{ST}$ is weighted by allele frequencies, but not by subpopulation size, on the grounds that the latter is usually unknown. $F_{IS}$ is the average over all subpopulations of the correlations between alleles in uniting gametes (individuals) sampled at random within subpopulations, and indicates the degree of substructuring within subpopulations resulting from nonrandom mating. $F_{IT}$ is the correlation between alleles of uniting gametes sampled at random relative to the distribution of alleles in the entire population.

Weir & Cockerham (1984) and Cockerham & Weir (1993) have argued that their theta ($\theta$) estimator is a more accurate descriptor of $F_{ST}$ because it includes corrections for sample size and the number of populations that were examined. Calculations of $\theta$ can be obtained using the FSTAT program (Goudet 1995).

Theoreticians have investigated different population structures based on three demographic approaches with the island, stepping-stone and isolation-by-distance models. In the island model, it is assumed that subpopulations are of equal size and that gene flow is symmetrical and can occur between them all. For many real world populations this model is unrealistic because gene flow will commonly be restricted to adjacent demes or subpopulations in a restricted part of the species total range. For these situations the stepping stone models or isolation-by-distance are more appropriate.

Allele frequency data from protein electrophoresis are extremely useful in obtaining an estimate of the amount of gene flow among subpopulations; for example, it is common practice to calculate gene flow for neutral genes by rearranging Wright's formula:

$$F_{ST} = 1/(1 + 4Nm),$$

to give $Nm$, the average number of migrants exchanged between demes per generation (Slatkin 1987). $N$ is the effective population size and $m$ is the migration rate. However, this formula is the equilibrium solution for the island model of population structure, and is strictly valid only in this case. Fortunately, theoretical studies suggest that this estimate is also a reasonable approximation for stepping-stone models (Crow & Aoki 1984; Slatkin 1987).

Another way to estimate gene flow from electrophoretic data is using the method based on alleles that occur in only one population, referred to as private alleles (Slatkin 1981,1985). The rationale for this approach is that private alleles will only exist in populations if there is little gene flow between them. Using computer simulations, Slatkin established that there is a log-linear relationship between the average frequency ($p_1$) of private alleles in populations and gene flow ($Nm$) between them as:

$$\ln(p_I^-) = -0.505\ln(Nm) - 2.44,$$

which can be rearranged to give:

$$Nm = \exp[\ln(p_I^-) + 2.44/(-0.505)].$$

Both methods of estimating gene flow assume that the populations are at equilibrium for migration and drift, and that there has been enough time for mutation to have generated new alleles in populations. This is clearly not the case in colonizing species that have relatively recently invaded part of their range (e.g. many northern species that have colonized regions exposed by the retreat of Pleistocene ice sheets).

*Effective population size and founder effect*

Allele frequency data from protein electrophoresis are commonly used to draw inferences about effective population size, which is a measure of the number of breeding adults in a population, and thus is clearly less than the total population size; for example, effective population sizes in birds are typically $10^2$–$10^3$ (Barrowclough 1980), often orders of magnitude less than total population size. Effective population size is affected by factors such as fluctuations in population size, unequal numbers of the sexes, unequal family size and age structure.

In the absence of selection and mutation, heterozygosity in a population will decline at a geometric rate as a result of random drift because:

$$H_t = (1 - 1/2N)^t H_0,$$

where $H_0$ and $H_t$ are the heterozygosities in the founder and $t$th generations, respectively, and $N$ is effective population size. If a population is founded with a very small number of individuals, then the rate of reduction of heterozygosity will increase substantially over that in a larger population. Thus a reduction in average heterozygosity is an indicator of founder effects, where populations are established with a small number of founders (Baker & Moeed 1987). Bottlenecks of small population size can also dramatically reduce heterozygosity in fluctuating populations, because long-term effective population size is approximately the harmonic mean of population size in each generation. Low levels of allozyme variation in most species of shorebirds (Baker & Strauch 1988), chaffinches (Baker *et al.* 1990), the elephant seal (Bonnell & Selander 1974), and the cheetah (O'Brien *et al.* 1983, 1987) have been attributed to bottlenecks.

*Gene duplication and polyploidy*

Ploidy levels in organisms have been determined based on the staining intensities of isozymes when subunit interactions are additive; for example, gene dosages in tree frogs (*Hyla versicolor*) and *Cnemidophorus* lizards were used to identify diploids, triploids and tetraploids (Danzmann & Bogart 1982; Dessauer & Cole 1984). In polyploid plants that have arisen by hybridization, isozymes have been used to infer the putative parental species (e.g. Roose & Gottlieb 1976; Holsinger & Gottlieb 1988). In some cases, autopolyploidy (having multiple chromosome sets from the same species) has been distinguished, and in others allopolyploidy (having chomosome sets from different pecies) has been inferred (Werth *et al.* 1985; Soltis & Rieseberg 1986). The inference of ploidy levels from isozymes is difficult, however, especially because polyploid genomes often undergo re-diploidization by gene silencing (Allendorf *et al.* 1975; Ferris & Whitt 1978).

### 4.6.5 Selective neutrality of protein polymorphisms and ecological genetics

The hypothesis of selective neutrality of allozyme polymorphisms states that neutral alleles are maintained in a population by a balance between mutation (which produces new alleles) and random drift (which eliminates existing ones) (Kimura 1968; King & Jukes 1969). This hypothesis predicts that the vast majority of mutations are effectively neutral, and thus are not subject to increase in frequency by positive natural selection. Most protein variants are thought to function equivalently in organisms, but deleterious ones are eliminated by purifying selection.

Allozyme data have been used extensively to test some predictions of this hypothesis. First, in a wide range of species there is a fairly good fit between the observed distributions of allele frequencies and those expected with the infinite alleles model of neutral mutation (Fig. 4.3). Second, the mean and variance of heterozygosity in 77 vertebrate species fit quite closely to a theoretical curve again derived using the infinite alleles model (Nei *et al.* 1976). Third, when gene identity ($F$) at various loci is plotted against the number of alleles in a large sample of organisms, the loci all fall within the 95% confidence regions predicted by the neutrality hypothesis (Whittam *et al.* 1983).

Despite this apparent concordance with theoretical predictions, the selective neutrality of allozymes is still controversial. Some loci appear to be associated with fitness, and have been the focus of intensive ecological genetic investigations. A well known example is the cline in allele frequency at *Lap*-1 (leucine aminopeptidase) in the mussel *Mytilus edulis* in Long Island Sound in the USA (Koehn & Hilbish 1987). In the high salinity oceanic populations an allele that regulates intracellular osmolarity is most frequent, pointing to selection as the cause of the cline.

Another example of a cline in allozyme allele frequencies is provided at the *Ldh*-B locus in the marine fish *Fundulus heteroclitus* along the eastern seaboard of the USA (Powers *et al.* 1983). In cooler northern latitudes the *b* allele predominates. Experiments showed that homozygotes for this allele have higher

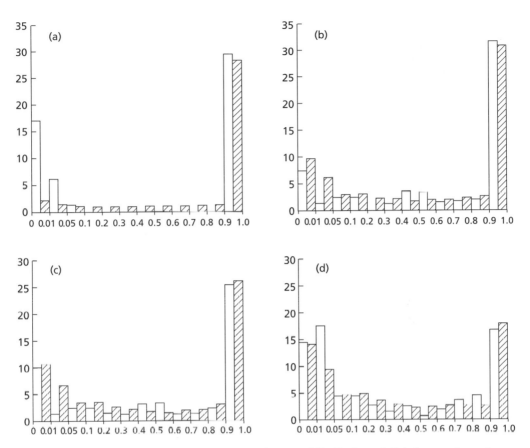

**Fig. 4.3** Observed (open columns) and expected distributions of allele frequencies under neutral mutation theory (cross hatched columns) for four animal species, (a) Japanese macaque (*Macaca fuscata*); (b) salamander (*Taricha rivularis*); (c) Eelpout (*Zoarces viviparus*); and (d) *Drosophila heteroneura*. With the exception of the Japanese macaque, which has an observed surplus of rare alleles, the two distributions are a good fit. (Redrawn from Chakraborty *et al.* 1980.)

activity in converting lactate to pyruvate at 10 °C than do *aa* homozygotes, and thus can sustain a much higher swimming speed at this temperature. No difference was observed in either enzyme kinetics or swimming ability of fish at 25 °C, and thus the *a* allele occurs at higher frequency in fish living in warmer waters.

Multilocus adaptation is suggested in other species. Growth rate has been shown to be correlated with the number of heterozygous loci in molluscs (Zouros *et al.* 1980; Koehn & Gaffney 1984; Koehn *et al.* 1988), salamanders (Mitton *et al.* 1986), white-tailed deer (Cothran *et al.* 1983), trembling aspen (Mitton & Grant 1980) and pitch pine (Ledig *et al.* 1983). An alternative explanation is inbreeding depression, because its effect would be inversely correlated with the number of heterozygous loci.

Another multilocus example involves the geographical uniformity of allozyme allele frequencies of populations of the American oyster (*Crassostrea virginica*) in the Gulf and Atlantic coasts of the USA. The lack of geographical variation in allozyme loci contrasts with the striking subdivision in both single-copy nuclear DNA markers (scnDNA) and mitochondrial DNA (mtDNA). This difference was attributed to balancing selection on the allozyme loci, rather than high gene flow that was homogenizing allozymes but not scnDNA and mtDNA (Karl & Avise 1992).

### 4.6.6 Among-species analysis

A frequent use of isozymes is in the determination of species limits and the construction of phylogenetic relationships among taxa. Fixed allele differences among populations usually warrant their recognition as separate species, as in Brown Kiwis (Baker *et al.* 1995). Phylogeny reconstruction based on isozyme data has been accompanied by methodological problems. Character-based methods use either the allele or locus as character, and employ different coding schemes to define character states (Buth 1984). Alternatively, distance methods are employed, but some algorithms assume that rates of molecular evolution are constant. However, all these methods are affected by sampling error and by homoplasy arising from the inability of protein electrophoresis to detect more than about 10% of the allelic variants that exist at the DNA level (Nei 1987).

Character-based parsimony methods can be used to depict branching patterns in the tree, and genetic distance analysis can be used to quantify the amount of divergence within and among different lineages. If large sample sizes are available for all taxa, then methods that minimize the total amount of change in allele frequencies over the tree can be used (Rogers 1984, 1986; Swofford & Berlocher 1987). For a comprehensive discussion of methods of phylogeny reconstruction see Swofford *et al.* (1996).

Isozyme data seem most appropriate in reconstructing relationships within and among genera in many groups of organisms (Murphy *et al.* 1996) because they have usually diverged long enough ago to allow the accumulation of phylogenetically informative mutations, but are not ancient enough to have evolved a large amount of homoplasy resulting from saturation at silent sites in codons; for example, the relationships of the Hawaiian honeycreepers (Fig. 4.4) have been elucidated using protein electrophoresis to resolve variation at 36 loci. In this group, monophyly dates to about 7–8 million years ago based on a molecular clock for the proteins (Johnson *et al.* 1989).

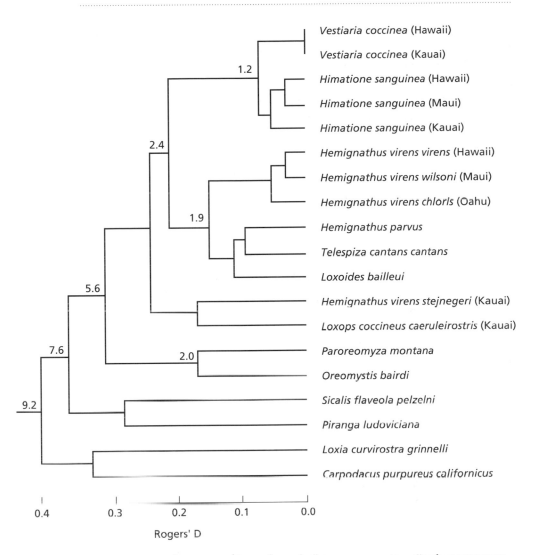

**Fig. 4.4** UPGMA phenogram of Rogers' genetic distances among Hawaiian honeycreepers, computed across 36 protein-encoding loci. The tree is a good fit to the genetic distance matrix from which it was computed; the matrix correlation between the two is 0.969. Numbers at the branching points are approximate times in millions of years for divergence of the taxa (redrawn from Johnson *et al.* 1989).

Although the newer techniques of restriction analysis of DNA and DNA sequencing have superseded protein electrophoresis, the latter remains an extremely useful adjunct in phylogeny reconstruction. Not only is it relatively cheap and technically undemanding, but it also provides an independent estimate of phylogeny to compare with the other gene trees, or for inclusion with sequence data in a total evidence approach (e.g. Friesen *et al.* 1996; Cohen *et al.* 1997).

Hybrid zones and the degree of introgression of alleles into the gene pools of the hybridizing species can often be detected by protein electrophoresis (e.g. Randi & Bernard-Laurent 1999). However, unless the hybridizing species are each fixed for different alleles at a number of loci, it is not possible to assess the directionality or extent of introgression (e.g. Gorman & Yang 1975; Patton *et al.* 1984; Szymura & Barton 1986), and thus faster evolving regions of DNA need to be studied. A classic case is provided by two species of minnows that hybridize in North America. Red-bellied (*Phoxinus eos*) and Fine Scale Dace (*P. neogaeus*) regularly hybridize and produce exclusively female offspring. The parental species are fixed for different alleles at seven allozyme loci, and thus hybrids can be definitively identified by their heterozygote constitution at these loci (Dawley *et al.* 1987). More sophisticated analyses using flow cytometry, however, determined that ploidy levels in the hybrids include diploids, triploids and diploid–triploid mosaics. Breeding experiments between parentals and hybrids along with tissue grafts showed that the third genome in triploids and mosaics was contributed by sperm from the parental species (Goddard & Dawley 1990).

Allozyme electrophoresis has been used successfully to infer the origin of parthenogenetic lizards in the genus *Cnemidophorus* (Wright 1978; Dessauer & Cole 1989; Sites *et al.* 1990); for example, in *C. lemniscatus* the unisexuals are all heterozygous at eight loci that are fixed or nearly fixed in the presumed parental species. The hybrid origin of the unisexuals was further confirmed by the observation that they were all heterozygous for a pericentric inversion that distinguishes the parentals (Sites *et al.* 1990).

The extent of introgression between two species of fire-bellied toads (*Bombina bombina* and *B. variegata*) in a hybrid zone in southern Poland has been estimated using five allozyme loci that are diagnostic for each species (Szymura & Barton 1986). For each locus, clines in allele frequencies are coincident and sharply stepped, with most change occurring over about 6 km. Selection against hybrids maintains a barrier to gene flow between the species, and gene flow appears to be limited to about 220 km.

Widths of clines separating hybridizing ecotypes of plants may also be influenced by their different breeding systems; for example, subspecies of the outcrossing *Gaillardia pulchella* (Compositae) are electrophoretically divergent at four allozyme loci. Recent spread of these subspecies along roadsides and in pastures has led to their hybridization, and the establishment of consistent clines in allele frequencies associated with edaphic ecotones (Heywood 1986). In the annual plant *Avena barbata*, however, the high frequency of self-fertilization not only limits gene flow (Loveless & Hamrick 1984) but also allows selection to affect all loci equally (Heywood 1986).

*Rates of evolution and molecular clocks*

The average rate of codon substitution that is detectable by protein electrophoresis is about $10^{-7}$ per locus per year (Nei 1987). Nei's genetic distance can be used to estimate time since divergence of population or species by noting that:

$$D = \ln I = 2\alpha t$$

where $\alpha$ is the rate of substitution. Thus the time $t$ can be estimated by substituting the empirical value of $\alpha$ in this equation and rearranging to give

$$t = 5 \times 10^6 D.$$

This formula, and related ones calibrated differently for different groups of organisms, are useful for obtaining rough estimates of evolutionary time. The question of rate constancy is controversial, and appears to vary among different classes of organisms. Estimates range from $0.8 \times 10^6$ in woodrats (Zimmerman & Nejtek 1977) to $18 \times 10^6$ in some fish and lizards (Adest 1977; Gorman & Kim 1977). Part of the reason for this rate variation is that different investigators use different protein loci, different conditions for electrophoresis, inadequate numbers of loci, or the populations and species have been bottlenecked.

Nei (1987) pointed out that the formula is best restricted to closely related species because when $D \geq 1$ its variance becomes very large and estimates are unreliable. The main problem with an electrophoretic protein clock is that reliable independent estimates of species or population divergence are largely lacking (Avise & Aquadro 1982), although some recent estimates of time of divergence of Hawaiian honeycreepers, for example, fit with the geological age of the islands (Johnson *et al.* 1989) (see Fig. 4.4).

## 4.7 Problems in interpretation

There are a number of problems to guard against in the interpretation of the results of protein electrophoresis. Banding patterns on gels can sometimes be complicated by the formation of conformational isozymes that are really multiple forms of the same gene product. These subbands have different mobilities because they differ in secondary or tertiary structure, or in thermal stability (Richardson *et al.* 1986). The electrophoretic mobilities of protein products can also be affected by polymorphic modifier loci, the physiological condition of the organism, ontogenetic changes, or by environmental effects (Conkle 1971; Harry 1983; Lebherz 1983; Womack 1983). Another common cause of mobility shifts in isozymes is the protease degradation accompanying repeated freezing and thawing of samples (Harris & Hopkinson 1976). Careful preservation of samples, correct choice of buffers and inheritance studies serve to minimize these problems.

Occasionally, products from different loci can form heteropolymers that mimic banding patterns expected for heterozygotes at one locus. In such cases the so-called heterozygotes will probably not be in Hardy–Weinberg proportions, and the staining intensities will vary from tissue to tissue. A more difficult problem is when subunits of multimeric enzymes do not associate randomly, as expected from their structure. Ferris & Whitt (1978) demonstrated that the heterodimer was not formed in heterozygotes at the *Ck-A* locus in teleost fishes. Inheritance data are again required to understand such phenomena.

The tissue-specific expression of isozymes such as *Ldh* can be confusing to inexperienced workers. Two loci of the tetrameric lactate dehydrogenase (*Ldh-A* and *Ldh-B*) are commonly expressed in vertebrate livers. Unequal expression in other tissues such as heart, brain and muscle can give one to four bands on gels for double homozygotes, and not the five expected by random association of subunits encoded by each locus. Banding patterns for heterozygotes can be even more confusing, especially if different tissue types are used for different individuals.

Alleles that do not code for proteins are called null alleles, and they can cause problems in interpreting zymograms unless they are recognized (Stoneking *et al.* 1981). The absence of any staining activity in a sample lane could be attributable to homozygosity for the null allele, but this must be confirmed by inheritance studies to differentiate it from sample degradation or loss. Failure to identify null alleles can lead to erroneous assignment of genotypes because heterozygotes with a coding allele will appear to be single-banded homozygotes in monomers. In multimeric proteins, null alleles are apparent when some bands on gels are missing.

Sometimes isozymes encoded by different loci overlap at the same location on gels, and are difficult to score, especially when one or all are polymorphic (e.g. peptidase A and C in many vertebrates). These isoloci can often be separated by altering electrophoretic conditions, such as buffers, run times and gel concentrations. Specific staining reactions for the products of each locus can usually assist in scoring the loci correctly.

Intragenic recombination can also generate confusing polymorphism in protein-encoding genes. This phenomenon has been observed in *Drosophila virilis* by Tsuno (1985, 1988), who identified 83 new allozymes among large numbers of progeny from flies heterozygous for the hypervariable esterase α locus.

## 4.8  Limitations of protein electrophoresis

A pervasive problem of protein electrophoresis is that it fails to detect even a majority of the genetic variation that exists at the nucleotide level, as only about one-third of nucleotide substitutions result in amino acid changes (of which about 25% are detectable by electrophoresis) (Nei 1987). This limitation will not matter where only relative indices of genetic variation are required, as in estimating heterozygosity of populations, but will be potentially much more serious in phylogenetic studies (but see Hackett 1989). Similarly, the lack of resolution of all genetic variants that exist at a locus and the slow rate of mutation of protein-encoding genes make paternity analysis very marginal with allozymes.

For populations and species that have diverged recently or that have undergone recent or recurrent bottlenecks, a major limitation of protein electrophoresis can be that insufficient polymorphism exists for an accurate assessment of population structure or for reconstruction of phylogenetic relationships. The slow rate of mutation means that populations can take a very long time (about the reciprocal of the mutation rate in generations) to recover genetic variability lost during a severe bottleneck, unless population growth is rapid afterwards (Nei et al. 1975). The genetic uniformity in a recently colonized part of a species range may be misinterpreted as being a result of homogenizing gene flow, when it really emanates from recent colonization by a genetically homogeneous ancestral stock (Larson et al. 1984).

At higher taxonomic levels, the opposite problem of too much allozyme divergence can arise, and it can be difficult or impossible to deduce phylogenies because of the inability to resolve loci for all taxa and because increased levels of homoplasy can be generated by extensive back mutation. Electrophoresis of 17 genera of turtles failed to resolve basal relationships in the tree because they are very divergent (Sites et al. 1984), whereas analysis of generic relationships in birds, with their more conserved proteins, is almost always phylogenetically informative (e.g. Zink 1982; Gill & Gerwin 1989; Johnson et al. 1989; Randi et al. 1991; Gerwin & Zink 1998).

## References

Adest, G.A. (1977) Genetic relationships in the genus *Uma* (Iguanidae). *Copeia*, **1977**, 47–52.

Allendorf, F.W., Utter, F.M. & May, B.P. (1975) Gene duplication within the family Salmonidae. Detection and determination of the genetic control of duplicate loci through inheritance studies and the examination of populations. In: *Isozymes IV: Genetics and Evolution.* (ed. Mairkert, C.L.) pp 415–432, Academic Press, New York.

Avise, J.C. & Aquadro, C.F. (1982) A comparative summary of genetic distances in the vertebrates: Patterns and correlations. *Evolutionary Biology* **15**, 151–185.

Baker, A.J. & Moeed, A. (1987) Rapid genetic differentiation and founder effect in colonizing populations of common mynas (*Acridotheres tristis*). *Evolution* **41**, 525–538.

Baker, A.J. & Strauch, J.G. (1988) Genetic variation and differentiation in shorebirds. *Acta
XIX Congress of International Ornithology (Ottawa)* **II**, 1639–1645.

Baker, A.J., Dennison, M.D., Lynch, A. & Le Grand, G. (1990) Genetic divergence in
peripherally isolated populations of chaffinches in the Atlantic islands. *Evolution* **44**,
981–999.

Baker, A.J., Daugherty, C.H., Colbourne, R. & McLennan, J.L. (1995). Flightless brown
Kiwis of New Zealand possess extremely subdivided population structure and cryptic
species like small mammals. *Proceedings of the National Academy of Sciences, USA*. **92**,
8254–8258.

Barrowclough, G.F. (1980) Gene flow, effective population sizes, and genetic variance
components in birds. *Evolution* **34**, 789–798.

Barrowclough, G.F. & Corbin, K.W. (1978) Genetic variation and differentiation in the
Parulidae. *Auk* **95**, 691–702.

Baverstock, P.R., Adams, M., Polkinghorne, R.W. & Gelder, M. (1982) The sex-linked
enzyme in birds—Z-chromosome conservation but no dosage compensation. *Nature* **296**,
763–766.

Bonhomme, F. & Selander, R.K. (1978) Estimating total genic diversity in the house mouse.
*Biochemical Genetics* **16**, 287–298.

Bonnell, M.T. & Selander, R.K. (1974) Elephant seals: genetic variation and near extinction.
*Science* **184**, 908–909.

Buth, D.G. (1984) The application of electrophoretic data in systematic studies. *Annual
Review of Ecology and Systematics* **15**, 501–522.

Chagala, E.M. (1991) *Genetic studies of five white pine species and their interspecific hybrids*. PhD
Thesis, University of Toronto, Toronto.

Chakraborty, R., Fuerst, P.A. & Nei, M. (1980) Statistical studies on protein polymorphism in
natural populations. III. Distribution of allele frequencies and the number of alleles per
locus. *Genetics* **94**, 1039–1063.

Chrambach, A. & Rodbard, D. (1971) Polyacrylamide gel electrophoresis. *Science* **172**,
440–451.

Cockerham, C. C. & Weir, B.S. (1993) Estimation of gene flow from *F*-statistics. *Evolution* **47**,
855–863.

Cohen, B.L., Baker, A.J., Blechschmidt, D.L. *et al.* (1997) Enigmatic phylogeny of skuas
(Aves: Stercorariidae). *Proceedings of the Royal Society* **B 264**, 181–190.

Cole, S.R. & Parkin, D.T. (1979) Enzyme polymorphisms in the house sparrow, *Passer domesticus*. *Biological Journal of the Linnean Society* **15**, 13–22.

Conkle, M.T. (1971) Inheritance of alcohol dehydrogenase and leucine aminopeptidase
isoenzymes in knobcone pine. *Forest Science* **17**, 190–195.

Cothran, E.G., Chesser, R.K., Smith, M.H. & Johns, P.E. (1983) Influences of genetic
variability and maternal factors on fetal growth in white-tailed deer. *Evolution* **37**,
282–291.

Crow, J.F. & Aoki, K. (1984) Group selection for a polygenic behavioral trait: estimating the
degree of population subdivision. *Proceedings of the National Academy of Sciences USA* **81**,
6073–6077.

Danzmann, R.G. & Bogart, J.P. (1982) Gene dosage effects on MDH isozyme expression in
diploid, triploid, and tetraploid treefrogs of the genus *Hyla*. *Journal of Heredity* **73**,
277–280.

Dawley, R.M., Schultz, R.J. & Goddard, K.A. (1987) Clonal reproduction and polyploidy in
unisexual hybrids of *Phoxinas eos* and *Phoxinas neogaeus*. *Copeia*, **1987**, 275–283.

Dessauer, H.C. & Cole, C.J. (1984) Influence of gene dosage on electrophoretic phenotypes
of proteins from lizards in the genus *Cnemidophorus*. *Comparative Biochemistry and Physiology* **77B**, 181–189.

Dessauer, H.C. & Cole, C.J. (1989) Diversity between and within nominal forms of unisexual teiid lizards. In: *Evolution and Ecology of Unisexual Vertebrates* (eds R. Dawley & J. Bogart), pp. 49–70. New York State Museum, Albany, New York.

Easteal, S. & Boussey, I.A. (1987) A sensitive and efficient isozyme technique for small arthropods and other invertebrates. *Bulletin of Entomolecular Research* **77**, 407–415.

Ferris, S.D. & Whitt, G.S. (1978) Genetic and molecular analysis of non-random dimer assembly of the creatine kinase isozymes of fishes. *Biochemical Genetics* **16**, 811–829.

Friesen, V.L., Baker, A.J. & Piatt, J.F. (1996) Phylogenetic relationships within the Alcidae (Charadriiformes: aves) inferred from total molecular evidence. *Molecular Biology and Evolution* **13**, 359–367.

Gerwin, J.A. & Zink, R.M. (1998) Phylogenetic patterns in the Trochilidae. *Auk* **115**, 105–118.

Gill, F.B. & Gerwin, J.A. (1989) Protein relationships among hermit hummingbirds. *Proceedings of the Academy of Natural Sciences of Philadelphia* **141**, 409–421.

Goddard, K.A. & Dawley, R.M. (1990) Clonal inheritance of a nuclear genome by a hybrid freshwater minnow (*Phoxinas eos-neogaeus*, Pisces: Cyprinidae). *Evolution* **44**, 1052–1065.

Gorman, G.C. & Kim, Y.J. (1977) Genotypic evolution in the face of phenotypic conservativeness: *Abudefduf* (Pomacentridae) from the Atlantic and Pacific sides of Panama. *Copeia* **1977**, 694–697.

Gorman, G.C. & Yang, S.Y. (1975) A low level of back-crossing between the hybridizing *Anolis* lizards of Trinidad. *Herpetologica* **31**, 196–198.

Goudet, J. (1995) *FSTAT, a program for IBM PC compatibles to calculate Weir and Cockerham's (1984) estimators of F-statistics*, v. 1.2.

Hackett, S.J. (1989) Effects of varied electrophoretic conditions on detection of evolutionary patterns in the Laridae. *Condor* **91**, 73–90.

Harris, H. (1966) Enzyme polymorphisms in man. *Proceedings of the Royal Society of London Series B* **164**, 298–310.

Harris, H. & Hopkinson, D.A. (1976) *Handbook of Enzyme Electrophoresis in Human Genetics*. North-Holland, Amsterdam.

Harry, D.E. (1983) Identification of a locus modifying the electrophoretic mobility of malate dehydrogenase isozymes in incense-cedar (*Calocedrus decurrens*), and its implications for population studies. *Biochemical Genetics* **21**, 417–434.

Hebert, P.D.N. & Beaton, M.J. (1989) *Methodologies for Allozyme Analysis Using Cellulose Acetate Electrophoresis A Practical Handbook*. Helena Laboratories, Beaumont, Texas.

Heywood, J.S. (1986) Clinal variation associated with edaphic ecotones in hybrid populations of *Gaillardia pulchella*. *Evolution* **40**, 1132–1140.

Holsinger, K.E. & Gottlieb, L.D. (1988) Isozyme variability in the tetraploid *Clarkia gracilis* (Onagraceae) and its diploid relatives. *Systematic Botany* **13**, 1–6.

Hubby, J.L. & Lewontin, R.C. (1966) A molecular approach to the study of genic heterozygosity in natural populations I The number of alleles at different loci in *Drosophila pseudoobscura*. *Genetics* **54**, 577–594.

Hunter, R.L. & Markert, C.L. (1957) Histochemical demonstration of enzymes separated by zone electrophoresis in starch gels. *Science* **125**, 1294–1295.

International Union of Biochemistry. Nomenclature Committee. (1984) *Enzyme Nomenclature, 1984*. Academic Press, Orlando, Florida.

Johnson, N.K., Marten, J.A. & Ralph, C.J. (1989) Genetic evidence for the origin and relationships of the Hawaiian honeycreepers (Aves: Fringillidae). *Condor* **91**, 379–396.

Karl, S.F. & Avise, J.C. (1992) Balancing selection at allozyme loci in oysters: implications from nuclear RFLPs. *Science* **256**, 100–102.

Keith, T.P. (1983) Frequency distribution of esterase 5 in two populations of *Drosophila pseudoobscura*. *Genetics* **105**, 135–155.

Kimura, M. (1968) Evolutionary rate at the molecular level. *Nature* **217**, 624–626.

King, J.L. & Jukes, T.H. (1969) Non-Darwinian evolution. *Science* **164**, 788–798.

Koehn, R.K. & Gaffney, P.M. (1984) Genetic heterozygosity and growth rate in *Mytilus edulis*. *Marine Biology* **82**, 1–7.

Koehn, R.K. & Hilbish, T.J. (1987) The adaptive importance of genetic variation. *American Scientist* **75**, 134–141.

Koehn, R.K., Diehl, W.J. & Scott, T.M. (1988) The differential contribution by individual enzymes of glycolysis and protein catabolism to the relationship between heterozygosity and growth rate in the coot clam *Mulinia lateralis*. *Genetics* **118**, 121–130.

Larson, A., Wake, D.B. & Yanev, K.P. (1984) Measuring gene flow among populations having high levels of genetic fragmentation. *Genetics* **106**, 293–308.

Lebherz, H.G. (1983) On epigenetically generated isozymes ('pseudoisozymes') and their possible biological relevance. In: *Isozymes: Current Topics in Biological and Medical Research*, vol. 7 (eds M.C. Rattazzi, J.G. Scandalios & G.S. Whitt), pp. 203–219. A. R. Liss, New York.

Ledig, F.T., Guries, R.P. & Bonefield, B.A. (1983) The relation of growth to heterozygosity in pitch pine. *Evolution* **37**, 1227–1238.

Lewontin, R.C. & Hubby, J.L. (1966) A molecular approach to the study of genic heterozygosity in natural populations. II. Amounts of variation and degree of heterozygosity in natural populations of *Drosophila pseudoobscura*. *Genetics* **54**, 595–609.

Loveless, M.D. & Hamrick, J.L. (1984) Ecological determinants of genetic structure in plant populations. *Annual Review of Ecology and Systematics* **15**, 65–95.

Mitton, J.B. & Grant, M.C. (1980) Observations on the ecology and evolution of quaking aspen, *Populus tremuloides*, in the Colorado front range. *American Journal of Botany* **67**, 202–209.

Mitton, J.B., Carey, C. & Kocher, T.D. (1986) The relation of enzyme heterozygosity to standard and active oxygen consumption and body size of tiger salamanders, *Ambystoma tigrinum*. *Physiological Zoology* **59**, 574–582.

Murphy, R.W., Sites, J.W., Buth, D.G. & Haufler, C.H. (1996) Proteins: isozyme electrophoresis. In: *Molecular Systematics*, 2nd edn (eds D.M. Hillis, C. Moritz & B.K. Mable), pp. 51–120. Sinauer Associates, Sunderland, Massachusetts.

Nei, M. (1975) *Molecular Population Genetics and Evolution*. North-Holland, Amsterdam.

Nei, M. (1978) Estimation of average heterozygosity and genetic distance from a small number of individuals. *Genetics* **89**, 583–590.

Nei, M. (1987) *Molecular Evolutionary Genetics*. Columbia University Press, New York.

Nei, M. & Roychoudhury, A.K. (1974) Sampling variance of heterozygosity and genetic distance. *Genetics* **76**, 379–390.

Nei, M., Maruyama, T. & Chakraborty, R. (1975) The bottleneck effect and genetic variability in populations. *Evolution* **29**, 1–10.

Nei, M., Fuerst, P.A. & Chakraborty, R. (1976) Testing the neutral mutation hypothesis by distribution of single locus heterozygosity. *Nature* **262**, 491–493.

O'Brien, S.J., Wildt, D.E., Goldman, D., Merril, C.R. & Bush, M. (1983) The cheetah is depauperate in genetic variation. *Science* **221**, 459–462.

O'Brien, S.J., Wildt, D.E., Bush, M. *et al.* (1987) East African cheetahs: Evidence for two population bottlenecks? *Proceedings of the National Academy of Sciences USA* **84**, 508–511.

Ota, T. (1993) *DISPAN: Distance analysis and phylogenetic analysis*. Institute of Molecular Evolutionary Genetics, Pennsylvania State University.

Patton, J.L., Smith, M.F., Price, R.D. & Hellenthal, R.A. (1984) Genetics of hybridization between the pocket gophers *Thomomys bottae* and *Thomomys townsendi* in northeastern California. *Great Basin Naturalist* **44**, 431–440.

Powers, D.A., DiMichele, L. & Place, A.R. (1983) The use of enzyme kinetics to predict differences in cellular metabolism, developmental rate and swimming performance between LDH-B genotypes of the fish *Fundulus heteroclitus*. Isozymes. *Current Topics in Biological and Medical Research* **10**, 147–170.

Ramshaw, J.A.M., Coyne, J.A. & Lewontin, R.C. (1979) The sensitivity of gel electrophoresis as a detector of genetic variation. *Genetics* **93**, 1019–1037.

Randi, E. & Bernard-Laurent, A. (1999) Population genetics of a hybrid zone between the red-legged and rock partridge. *Auk* **116**, 324–337.

Randi, E., Fusco, G., Lorenzini, R. & Crowe, T.M. (1991) Phylogenetic relationships and rates of allozyme evolution within the Phasianidae. *Biochemical Systematics and Ecology* **19**, 213–221.

Raymond, M. & Rousset, F. (1995) GENEPOP (v. 1.2): a population genetics software for exact tests and ecumenicism. *Journal of Heredity* **86**, 248–249.

Richardson, B.J., Baverstock, P.R. & Adams, M. (1986) *Allozyme Electrophoresis: A Handbook for Animal Systematics and Population Structure*. Academic Press, Sydney.

Rockwell, R.F. & Barrowclough, G.F. (1987) Gene flow and the genetic structure of populations. In: *Avian Genetics: a Population and Ecological Approach* (eds F. Cooke & P.A. Buckley), pp. 223–255. Academic Press, London.

Rogers, J.S. (1972) Measures of genetic similarity and genetic distance. *Studies in Genetics. VII* University of Texas Publications 7213, 145–153.

Rogers, J.S. (1984) Deriving phylogenetic trees from allele frequencies. *Systematic Zoology* **33**, 52–63.

Rogers, J.S. (1986) Deriving phylogenetic trees from allele frequencies: a comparison of nine genetic distances. *Systematic Zoology* **35**, 297–310.

Roose, M.L. & Gottlieb, L.D. (1976) Genetic and biochemical consequences of polyploidy in *Tragopogon*. *Evolution* **30**, 818–830.

Singh, R.S., Lewontin, R.C. & Felton, A.A. (1976) Genetic heterogeneity within electrophoretic 'alleles' of xanthine dehydrogenase in *Drosophila pseudoobscura*. *Genetics* **84**, 609–629.

Sites, J.W., Bickham, J.W., Pytel, B.A., Greenbaum, I.F. & Bates, B.A. (1984) Biochemical characters and the reconstruction of turtle phylogenies: relationships among batagurine genera. *Systematic Zoology* **33**, 137–158.

Sites, J.W., Peccini-Seale, D.N., Moritz, C., Wright, J.W. & Brown, W.M. (1990) The evolutionary history of parthenogenetic *Cnemidophorus lemniscatus* (Sauria, Teiidae). I. Evidence for a hybrid origin. *Evolution* **44**, 906–921.

Slatkin, M. (1981) Estimating levels of gene flow in natural populations. *Genetics* **99**, 323–335.

Slatkin, M. (1985) Rare alleles as indicators of gene flow. *Evolution* **39**, 53–65.

Slatkin, M. (1987) Gene flow and the geographic structure of natural populations. *Science* **236**, 787–792.

Smithies, O. (1955) Zone electrophoresis in starch gels: group variations in the serum proteins of normal individuals. *Biochemical Journal* **61**, 629–641.

Sokal, R.R. & Rohlf, F.J. (1981) *Biometry*, 2nd edn. W. H. Freeman, San Francisco.

Soltis, D.E. & Rieseberg, L.J. (1986) Autopolyploidy in *Tolmeia menziesii* (Saxifragaceae). Genetic insights from enzyme electrophoresis. *American Journal of Botany* **73**, 310–318.

Stoneking, M., May, B. & Wright, J. (1981) Loss of duplicate gene expression in salmonids: Evidence for a null allele polymorphism at the duplicate aspartate aminotransferase loci in brook trout (*Salvelinus fontinalis*). *Biochemical Genetics* **19**, 1063–1077.

Swofford, D.L. & Berlocher, S.H. (1987) Inferring evolutionary trees from gene frequency data under the principle of maximum parsimony. *Systematic Zoology* **36**, 293–325.

Swofford, D.L. & Selander, R.B. (1981) BIOSYS-1: a FORTRAN program for the comprehensive analysis of electrophoretic data in population genetics and systematics. *Journal of Heredity* **72**, 281–283.

Swofford, D.L., Olsen, G.J., Wadell, P.J. & Hillis, D.M. (1996) Phylogenetic inference. In: *Molecular Systematics*, 2nd edn (eds D.M. Hillis, C. Moritz & B.K. Mable), pp. 407–514. Sinauer Associates, Sunderland, Massachusetts.

Szymura, J.M. & Barton, N.H. (1986) Genetic analysis of a hybrid zone between fire-bellied toads, *Bombina bombina* and *B. variegata*, near Cracow in southern Poland. *Evolution* **40**, 1141–1159.

Tsuno, K. (1985) Studies on mutation at esterase loci in Drosophila virilis. III. Genetic variation in Est-α alleles produced by mutation experiments. *Japanese Journal of Genetics* **60**, 103–118.

Tsuno, K. (1988) Contribution of recombinants produced by female flies heterozygous for Est-α alleles to genetic variation of *Drosophila virilis*. *Genetics Research* **51**, 217–222.

Weir, B.S. & Cockerham, C.C. (1984) Estimating F-statistics for the analysis of population structure. *Evolution* **38**, 1355–1370.

Werth, C.R., Guttman, S.I. & Eshbaugh, W.H. (1985) Recurring origins of allopolyploid species in. *Asplenium Science* **228**, 731–733.

Whittam, T.S., Ochman, H. & Selander, R.K. (1983) Geographic components of linkage disequilibrium in natural populations of *Escherichia coli*. *Molecular Biology and Ecology* **1**, 67–83.

Womack, J.E. (1983) Posttranslational modification of enzymes: Processing genes. In: *Isozymes: Current Topics in Biological and Medical Research*, vol. 7 (eds M.C. Rattazzi, J.C. Scandalios & G.S. Whitt), pp. 175–186. A. R. Liss, New York.

Wright, J.W. (1978) Parthenogenetic lizards. *Science* **201**, 1152–1154.

Wright, S. (1965) The interpretation of population structure by F-statistics with special regard to systems of mating. *Evolution* **9**, 395–420.

Wright, S. (1978) *Evolution and the Genetics of Populations IV. Variability Within and Among Natural Populations*. University of Chicago Press, Chicago, Illinois.

Zimmerman, E.G. & Nejtek, M.E. (1977) Genetics and speciation of three species of *Neotoma*. *Journal of Mammalogy* **58**, 391–402.

Zink, R.M. (1982) Patterns of genic and morphologic variation among sparrows in the genera *Zonotrichia, Melospiza, Junco*, and *Passerella*. *Auk* **99**, 632–649.

Zouros, E., Singh, S.M. & Miles, H.E. (1980) Growth rate in oysters: an overdominant phenotype and its possible explanations. *Evolution* **34**, 856–867.

# CHAPTER 5

# Solution DNA–DNA Hybridization

## ANTHONY H. BLEDSOE AND FREDERICK H. SHELDON

## 5.1 Introduction

Solution-based DNA–DNA hybridization is a technique for locating homologous DNA sequences in solution and measuring their overall basepair differences. As with *in situ* hybridization, single-stranded DNA sequences from different individuals or species are encouraged to reassociate into hybrid duplexes. These duplexes are then dissociated by heating, and the amount of energy required to 'melt' them is proportional to their basepair similarity. It is a fast and versatile technique that can be applied at a variety of levels to individual gene sequences, sets of coding genes, small genomes (e.g. mitochondrial DNA (mtDNA) or chloroplast DNA (cpDNA)), or the entire single-copy nuclear genome. The data produced by solution DNA hybridization are dissimilarity values. In systematic studies, they are fitted to branching patterns to depict phylogenetic trees, and they are often compared with other dissimilarity measures to uncover evolutionary rate and biogeographic patterns. DNA hybridization estimates of phylogeny have proved valuable for a variety of ecological investigations, from studies of the ecological basis of adaptive radiation to identification and analysis of ecologically important microorganisms.

Our main goal in this chapter is to provide an overview of the utility of DNA hybridization in ecological studies and a critical analysis of DNA hybridization methods. Because many of the pertinent ecological applications involve the use of phylogenies generated from DNA hybridization data, we have emphasized the features that we believe are crucial for success in using DNA hybridization to estimate phylogeny.

### 5.1.1 DNA hybridization and ecological inference

The most common form of DNA hybridization in systematic studies is single-copy nuclear DNA (scnDNA) hybridization. Because scnDNA tends to evolve relatively slowly (Helm-Bychowski & Wilson 1986; Vawter & Brown 1986), scnDNA hybridization has been applied mainly to phylogenetic problems that bear on major features of evolution. In plants and animals, taxa below the species level are not often compared by scnDNA hybridization because their

DNAs have not diverged adequately to permit resolution of genetic distances. Even so, DNA hybridization can serve as a powerful tool for ecological inference. The understanding of many ecological issues depends upon a broad historical context, which DNA hybridization can provide (e.g. Ricklefs 1987). In addition, there are numerous ways to increase close-range resolving power of DNA hybridization to permit the examination of problems at and below the species level.

DNA hybridization has supplied phylogenetic perspectives to ecological, behavioural and coevolutionary patterns in many groups of organisms. Sibley & Ahlquist (1990) published a thorough listing of the studies prior to 1990, and Sibley (1997) provided a historical overview of DNA hybridization. Some recent examples that are not included in Sibley and Ahlquist's book are: fish and their stomach parasites (Verneau *et al.* 1991); hermit crabs and their symbiotic hybrids (Cunningham *et al.* 1991); drosophilids (Caccone *et al.* 1996); sand dollars (Marshall 1992; Marshall & Swift 1992); marsupials (Springer & Kirsch 1991; Kirsch *et al.* 1990, 1992, 1997); artiodactyls (Douzery *et al.* 1995); megachiropterans (Springer *et al.* 1995b); and birds (Sheldon *et al.* 1992; Sheldon & Winkler 1993; Sheldon & Gill 1996; Slikas *et al.* 1996; Bleiweiss *et al.* 1997). Ecological insight from these studies was derived mainly as a by-product of investigations intended to solve phylogenetic problems; for example, in one of the bird studies (Sheldon *et al.* 1992; Slikas *et al.* 1996), scnDNA hybridization of well-diverged *Parus* species was used to complement allozyme and mtDNA restriction fragment-length polymorphism (RFLP) analyses of populations and closely related species (Gill & Slikas 1992; Gill *et al.* 1992). The allozyme and RFLP data alone were unable to distinguish the deeper branches within the genus. However, DNA hybridization data resolved quite clearly a deep bifurcation in the titmouse tree, and the bifurcation correlated exactly with well-studied patterns of hippocampal development, spatial memory and associated food-caching and flocking behaviour (e.g. Ekman 1989; Sherry 1989; Krebs *et al.* 1990). In another example, McCracken & Sheldon (1997) used a DNA hybridization phylogeny of herons to identify convergent vocal traits and specify their relationship to vegetation characteristics. Similarly, Winkler & Sheldon (1993) analysed nest evolution and its association with habitat and coloniality in swallows, again using DNA hybridization evidence as a basis for ecological inference. Primmer *et al.* (1996) related genetic distance, measured by DNA hybridization, to the taxonomic range of birds across which microsatellites can be amplified with a single set of primers. Such results have important practical applications for the use of microsatellites in studies of avian ecology. A final example serves to indicate the breadth of utility of DNA hybridization phylogenies. Krajewski (1994) compared seven phylogenetic measures of biodiversity, using a DNA hybridization phylogeny of cranes. His results identified several specific measures that most reflect the relative contribution of particular species to biodiversity. Such results have potential application well beyond the goals of phylogeny reconstruction *per se*.

DNA hybridization techniques provide a critical tool at a more basic taxonomic level in bacteriology, virology and mycology. DNA reassociation frequently provides the benchmark for designating bacterial forms at the strain-level, where DNA–DNA hybridization is still viewed as superior to, for example, 16s rRNA sequencing (Stackebrandt & Goebel 1994). Such work plays a crucial role in medically important areas, such as the study of infection-causing strains of *Mycobacterium* (Springer *et al.* 1995a). Basic and applied ecological studies of economically important forms such as *Rhizobium* bacteria also rely heavily on DNA hybridization data (e.g. Laguerre *et al.* 1993; Oyaizu *et al.* 1993). Recent DNA-hybridization-based studies of potential tools for pollution control, such as aromatic-degrading bacteria (e.g. Balkwill *et al.* 1997), further demonstrate the value of DNA hybridization techniques in ecological applications of microbiology.

Many DNA hybridization studies have taken advantage of idiosyncrasies of DNA evolution to increase close-range resolving power, even in plants and animals. When rates of scnDNA change are rapid, species-level problems may be routinely addressed, as in the cases of cave crickets (Caccone & Powell 1987), *Drosophila* (Hunt *et al.* 1981; Goddard *et al.* 1990; Caccone *et al.* 1987, 1988a, 1996), and sea urchins (e.g. Palumbi & Metz 1991). With the help of special buffers, DNA hybridization has also been able to distinguish individual variation in some of these invertebrate groups (e.g. Grula *et al.* 1982; Caccone *et al.* 1987, 1988a). Alternatively, the selection of fast or slowly evolving genes or genomes may enlarge the range of DNA hybridization resolution. Hybridization of vertebrate mtDNAs, for example, takes advantage of the fast rate of their evolution to compare closely related taxa (Brown *et al.* 1979). Hybridization of complementary DNA (cDNA) produced from protein-coding regions of messenger RNA (mRNA) permits the comparison of sequences of which the evolutionary rate of change has been slowed by natural selection. Thus, highly diverged taxa may be compared (Roberts *et al.* 1985; Caccone *et al.* 1992).

DNA hybridization also offers the possibility of palaeo-ecological inference. Genetic dissimilarities among taxa that are well diverged can be compared with fossil and biogeographical dates to provide approximate locations in time. These, in turn, can be used to infer the existence of taxa at ages for which there is no direct (fossil) evidence.

### 5.1.2  The state of DNA hybridization practice and theory

Phylogenetic reconstruction by DNA hybridization is controversial for a number of theoretical and practical reasons. Philosophically, DNA hybridization has been at odds with cladistic analysis, which is the predominant mode of phylogenetic inference. As we will explain, much of the tension between the two approaches stems from a misunderstanding of how DNA hybridization operates. However, some practitioners of DNA hybridization, in particular Sibley and Ahlquist (e.g. Sibley & Ahlquist 1990), exacerbated concerns about the technique by overstating its power and abilities and inadequately preparing experiments or analysing data. The combination of philosophical differences and practical problems has led to a series of critiques of DNA hybridization and its data. Theoretical concerns tend to focus on the properties of dissimilarity data and phylogenies derived from them: Are dissimilarity values metric (i.e. distances) and additive (e.g. Thorpe 1982; Farris 1981, 1985, 1986)? Are they distorted by variable rates of evolution (e.g. Carpenter 1985)? And is the derived information they contain overwhelmed or biased by primitive noise (e.g. Platnick 1985; Andrews 1986)? Other critiques have been aimed mainly at the application of DNA hybridization by Sibley and Ahlquist, but they also raise many theoretical and statistical questions (Templeton 1985, 1986; Cracraft 1987; Houde 1987a,b; Marks *et al.* 1988; Sarich *et al.* 1989; Schmid & Marks 1990; Lanyon 1992). We will discuss most of the issues raised in these critiques in the course of this chapter.

Recently, and partly in response to these criticisms, practitioners of DNA hybridization have attempted to come to grips with the operation, assumptions, strengths and weaknesses of the technique (Ahlquist *et al.* 1987; Bledsoe 1987a,b; Sheldon 1987a; Bledsoe & Sheldon 1989, 1990; Caccone & Powell 1989; Springer & Krajewski 1989a,b; Powell & Caccone 1990; Sibley *et al.* 1990; Sheldon & Bledsoe 1989, 1993; Sheldon & Kinnarney 1993). These efforts have been accompanied by improvements in experimental design and data analysis (Krajewski & Dickerman 1990; Blackstone & Sheldon 1991; Dickerman 1991; Marshall 1991; Bleiweiss & Kirsch 1993a,b). Previously, there existed only a vague sense that DNA hybridization operated as an ultimate form of numerical taxonomy—arriving at phylogeny by virtue of comparing huge numbers of basepair characters (Sibley & Ahlquist 1983; Sibley *et al.* 1987)—but this view did not explain why DNA hybridization should succeed where numerical taxonomy had failed; for example, why it should yield a nested phylogenetic hierarchy independent of clustering methods, or how it could possibly work in the face of variable rates of evolution. The answer is that DNA hybridization is not an ultimate form of numerical taxonomy. Instead, as reviewed by Bledsoe & Sheldon (1990), it operates in a fashion analogous to cladistics (see Fig. 5.1). DNA hybridization measures the sum of changes that have occurred between two species since they diverged from a common ancestor. When several taxa are compared and an outgroup is specified, phylogenetic hierarchy is provided by the pattern of distances in a

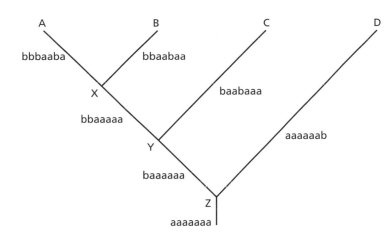

**Fig. 5.1** A hypothetical tree demonstrating the operation of DNA hybridization for three taxa (A, B, and C) and an outgroup taxon (D). Each lineage is accompanied by a hypothetical sequence of seven sites, each with two possible character states (a and b). At the internal nodes (X and Y), the sequences combine apomorphic states (i.e. derived in the lineage leading up to the node) and plesiomorphic states (i.e. derived prior to the lineage leading up to the node). The pattern of possession of these apomorphic characters delimits the monophyletic group consisting of A and B. The changes in character states that occur leading to terminal nodes are autapomorphous. Genetic distances among the taxa are determined by counting character state differences: A to B = 3, A to C = 4, B to C = 3, A to D = 5, B to D = 4, C to D = 3. Analysis of this distance matrix using a pairwise method and least-squares measure of fit (Cavalli-Sforza & Edwards 1967) retrieved the correct branching pattern when rooted at D. If, however, site 7 in the lineage leading from X to A were to evolve state b, nonhomologous identity between A and D would exist at this site. Analysis of the resulting distance matrix using the same method as above yielded an incorrect tree that groups B and C as monophyletic. (Adapted from Bledsoe & Sheldon 1990.)

matrix of pairwise comparisons of taxa. This pattern results from the underlying pattern of synapomorphic substitutions on the subterminal portions of the tree. Primitive homology plays a role in the formation of hybrid duplexes and in the automatic congruence assessment of basepair homology but does not contribute directly to dissimilarity that reflects underlying synapomorphies. This model requires that genetic change occurs and is forward (back-mutations are overwhelmed), that trees are rooted by outgroups determined by data other than that contained in the matrix and that dissimilarity values are fitted to trees by an algorithm that does not assume constant rates of evolution.

### 5.1.3 The effectiveness of DNA hybridization: phylogenetic congruence

Proponents of DNA hybridization have argued that the advantage of the technique is that it is a highly efficient averaging method of which the accuracy is ensured, when nuclear genomes are compared, by the input of information from millions of basepairs. Testing this proposition is problematic, because DNA hybridization produces dissimilarity values and bypasses characters. Although the relationship between dissimilarity values and simple basepair differences is well understood (e.g. Springer *et al.* 1992), the exact nucleotide changes are unknown. It is thus impossible to perform character tests like those used in cladistic analyses.

Andrews (1987) argued that this inability compromises the validity of the technique, and Zink & Weller (1991) suggested that, as a result, DNA hybridization should not be used as a primary method for phylogenetic analysis. However, comparison of phylogenetic estimates produced by different techniques indicates that DNA hybridization methods regularly yield phylogenies that are congruent with those based on independent data sets and methods of analysis, including cladistic methods (Bledsoe & Raikow 1990; Sheldon & Bledsoe 1993). Perhaps the best example of this is the hominoid case, in which a series of independent DNA hybridization efforts (Hoyer *et al.* 1972; Benveniste & Todaro 1976; Caccone & Powell 1989; Sibley *et al.* 1990) yielded measurements of dissimilarity and phylogeny consistent with the bulk of other molecular data, including extensive sequencing (e.g. Holmquist *et al.* 1988a,b; Goodman *et al.* 1990; Miyamoto & Goodman 1990). Likewise, Krajewski & Fetzner (1994) analysed mitochondrial cytochrome *b* DNA sequences of cranes, using discrete-character maximum parsimony methods, and obtained a phylogeny highly congruent with Krajewski's (1989) DNA-hybridization estimate for the same group. In other instances (e.g. megachiropterans; Springer *et al.* 1995), DNA hybridization and morphologically based maximum-parsimony results are strikingly discordant. Incongruence is particularly difficult to interpret with regard to the results of Sibley & Ahlquist (1990), who made various mistakes in the application of DNA hybridization methods (Lanyon 1992; Sheldon & Bledsoe 1993). Some instances of incongruence may merely reflect that Sibley and Ahlquist applied DNA hybridization methods incorrectly, not that the methods *per se* are invalid.

## 5.2 Biochemistry of DNA–DNA hybridization

The classic review of solution-based DNA–DNA hybridization biochemistry is by Britten *et al.* (1974). Sibley & Ahlquist (1981, 1990) and Powell & Caccone (1990) described the application of traditional and TEACL-based DNA hybridization techniques, respectively, to systematic problems. Sibley & Ahlquist (1986) provided a readable review of methods with excellent figures describing the chemistry of hybridization and the sequence of biochemical procedures used to obtain DNA hybridization data.

The three basic steps of DNA hybridization are:

1    the extraction and preparation of DNA;
2    the reassociation of single-stranded DNA into hybrids; and
3    the dissociation of the hybrids to determine their stability.

Werman *et al.* (1996) provided an overview, which contains a complete description of all current methods and a set of modern protocols. They cover in detail the preparation of 'driver' and 'tracer' DNAs, isolation of the single-copy sequences, preparation of homoduplex hybrids (formed from tracer and driver DNA of a single individual) and heteroduplex hybrids (formed from tracer and driver DNAs from different individuals, including different species), and determination of hybrid reassociation and stability. The degree of basepair similarity in a heteroduplex hybrid is proportional to the reduction in median melting temperature of that heteroduplex compared with its homoduplex counterpart. It is this reduction that forms the basis for calculating the measures of dissimilarity discussed in section 5.3.2.

## 5.3  Data analysis

The fundamental steps of DNA-hybridization data analysis are the computation of hybrid reassociation, dissociation and dissimilarity values, and the estimation of phylogenetic trees. In this section, we describe these steps and discuss various properties and potential problems of the data in terms of their metricity, additivity, error and biases.

### 5.3.1  Indexes to reassociation and dissociation

The raw data of DNA hybridization are radioactive counts corresponding to thermal fractionation temperatures (see Table 5.1). These counts are transformed into frequency distributions that supply indexes of percentage reassociation and median or model melting temperatures.

The calculation of indexes from hydroxyapatite-fractionated hybrids has been reviewed by Sheldon & Bledsoe (1989). The counts recorded at and below the incubation temperature, minus counts contributed by unbound isotopes, reflect the proportion of sequences that failed to hybridize. The total counts above the incubation temperature indicate the proportion of DNA that reassociated. A derivative plot is constructed from the percentage of counts recorded at each fractionation temperature (Fig. 5.2a) and from this, modes are estimated by least-squares fitting. Cumulative curves are constructed by adding the derivative plot percentages to one another sequentially (Fig. 5.2b). $T_m$ is the median of a cumulative curve. $T_{50}H$ ($T_mR$) is the median of a cumulative curve that has been corrected for reduced percentage reassociation (Fig. 5.2c) and has often been used to estimate large dissimilarity values

**Table 5.1** Sample data from an experiment in which Great Blue Heron (*Ardea herodias*) is the homoduplex control; American Bittern (*Botaurus lentiginosus*) and Lesser Yellowlegs (*Tringa flavipes*) are heteroduplexes. Hybrids were incubated at 60°C. (Adapted from Sheldon & Bledsoe 1989.)

| Fractionation temperature (°C) | *Ardea herodias* | | | | *Botaurus lentiginosus* | | | | *Tringa flavipes* | | | |
|---|---|---|---|---|---|---|---|---|---|---|---|---|
| | Counts | Mode* 87.1 | $T_m$† 84.1 | $T_{50}$H‡ 84.1 | Counts | Mode* 81.0 | $T_m$† 77.8 | $T_{50}$H‡ 75.8 | Counts | Mode* 71.7 | $T_m$† 71.1 | $T_{50}$H‡ 65.5 |
| 55.0 | 93344 | | | | 141252 | | | | 175274 | | | |
| 57.5 | 1921 | | | | 3495 | | | | 4671 | | | |
| 60.0 | 2260 | 0.00 | 0.00 | 0.00 | 4656 | 0.00 | 0.00 | 17.48 | 8240 | 0.00 | 0.00 | 36.39 |
| 62.5 | 3376 | 1.75 | 1.75 | 1.75 | 6308 | 3.48 | 3.48 | 20.35 | 11729 | 8.51 | 8.51 | 41.80 |
| 65.0 | 3476 | 1.80 | 3.55 | 3.55 | 7195 | 3.97 | 7.45 | 23.63 | 14255 | 10.35 | 18.86 | 48.38 |
| 67.5 | 3748 | 1.94 | 5.49 | 5.49 | 8172 | 4.51 | 11.95 | 27.34 | 16475 | 11.96 | 30.81 | 55.99 |
| 70.0 | 4382 | 2.27 | 7.76 | 7.76 | 10313 | 5.69 | 17.64 | 32.04 | 18020 | 13.08 | 43.89 | 64.31 |
| 72.5 | 5409 | 2.80 | 10.56 | 10.56 | 14133 | 7.79 | 25.43 | 38.47 | 20021 | 14.53 | 58.42 | 73.55 |
| 75.0 | 6919 | 3.58 | 14.14 | 14.14 | 17442 | 9.62 | 35.05 | 46.40 | 16834 | 12.22 | 70.64 | 81.32 |
| 77.5 | 9730 | 5.04 | 19.18 | 19.18 | 23724 | 13.08 | 48.13 | 57.20 | 14148 | 10.27 | 80.91 | 87.85 |
| 80.0 | 14182 | 7.34 | 26.52 | 26.52 | 26773 | 14.76 | 62.89 | 69.38 | 10016 | 7.27 | 88.17 | 92.48 |
| 82.5 | 26417 | 13.68 | 40.20 | 40.20 | 28193 | 15.54 | 78.43 | 82.20 | 6810 | 4.94 | 93.12 | 95.62 |
| 85.0 | 30189 | 15.63 | 55.83 | 55.83 | 19890 | 10.97 | 89.40 | 91.25 | 3930 | 2.85 | 95.97 | 97.44 |
| 87.5 | 35057 | 18.15 | 73.98 | 73.98 | 12171 | 6.71 | 96.11 | 96.79 | 2625 | 1.91 | 97.87 | 98.65 |
| 90.0 | 26653 | 13.80 | 87.78 | 87.78 | 4961 | 2.74 | 98.85 | 99.05 | 1848 | 1.34 | 99.22 | 99.50 |
| 92.5 | 16195 | 8.38 | 96.16 | 96.16 | 1584 | 0.87 | 99.72 | 99.77 | 769 | 0.56 | 99.77 | 99.86 |
| 95.0 | 7414 | 3.84 | 100.00 | 100.00 | 508 | 0.28 | 100.00 | 100.00 | 312 | 0.23 | 100.00 | 100.00 |

* Values in the modal column are percentages computed by dividing the total counts from 60 to 95°C by the counts at a given temperature.

† Values in the $T_m$ column are cumulative percentages added from the adjacent modal column.

‡ Values in the $T_{50}$H column are percentages that have been normalized against data of the homoduplex. The first value in the $T_{50}$H column is the difference between the normalized percentage reassociation of the homoduplex (assumed to be 100%) and the heteroduplex. The rest of the values are normalized so that they add up to 100% (see Sheldon & Bledsoe 1989, for equations and details).

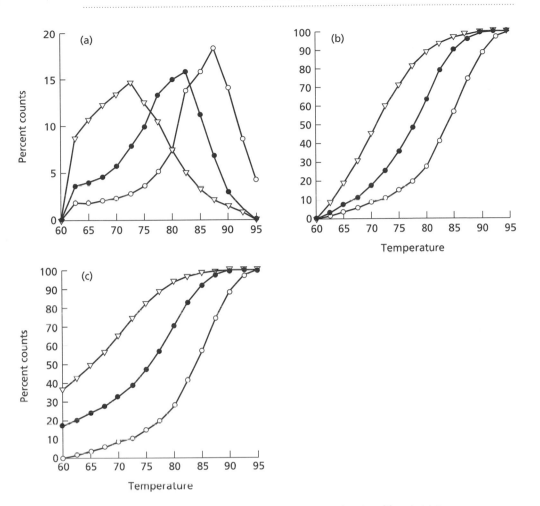

**Fig. 5.2** Graphical representations of the dissociation data in Table 5.1. (a) Frequency distributions for modal values. (b) Cumulative distributions for median values ($T_m$s). (c) Normalized cumulative distributions for median values corrected for reduced percentage reassociation ($T_{50}$Hs). Open circles denote data for homoduplex DNA of *Ardea herodias*. Closed circles denote data from heteroduplex *A. herodias*/*Botaurus lentiginosus* DNA. Triangles denote data from heteroduplex *A. herodias*/*Tringa flavipes* DNA. (Adapted from Sheldon & Bledsoe 1989.)

(e.g. Kohne 1970; Benveniste & Todaro 1976; Sibley & Ahlquist 1983). Bleiweiss & Kirsch (1993a,b) have provided a variance analysis for $T_{50}$H and other measures of central tendency derived from thermal elution curves.

The main difference in computation of hydroxyapatite and S1-nuclease fractionated hybrids is that in the latter, reassociation is determined by digested (unreassociated) and undigested (reassociated) DNA. The advantage of this approach is that single-stranded tails on otherwise duplex molecules are digested along with completely single-stranded DNAs. Therefore, percentage reassociation is more accurately assessed (Benveniste 1985; Caccone *et al.* 1988b).

### 5.3.2 Dissimilarity values and phylogenetic trees

Dissimilarity (also called delta, Δ) values are computed by subtracting heteroduplex from homoduplex values. A phylogeny is then estimated by evaluating the fit of dissimilarity values to trees with alternative branching patterns. The tree whose branching pattern and branch lengths correspond best with the DNA hybridization dissimilarities, based on various statistics of fit, is presumed to reflect the best estimate of phylogeny. The main fitting method is least-squares. In addition, two algorithmic methods—distance-Wagner (Farris 1972) and neighbour-joining (Saitou & Nei 1987)—can be used to obtain estimates of phylogeny. Of these, least-squares is the most commonly used. Felsenstein's (1995) computer-based phylogenetic inference package, PHYLIP, uses an alternating least-squares approach (Felsenstein 1997) that permits fitting by weighted (Fitch & Margoliash 1967) and unweighted (Cavalli-Sforza & Edwards 1967) least-squares, and parameters can be changed so that sister branches are either equal in length (i.e. evolutionary rates are assumed to be constant) or they may vary in length (constant rates are not assumed). The computer program PAUP*, written by David Swofford and commercially available through Sinauer Associates (Sunderland, Massachusetts), also provides options for the analysis of distance data using a variety of methods, including neighbour-joining (Saitou & Nei 1987). A comparison of neighbour-joining and weighted least-squares methods (Kuhner & Felsenstein 1994) suggests that the latter are slightly superior in certain circumstances. These methods do not assume constant rates of molecular change and hence are usually preferable to methods such as UPGMA, which relies on constant rates (e.g. Sneath & Sokal 1973).

### 5.3.3 Measurement error in DNA-hybridization studies

*Sources of error*

The sources of measurement error in DNA-hybridization studies can be either experimental or biological (e.g. Springer & Krajewski 1989a). Experimental error, above and beyond the usual laboratory sources (e.g. variations in buffer preparation), may result from contamination of DNA samples by substances that either hinder hybridization or compete with DNA in the hybridization process. Such substances can be removed or avoided by careful extraction and, if necessary, by column, filter or glass-emulsion purification. Experimental error also arises from DNA fragment length variations, which alter hybrid duplex melting properties (Crothers *et al.* 1965; Hayes *et al.* 1970). Finally, experimental error is introduced during thermal fractionation by variations in such factors as temperature, elution volume, S1 digestion and scintillation-cocktail quenching. Bleiweiss & Kirsch (1993a,b) found that preparation error was the single most consistent effect on tracer performance and that effects associated with spacing of thermal elution columns within their experimental device were not statistically significant.

Of particular concern are potential differences in AT:GC ratios and genome sizes among the taxa under comparison. Such differences may bias dissimilarity measurements substantially and therefore consistently affect measurement accuracy. Usually, researchers simply assume that AT:GC ratios and genome sizes are the same among similar taxa. Fortunately, DNA-hybridization data can provide clues to genomic differences. Relative AT:GC values are indicated in the width of melting profiles and by the melting temperature ($T_m$) of homoduplex hybrids (Marshall & Swift 1992). Relative amounts of scnDNA are determined routinely during their preparation, by virtue of the extent of reassociation at a given value of $C_0t$ (initial single stranded DNA concentration multiplied by reassociation time) value (Britten *et al.* 1974; Werman *et al.* 1996).

*Detecting error*

DNA hybridization error is determined principally from the variance in replicate hybrid indexes (e.g. Werman *et al.* 1996). Unfortunately, replicate error distributions tend to differ because of variations in DNA preparation and experimentation (Bleiweiss & Kirsch 1993a,b), and they are difficult to predict. Overall standard deviations for $\Delta T_m$ in major bird studies, for example, have ranged from 0.20 to 0.48 (Sheldon 1987b; Krajewski 1989). Systematic error may be detected by comparison of reciprocal dissimilarity values (radio-labelled DNA of taxon A to driver DNA of taxon B, and vice versa). If one sample or species consistently yields asymmetric reciprocals, then the problem is systematic (Bledsoe & Sheldon 1989; Springer & Krajewski 1989a), and steps must be taken to determine the cause and correct it. Otherwise, the phylogeny based on those data is likely to be distorted.

*Individual variation*

Genetic differences can be measured between strict homoduplexes (label and driver DNA from a single individual) and intraspecific heteroduplexes (label and driver DNAs from different individuals of the same species). Among verte-brates, these differences are typically less than $\Delta T_m = 0.5$ (Sheldon 1987b;

Werman *et al.* 1996), but much higher levels have been found in invertebrate groups such as sea urchins, for which $\Delta T_m$ values of 2–4 have been obtained (Britten *et al.* 1978; Palumbi & Metz 1991). Despite such high levels, individual variation should not compromise comparisons between species, unless there are variable rates of evolution (or other major genomic differences) among individuals within species. In other words, distances from a labelled DNA of one species to driver DNA of several individuals of another species is expected to be equal unless those individuals have evolved at different rates.

### 5.3.4  Tree-building problems and solutions

*Data metricity and additivity*

Many criticisms of DNA hybridization phylogenies (e.g. Farris 1981, 1985, 1986) focus on whether DNA-hybridization dissimilarity values meet the requirements of algorithms (section 5.3.2) used to obtain phylogenetic esti-mates. In basic terms, a dissimilarity measure must be metric (that is, have the properties of a distance) and additive (that is, the specific values of the measure must be consistent with one another throughout the matrix). Two commonly used measures, $\Delta T_m$ and $\Delta$mode, are metrics in expectation (Bledsoe & Sheldon 1989). In contrast, two other measures, $\Delta T_{50}H$ and $\Delta$NPR (normalized percentage hybridization), are not certain to be metric, because they are based on the extent of nonhybridizing sequences not directly com-pared by thermal stability analysis. Thus, when single-copy genome sizes vary, reciprocal measurements of $\Delta T_{50}H$ and $\Delta$NPR are expected to differ. With regard to additivity, no DNA hybridization dissimilarity measure is additive in expectation. Nonadditivity can be caused by multiple mutations at single nucleotide sites, homoplasy induced by multiple mutations or sequence rearrangements, variation in single-copy genome size, and measurement error that is systematic in its effects. Researchers have responded to these problems by avoiding measures that are not metric in expectation, improving biochemical procedures and transforming data based on models of DNA evo-lution. These steps are essential to obtain a matrix of distances that is additive in expectation.

*Correcting and transforming data*

Problems such as nonadditivity have prompted researchers to apply a variety of transformations and corrections to improve data. As long as these manipu-lations are based on statistical analyses of hybrid behaviour or on evolutionary models, they are reasonable and can be helpful; for example, correcting dis-similarity values for measured differences in hybrid length is appropriate and often necessary. On the other hand, subjective manipulations based largely on experience or *a posteriori* criteria of what constitutes 'good' or 'bad' data are not acceptable. These kinds of 'corrections' inject an amount of subjectivity into a data set, the main strength of which is its objectivity (e.g. Sarich *et al.* 1989; Gill & Sheldon 1991).

*Correcting for size variation*. This is the most common form of correction based on empirical evidence (e.g. Hunt *et al.* 1981; Powell & Caccone 1990). Duplex melting temperature decreases with a decrease in fragment length. Thus, in the fractionation step, size changes can confound the measurement of differences in base-pair complementarity.

*Correcting for reciprocal asymmetry*. When believed to be caused by experimental problems, asymmetry has been corrected by a procedure similar to that used in immunological comparisons (Sarich & Cronin 1976; Springer & Kirsch 1989; Krajewski & Dickerman 1990). Asymmetry can also be caused by hybrid fragment-length differences and corrected accordingly.

*Transforming data to improve additivity*. Jukes & Cantor (1969), Palumbi (1989) and colleagues have suggested methods to correct dissimilarity values for multiple mutations at single base sites. The rationale of these corrections as applied in DNA hybridization studies has been discussed by Springer & Kirsch (1989) and Springer & Krajewski (1989a,b), and they have been used by several researchers (e.g. Catzeflis *et al.* 1987; Bledsoe 1988; Sheldon *et al.* 1992).

### Strength of support for phylogenies

Support for phylogenetic trees may be tested by bootstrapping (Krajewski & Dickerman 1990; Marshall 1991) or jack-knifing (Lanyon 1985). Bootstrapping assigns quantitative levels of stability to branching points by sampling different sets of replicate dissimilarity measurements. Jack-knifing, i.e. estimating trees with different numbers of taxa, assays the effects that taxa have on tree topology. Both methods have proved useful, for example in testing controversial aspects of the Sibley & Ahlquist (1990) phylogenetic proposals (Bleiweiss *et al.* 1995). Felsenstein (1987) presented a mixed-model ANOVA to test DNA-hybridization tree hypotheses. Other methods based on Pielou's Q-statistic (Templeton 1985) or the Mann–Whitney $U$-test (Fitch 1986) have not been found valid upon scrutiny (Fitch 1986; Saitou 1986; Marshall 1991).

### 5.3.5  Testing rates of evolution

Lineage-based variable rates have been discovered several times in DNA hybridization studies (e.g. Laird *et al.* 1969; Kohne *et al.* 1972; Bonner *et al.* 1980,1981; Brownell 1983; Britten 1986; Catzeflis *et al.* 1987; Sheldon 1987a; Springer & Kirsch 1989; Caccone & Powell 1990; Goddard *et al.* 1990; Krajewski 1990). Variable rates may be detected indirectly by calibrating nucleotide change against absolute time (see section 5.5.3), or directly by comparing dissimilarity values. Direct rate testing may be accomplished in two ways: Sarich & Wilson's (1967) relative rate and Felsenstein's (1984) comparative tree test.

The relative rate test compares outgroup to ingroup distances. If rates are constant, distances from an outgroup to different ingroup members will be the same. DNA hybridization data have been examined in this way many times using a variety of statistical tests, e.g. chi-square and SNK (Bledsoe 1987a; Houde 1987a), and ANOVA and *t*-tests (e.g. Krajewski 1990; Sheldon *et al.* 1992). Felsenstein's comparative tree test uses an *F*-statistic to compare trees built with and without the assumption of a constant rate of evolution (e.g. Bledsoe 1987a; Sheldon 1987a; Krajewski 1990). A problem with all of these tests is that different dissimilarity values are based on a single average homoduplex value and thus are not independent. Independence may be improved, however, by bootstrapping dissociation values (including homoduplex values) and then computing dissimilarities (e.g. Krajewski & Dickerman 1990; Dickerman 1991).

## 5.4  Experimental design

DNA hybridization studies must be structured to permit a statistical assessment of data metricity, measurement error and rate variability. They must also accommodate the operational model of DNA hybridization presented in section 5.1.2 and the assumptions of tree-building methods. In theory, this may be difficult, especially because of the additivity requirement (section 5.3.4). The practical requirements, however, are straightforward. To detect the pattern of unique and shared basepair differences among the study group taxa, all pairwise distances must be measured (or, if not possible, at least estimated; Landry *et al.* 1996) and at least one taxon must be an outgroup. The outgroup must be designated based on data other than genetic dissimilarity. The use of more than one outgroup helps to corroborate tree rooting. It also increases the number of possible statistical tests of rates (Bledsoe 1987a; Houde 1987a). To determine error structure and data metricity, replicate and reciprocal distances need to be measured. Finally, distances must be converted into trees via a branch-fitting method that does not assume constant rates of evolution.

## 5.5 The application of DNA–DNA hybridization: strengths and limitations

### 5.5.1 Distances versus characters

Unlike most other molecular methods, DNA hybridization can only produce distances, which precludes study of molecular evolutionary patterns (e.g. in transition/transversion ratios), that require direct sequence information. In addition, every time a taxon is added to a DNA-hybridization matrix, it must be hybridized with every other taxon to obtain a complete set of dissimilarity values. As a result, the number of required comparisons increases geometrically as taxa are added to the matrix. In contrast, once a sequence has been characterized, no additional work is needed to include it in an expanded analysis of taxa.

What then are the advantages of DNA hybridization? First, a large amount of DNA can be compared quickly, and the resultant dissimilarity values are expected to be reliable because they are based on a large number of underlying character changes (Sibley & Ahlquist 1983). Second, homoplasy resulting from nonhomologous nucleotide identity is minimized by a form of congruence assessment embedded in the reassociation process (Bledsoe & Sheldon 1990). That is, the reassociation step sorts out sequences that are at least 75–80% identical in basepair complementarity (e.g. Bonner et al. 1973). This high level of identity strengthens the hypothesis of nucleotide homology at any given site within a sequence, by virtue of the congruence test (see Riedl 1978; Patterson 1988). Third, this congruence assessment becomes particularly important when comparing outgroups to ingroups. Outgroup sequences are likely to contain many sites for which identity to ingroup members is nonhomologous (Sober 1983; Wheeler 1990). Fourth, DNA hybridization provides a genome-wide measure of dissimilarity, against which the divergence of specific sequences can be assessed. Other methods lack the ability to ascertain the distribution of sequence divergence values without resort to complete or extensive sequencing of genomes. Finally, a great potential advantage of accurate estimates of dissimilarity is that they can be calibrated against time or compared with other dissimilarity measures, such as geographical distances or independently produced genetic distances, to establish rates of evolution and movement. Moreover, distance correlation can be tested by various statistical methods (e.g. the Mantel test for matrices; Rohlf 1990). Although the establishment of absolute rates remains challenging, because accurate dating of fossils is often difficult and variable rates of molecular change can undermine extrapolation from one set of lineages to another, DNA-hybridization phylogenies nonetheless form a powerful tool in the

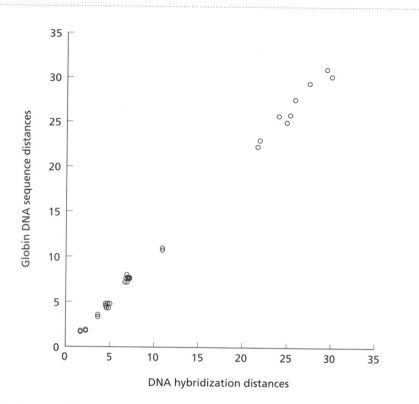

**Fig. 5.3** A comparison of DNA-hybridization $T_{50}H$ distances and distances derived from DNA sequencing of noncoding β-globin genes, including 10.8 kb of the psi-eta locus. The graph is based on values summarized by Goodman *et al.* (1990, Fig. 3). The DNA-hybridization distances were taken by Goodman and colleagues from Sibley & Ahlquist (1987) and Bonner *et al.* (1980).

phylogenetic arsenal available to researchers interested in ecological studies. DNA hybridization yields accurate measures of dissimilarity which, when compared with large sequence data sets, are remarkably similar to those computed from direct sequence information (see Fig. 5.3).

### 5.5.2 Relative dissimilarity and absolute difference

The relationship between DNA-hybridization dissimilarity values and nucleotide differences has been calibrated by hybridizing DNAs of known base-pair composition. Although estimates for $\Delta T_m$ value of 1.0 °C have ranged between 0.7% and 1.7%, depending upon which sequences were hybridized (Bautz & Bautz 1964; Britten *et al.* 1974; Caccone *et al.* 1988a), it appears that the best calibration is that of Springer *et al.* (1992). They measured melting temperature depression for primate eta-globin pseudogenes that ranged from 1.6%

difference (human and chimpanzee) to nearly 11% difference (human and spider monkey). Depending upon regression criteria, they found that a $\Delta T_m$ value of 1.0 °C corresponds to a 1.14–1.22 nucleotide difference, and the relationship is linear throughout the range. Unlike $\Delta T_m$, $\Delta$mode and $\Delta T_{50}H$ have never been calibrated. Their approximate relationship to percentage difference, however, has been demonstrated indirectly by comparing them with $\Delta T_m$; for example, Springer et al. (1990) found $\Delta T_{50}H = 1.25$ ($T_m^{1.04}$), and Sheldon et al. (1992) found $\Delta T_{50}H = 1.08$ ($\Delta T_m$) $+0.007$ ($\Delta T_m^2$). Because $\Delta T_m$ and $\Delta$mode are essentially equivalent for most taxa and $\Delta T_{50}H$ is well-correlated with them up to approximately $\Delta 10$–12 (Bledsoe & Sheldon 1989; Sheldon et al. 1992), the calculation of percent difference among them usually presents no problem. Even so, some caution is required with $\Delta T_{50}H$, which depends upon the assumption that percentage reassociation decreases linearly with duplex melting temperature. This assumption does not hold true for invertebrate groups that have large sections of extremely rapidly evolving sequences (e.g. Caccone et al. 1988a). Beyond approximately $\Delta 12$, $T_m$ becomes highly compressed, but $\Delta$mode and $\Delta T_{50}H$ are well-correlated up to approximately $\Delta 18$–20 (Sibley & Ahlquist 1983; Sheldon & Bledsoe 1989). Beyond a $\Delta$mode value of 20, only $\Delta T_{50}H$ can be used to measure dissimilarity.

### 5.5.3 Divergence dating and absolute rate calibrations

The existence of evolutionary rate differences (section 5.3.5) requires the careful calibration of distance with time and extrapolation of rates and dates from one lineage to another. In all but those taxa for which the fossil record is extremely good, the determination of absolute rates of evolution is difficult, because four independent dates are necessary (Gingerich 1986; Moritz et al. 1987). Even hominoid datings, which are based on a good fossil record, are sloppy enough to introduce tremendous variation in estimated rates; for example, Sibley & Ahlquist (1987, p. 118) found that distance-to-time for the apes ranged from $\Delta T_{50}H$ 1.0 = 3.42 to $\Delta T_{50}H = 4.76$ because of the uncertainty in Orangutan-African ape divergence date.

### 5.6 Summary

Solution DNA–DNA hybridization provides a powerful technique for estimating phylogenetic relationships, and hence for providing ecologists with information on evolutionary history that is essential for testing many ecological hypotheses. The state of DNA hybridization practice and theory is sufficiently well-advanced that the technique, when applied phylogenetically, routinely gives results concordant with other techniques (e.g. cladistic analysis). Recent studies of measurement error in DNA-hybridization data have defined the level of resolution, in terms of percentage sequence divergence, that the technique offers, and biochemical advances have narrowed that level compared to previous studies (e.g. Sibley & Ahlquist 1990). Solution DNA–DNA hybridization offers the opportunity to detect changes in the rate of DNA among lineages, and well-supported analytical procedures exist by which to reconstruct phylogeny, despite rate variation. The technique has the limitation that the number of required pairwise measurements increases with the square of the number of species studied, making the technique inappropriate for large groups. Nonetheless, the technique gives good measures of genome-wide genetic distance, and hence has applications that are not possible in the absence of extensive direct sequence data.

## References

Ahlquist, J.E., Bledsoe, A.H., Sheldon, F.H. & Sibley, C.G. (1987) DNA hybridization and avian systematics. *Auk* **104**, 556–563.

Andrews, P. (1986) Molecular evidence for catarrhine evolution. In: *Major Topics in Primate and Human Evolution* (eds B. Wood, L. Martin & P. Andrews), pp. 107–129. Cambridge University Press, Cambridge.

Andrews, P. (1987) Aspects of hominoid phylogeny. In: *Molecules and Morphology in Evolution: Conflict or Compromise* (ed. C. Patterson), pp. 23–53. Cambridge University Press, Cambridge.

Balkwill, D.L., Drake, G.R., Reeves, R.H., *et al.* (1997) Taxonomic study of aromatic-degrading bacteria from deep-terrestrial-subsurface sediments and description of *Sphingomonas aromaticivorans* sp. nov., *Sphingomonas subterranea* sp. nov. & *Sphingomonas stygia* sp. nov. *International Journal of Systematic Bacteriology* **47**, 191–201.

Bautz, E.K. & Bautz, F.A. (1964) The influence of noncomplementary bases on the stability of ordered polynucleotides. *Proceedings of the National Academy of Sciences USA* **52**, 1476–1481.

Benveniste, R.E. (1985) The contributions of retroviruses to the study of mammalian evolution. In: *Molecular Evolutionary Genetics* (ed. R.J. MacIntyre), pp. 359–417. Plenum Press, New York.

Benveniste, R.E. & Todaro, G.J. (1976) Evolution of type C viral genes: evidence for an Asian origin of man. *Nature* **261**, 101–108.

Blackstone, N.W. & Sheldon, F.H. (1991) The relationship between hetero- and homoduplex melting temperatures in studies of DNA–DNA hybridization. *Systematic Zoology* **40**, 89–95.

Bledsoe, A.H. (1987a) DNA evolutionary rates in nine-primaried passerine birds. *Molecular Biology and Evolution* **4**, 559–571.

Bledsoe, A.H. (1987b) Estimation of phylogeny from molecular distance data: the issue of variable rates. *Auk* **104**, 563–565.

Bledsoe, A.H. (1988) Nuclear DNA evolution and phylogeny of the New World nine-primaried oscines. *Auk* **105**, 504–515.

Bledsoe, A.H. & Raikow, R.J. (1990) A quantitative assessment of congruence between molecular and nonmolecular estimates of phylogeny. *Journal of Molecular Evolution* **30**, 247–259.

Bledsoe, A.H. & Sheldon, F.H. (1989) The metric properties of DNA–DNA hybridization dissimilarity measures. *Systematic Zoology* **38**, 93–105.

Bledsoe, A.H. & Sheldon, F.H. (1990) Molecular homology and DNA hybridization. *Journal of Molecular Evolution* **30**, 425–433.

Bleiweiss, R. & Kirsch, J.A.W. (1993a) Experimental analysis of variance for DNA hybridization: I. Accuracy. *Journal of Molecular Evolution* **37**, 504–513.

Bleiweiss, R. & Kirsch, J.A.W. (1993b) Experimental analysis of variance for DNA hybridization: II. Precision. *Journal of Molecular Evolution* **37**, 514–524.

Bleiweiss, R., Kirsch, J.A.W. & Shafi, N. (1995) Confirmation of a portion of the Sibley-Ahlquist 'tapestry'. *Auk* **112**, 87–97.

Bleiweiss, R., Kirsch, J.A.W. & Matheus, J.C. (1997) DNA hybridization evidence for the principal lineages of hummingbirds. *Molecular Biology and Evolution* **14**, 325–343.

Bonner, T.I., Brenner, D.J., Neufeld, B.R. & Britten, R.J. (1973) Reduction in the rate of DNA reassociation by sequence divergence. *Journal of Molecular Biology* **81**, 123–135.

Bonner, T.I., Heinemann, R. & Todaro, G.J. (1980) Evolution of DNA sequences has been retarded in Malagasy primates. *Nature* **286**, 420–423.

Bonner, T.I., Heinemann, R. & Todaro, G.J. (1981) A geographical factor involved in the evolution of the single copy DNA sequences of primates. In: *Evolution Today* (eds G.G.E. Scudder & J.L. Reveal), pp. 293–300. Carnegie-Mellon University, Pittsburgh, Pennsylvania.

Britten, R.J. (1986) Rates of DNA sequence evolution differ between taxonomic groups. *Science* **231**, 1393–1398.

Britten, R.J., Graham, D.E. & Neufeld, B.R. (1974) Analysis of repeating DNA sequences by reassociation. *Methods in Enzymology* **29**, 363–418.

Britten, R.J., Cetta, A. & Davidson, E.H. (1978) The single-copy DNA sequence polymorphism of the sea urchin *Strongylocentrotus purpuratus*. *Cell* **15**, 1175–1186.

Brown, W.M., George, M. & Wilson, A.C. (1979) Rapid evolution of animal mitochondrial DNA. *Proceedings of the National Academy of Sciences USA* **76**, 1967–1971.

Brownell, E. (1983) DNA/DNA hybridization studies of muroid rodents: symmetry and rates of molecular evolution. *Evolution* **37**, 1034–1051.

Caccone, A. & Powell, J.R. (1987) Molecular evolutionary divergence among North American cave crickets. II. DNA–DNA hybridization. *Evolution* **41**, 1215–1238.

Caccone, A. & Powell, J.R. (1989) DNA divergence among hominoids. *Evolution* **43**, 925–942.

Caccone, A. & Powell, J.R. (1990) Extreme rates and heterogeneity in insect DNA evolution. *Journal of Molecular Evolution* **30**, 273–280.

Caccone, A., Amato, G.D. & Powell, J.R. (1987) Intraspecific DNA divergence in *Drosophila*: a study on parthenogenetic *D. mercatorum*. *Molecular Biology and Evolution* **4**, 343–350.

Caccone, A., Amato, G.D. & Powell, J.R. (1988a) Rates and patterns of scnDNA and mtDNA divergence within the *Drosophila melanogaster* subgroup. *Genetics* **118**, 671–783.

Caccone, A., DeSalle, R. & Powell, J.R. (1988b) Calibration of the change in thermal stability of DNA duplexes and degree of base pair mismatch. *Journal of Molecular Evolution* **27**, 212–216.

Caccone, A., Gleason, J.M. & Powell, J.R. (1992) Complementary DNA–DNA hybridization in *Drosophila*. *Journal of Molecular Evolution* **34**, 130–140.

Caccone, A., Moriyama, E.N., Gleason, J.M., Nigro, L. & Powell, J.R. (1996) A molecular phylogeny for the *Drosophila melanogaster* subgroup and the problem of polymorphism data. *Molecular Biology and Evolution* **13**, 1224–1232.

Carpenter, J.M. (1985) The clock of evolution needs repair. *Natural History* **94** (6), 6.

Catzeflis, F.M., Sheldon, F.H., Ahlquist, J.E. & Sibley, C.G. (1987) DNA–DNA hybridization evidence of the rapid rate of muroid rodent DNA evolution. *Molecular Biology and Evolution* **4**, 242–253.

Cavalli-Sforza, L.L. & Edwards, A.W.F. (1967) Phylogenetic analysis: models and estimation procedures. *Evolution* **21**, 550–570.

Cracraft, J. (1987) DNA hybridization and avian phylogenetics. *Evolutionary Biology* **21**, 47–96.

Crothers, D.M., Kallenback, N.R. & Zimm, B.H. (1965) The melting transition of low-molecular-weight DNA: theory and experiment. *Journal of Molecular Biology* **11**, 802–820.

Cunningham, C.W., Buss, L.W. & Anderson, C. (1991) Molecular and geologic evidence of shared history between hermit crabs and the symbiotic genus *Hydractinia*. *Evolution* **45**, 1301–1316.

Dickerman, A.W. (1991) Among-run artifacts in DNA hybridization. *Systematic Zoology* **40**, 494–499.

Douzery, E., Lebreton, J.D. & Catzeflis, F.M. (1995) Testing the generation time hypothesis using DNA/DNA hybridization between artiodactyls. *Journal of Evolutionary Biology* **8**, 511–529.

Ekman, J. (1989) Ecology of non-breeding social systems of *Parus. Wilson Bulletin* **101**, 263–288.

Farris, J.S. (1972) Estimating phylogenetic trees from distance matrices. *American Naturalist* **106**, 645–668.

Farris, J.S. (1981) Distance data in phylogenetic analysis. *Advances in Cladistics* **1**, 3–23.

Farris, J.S. (1985) Distance data revisited. *Cladistics* **1**, 67–85.

Farris, J.S. (1986) Distances and statistics. *Cladistics* **2**, 144–157.

Felsenstein, J. (1984) Distance methods for inferring phylogenies: a justification. *Evolution* **38**, 16–24.

Felsenstein, J. (1987) Estimation of hominoid phylogeny from a DNA hybridization data set. *Journal of Molecular Evolution* **26**, 123–131.

Felsenstein, J. (1995) *PHYLIP—phylogeny inference package*, v. 3.5c. Department of Genetics, University of Washington, Seattle, Washington.

Felsenstein, J. (1997) An alternating least squares approach to inferring phylogenies from pairwise distances. *Systematic Biology* **46**, 101–111.

Fitch, W.M. (1986) Commentary. *Molecular Biology and Evolution* **3**, 296–298.

Fitch, W.M. & Margoliash, E. (1967) Construction of phylogenetic trees. *Science* **155**, 279–284.

Gill, F.B. & Sheldon, F.H. (1991) The birds reclassified. [Review of] *Phylogeny and classification of birds*, by C. G. Sibley and J. E. Ahlquist. *Science* **252**, 1003–1005.

Gill, F.B. & Slikas, B. (1992) Patterns of mitochondrial DNA divergence in North American crested titmice. *Condor* **94**, 20–28.

Gill, F.B., Mostrom, A.M. & Mack, A.L. (1992) Speciation in North American chickadees: patterns of genetic divergence. *Evolution* **47**, 195–212.

Gingerich, P.D. (1986) Temporal scaling of molecular evolution in primates and other mammals. *Molecular Biology and Evolution* **3**, 205–221.

Goddard, K., Caccone, A. & Powell, J.R. (1990) Evolutionary implications of DNA divergence in the *Drosophila obscura* group. *Evolution* **44**, 1656–1670.

Goodman, M., Tagle, D.A., Fitch, D.H.A. *et al.* (1990) Primate evolution at the DNA level and a classification of hominoids. *Journal of Molecular Evolution* **30**, 260–266.

Grula, J.W., Hall, T.J., Hunt, J.A. *et al.* (1982) Sea urchin DNA sequence variation and reduced interspecies differences of the less variable DNA sequences. *Evolution* **36**, 665–676.

Hayes, F.N., Lilly, E.H., Ratliff, R.L., Smith, D.A. & Williams, D.L. (1970) Thermal transitions in mixtures of polydeoxynucleotides. *Biopolymers* **9**, 1105–1117.

Helm-Bychowski, K.M. & Wilson, A.C. (1986) Rates of nuclear DNA evolution in pheasant-like birds: evidence from restriction maps. *Proceedings of the National Academy of Sciences USA* **83**, 688–692.

Holmquist, R., Miyamoto, M.M. & Goodman, M. (1988a) Higher-primate phylogeny—why can't we decide? *Molecular Biology and Evolution* **5**, 201–216.

Holmquist, R., Miyamoto, M.M. & Goodman, M. (1988b) Analysis of higher-primate phylogeny from transversion differences in nuclear and mitochondrial DNA by Lake's methods of evolutionary parsimony and operator metrics. *Molecular Biology and Evolution* **5**, 217–236.

Houde, P. (1987a) Critical evaluation of DNA hybridization studies in avian systematics. *Auk* **104**, 17–32.

Houde, P. (1987b) Response to A. H. Bledsoe and J. F. Ahlquist *et al. Auk* **104**, 566–568.

Hoyer, B.H., van de Velde, N.W., Goodman, M. & Roberts, R.B. (1972) Examination of hominid evolution by DNA sequence homology. *Journal of Human Evolution* **1**, 645–649.

Hunt, J.A., Hall, T.J. & Britten, R.J. (1981) Evolutionary distance in Hawaiian Drosophila measured by DNA reassociation. *Journal of Molecular Evolution* **17**, 361–367.

Jukes, T.H. & Cantor, C.R. (1969) Evolution of protein molecules. In: *Mammalian Protein Metabolism.* (ed. H.N. Munro), pp. 21–123. Academic Press, New York.

Kirsch, J.A.W., Springer, M.S., Krajewski, C., Archer, M., Aplin, K. & Dickerman, A.W. (1990) DNA/DNA hybridization studies of the carnivorous marsupials. I: The intergeneric relationships of bandicoots (Marsupialia: Perameloidea). *Journal of Molecular Evolution* **30**, 434–448.

Kirsch, J.A.W., Dickerman, A.W., Reig, O.A. & Springer, M.S. (1992) DNA hybridization evidence for the Australasian affinity of the American marsupial *Dromiciops australis*. *Proceedings of the National Academy of Sciences USA* **88**, 10465–10469.

Kirsch, J.A.W., Lapointe, F.J. & Springer, M.S. (1997) DNA-hybridisation studies of marsupials and their implications for metatherian classification. *Australian Journal of Zoology* **45**, 211–280.

Kohne, D.E. (1970) Evolution of higher-organism DNA. *Quarterly Review of Biophysics* **3**, 327–375.

Kohne, D.E., Chiscon, J.A. & Hoyer, B.H. (1972) Evolution of primate DNA sequences. *Journal of Human Evolution* **1**, 627–644.

Krajewski, C. (1989) Phylogenetic relationships among cranes (Gruiformes: Gruidae) based on DNA hybridization. *Auk* **106**, 603–618.

Krajewski, C. (1990) Relative rates of single-copy DNA evolution in cranes. *Molecular Biology and Evolution* **7**, 65–73.

Krajewski, C. (1994) Phylogenetic measures of biodiversity: a comparison and critique. *Biological Conservation* **69**, 33–39.

Krajewski, C. & Dickerman, A.W. (1990) Bootstrap analysis of phylogenetic trees derived from DNA hybridization distances. *Systematic Zoology* **39**, 383–390.

Krajewski, C. & Fetzner, J.W. Jr (1994) Phylogeny of cranes (Gruiformes: Gruidae) based on cytochrome-B DNA sequences. *Auk* **111**, 351–365.

Krebs, J.R., Healy, S.D. & Shettleworth, S.J. (1990) Spatial memory of Paridae: comparison of a storing and a non-storing species, the Coal Tit, *Parus ater*, and the Great Tit, *P. major*. *Animal Behavior* **39**, 1127–1137.

Kuhner, M.K. & Felsenstein, J. (1994) A simulation comparison of phylogeny algorithms under equal and unequal evolutionary rates. *Molecular Biology and Evolution* **11**, 459–468.

Laguerre, G., Fernandez, M.P., Edel, V., Normand, P. & Amarger, N. (1993) Genomic heterogeneity among French *Rhizobium* strains isolated from *Phaseolus vulgaris* L. *International Journal of Systematic Bacteriology* **43**, 761–767.

Laird, C.D., McConaughy, B.L. & McCarthy, B.J. (1969) Rate of fixation of nucleotide substitutions in evolution. *Nature* **224**, 149–154.

Landry, P.A., Lapointe, F.J. & Kirsch, J.A.W. (1996) Estimating phylogenies from lacunose distance matrices: additive is superior to ultrametric estimation. *Molecular Biology and Evolution* **13**, 818–823.

Lanyon, S.M. (1985) Detecting internal inconsistencies in distance data. *Systematic Zoology* **34**, 397–403.

Lanyon, S.M. (1992) [Review of] 'Phylogeny and classification of birds: a study in molecular evolution', by C. G. Sibley and J. E. Ahlquist. *Condor* **94**, 304–307.

McCracken, K.G. & Sheldon, F.H. (1997) Avian vocalizations and phylogenetic signal. *Proceedings of the National Academy of Sciences USA* **94**, 3833–3836.

Marks, J., Schmid, C.W. & Sarich, V.M. (1988) DNA hybridization as a guide to phylogenies: relations of the Hominoidea. *Journal of Human Evolution* **17**, 769–786.

Marshall, C.R. (1991) Statistical tests and bootstrapping: assessing the reliability of phylogenies based on distance data. *Molecular Biology and Evolution* **8**, 386–391.

Marshall, C.R. (1992) Character analysis and the integration of molecular and morphological data in an understanding of sand dollar phylogeny. *Molecular Biology and Evolution* **9**, 309–322.

Marshall, C.R. & Swift, H. (1992) DNA–DNA hybridization phylogeny of sand dollars and highly reproducible extent of hybridization values. *Journal of Molecular Evolution* **34**, 31–44.

Miyamoto, M.M. & Goodman, M. (1990) DNA systematics and evolution of primates. *Annual Review of Ecology and Systematics* **21**, 197–220.

Moritz, C., Dowling, T.E. & Brown, W.M. (1987) Evolution of animal mitochondrial DNA: Relevance for population biology and systematics. *Annual Review of Ecology and Systematics* **18**, 269–292.

Oyaizu, H., Matsumoto, S., Minamisawa, K. & Gamou, T. (1993) Distribution of rhizobia in leguminous plants surveyed by phylogenetic identification. *Journal of General and Applied Microbiology* **39**, 339–354.

Palumbi, S.R. (1989) Rates of molecular evolution and the fraction of nucleotide positions free to vary. *Journal of Molecular Evolution* **29**, 180–187.

Palumbi, S.R. & Metz, E.C. (1991) Strong reproductive isolation between closely related tropical sea urchins (genus *Echinometra*). *Molecular Biology and Evolution* **8**, 227–239.

Patterson, C. (1988) Homology in classical and molecular biology. *Molecular Biology and Evolution* **5**, 603–625.

Platnick, N.I. (1985) More clock of evolution. *Natural History* **94** (8), 4.

Powell, J.R. & Caccone, A. (1990) The TEACL method of DNA–DNA hybridization: technical considerations. *Journal of Molecular Evolution* **30**, 267–272.

Primmer, C.R., Moller, A.P. & Ellegren, H. (1996) A wide-range survey of cross-species microsatellite amplification in birds. *Molecular Ecology* **5**, 365–378.

Ricklefs, R.E. (1987) Community diversity: relative roles of local and regional processes. *Science* **235**, 167–171.

Riedl, R.E. (1978) *Order in Living Organisms*. J. Wiley, Chichester.

Roberts, J.W., Johnson, S.A., Kier, P., Hall, T.J., Davidson, E.H. & Britten, R.J. (1985) Evolutionary conservation of DNA sequences expressed in sea urchin eggs and early embryos. *Journal of Molecular Evolution* **22**, 99–107.

Rohlf, F.J. (1990) *NTSYS-pc. Numerical taxonomy and multivariate analysis system*, v. 1.60. Exeter Software, Setauket, New York.

Saitou, N. (1986) On the delta Q-test of Templeton. *Molecular Biology and Evolution* **3**, 282–284.

Saitou, N. & Nei, M. (1987) The neighbor-joining method: a new method for reconstructing phylogenetic trees. *Molecular Biology and Evolution* **4**, 406–425.

Sarich, V.M. & Cronin, J.E. (1976) Molecular systematics of the primates. In: *Molecular Anthropology* (eds M. Goodman & R.E. Tashian), pp. 141–170. Plenum Press, New York.

Sarich, V.M. & Wilson, A.C. (1967) Immunological time scale for hominoid evolution. *Science* **158**, 1200–1203.

Sarich, V.M., Schmid, C.W. & Marks, J. (1989) DNA hybridization as a guide to phylogenies: a critical analysis. *Cladistics* **5**, 3–32.

Schmid, C.W. & Marks, J. (1990) DNA hybridization as a guide to phylogeny: chemical and physical limits. *Journal of Molecular Evolution* **30**, 237–246.

Sheldon, F.H. (1987a) Rates of single-copy DNA evolution in herons. *Molecular Biology and Evolution* **4**, 56–59.

Sheldon, F.H. (1987b) Phylogeny of herons estimated from DNA–DNA hybridization data. *Auk* **104**, 97–108.

Sheldon, F.H. & Bledsoe, A.H. (1989) Indexes to the reassociation and stability of solution DNA hybrids. *Journal of Molecular Evolution* **29**, 328–343.

Sheldon, F.H. & Bledsoe, A.H. (1993) Avian molecular systematics, 1970s to 1990s. *Annual Review of Ecology and Systematics* **24**, 243–278.

Sheldon, F.H. & Gill, F.B. (1996) A reconsideration of songbird phylogeny, with emphasis on the evolution of titmice and their sylvioid relatives. *Systematic Biology* **45**, 473–495.

Sheldon, F.H. & Kinnarney, M. (1993) The effects of sequence removal on DNA-hybridization estimates of distance, phylogeny, and rates of evolution. *Systematic Biology* **42**, 32–48.

Sheldon, F.H. & Winkler, D.W. (1993) Intergeneric phylogenetic relationships of swallows estimated by DNA–DNA hybridization. *Auk* **110**, 798–824.

Sheldon, F.H., Slikas, B., Kinnarney, M., Gill, F.B., Zhao, E. & Silverin, B. (1992) DNA–DNA hybridization evidence of phylogenetic relationships among major lineages of *Parus*. *Auk* **109**, 173–185.

Sherry, D.F. (1989) Food storing in the Paridae. *Wilson Bulletin* **101**, 289–304.

Sibley, C.G. (1997) Proteins and DNA in systematic biology. *Trends in Biochemical Science* **22**, 364–367.

Sibley, C.G. & Ahlquist, J.E. (1981) The phylogeny and relationships of the ratite birds as indicated by DNA–DNA hybridization. In: *Evolution Today* (eds G.G.E. Scudder & J.L. Reveal), pp. 301–335. Carnegie-Mellon University, Pittsburgh, Pennsylvania.

Sibley, C.G. & Ahlquist, J.E. (1983) Phylogeny and classification of birds based on the data of DNA–DNA hybridization. *Current Ornithology* **1**, 245–292.

Sibley, C.G. & Ahlquist, J.E. (1986) Reconstructing bird phylogeny and comparing DNAs. *Scientific American* **254** (2), 82–92.

Sibley, C.G. & Ahlquist, J.E. (1987) DNA hybridization evidence of hominoid phylogeny: results from an expanded data set. *Journal of Molecular Evolution* **26**, 99–121.

Sibley, C.G. & Ahlquist, J.E. (1990) *Phylogeny and Classification of Birds*. Yale University Press, New Haven, Connecticut.

Sibley, C.G., Ahlquist, J.E. & Sheldon, F.H. (1987) DNA hybridization and avian phylogenetics: reply to Cracraft. *Evolutionary Biology* **21**, 97–125.

Sibley, C.G., Comstock, J.A. & Ahlquist, J.E. (1990) DNA hybridization evidence of hominoid phylogeny: a reanalysis of the data. *Journal of Molecular Evolution* **30**, 202–236.

Slikas, B., Sheldon, F.H. & Gill, F.B. (1996) Phylogeny of titmice (Paridae): I. Estimate of relationships among subgenera based on DNA–DNA hybridization. *Journal of Avian Biology* **27**, 70–82.

Sneath, P.H.A. & Sokal, R.R. (1973) *Numerical Taxonomy*. W. H. Freeman & Co, San Francisco.

Sober, E. (1983) Parsimony in systematics: philosophical issues. *Annual Review of Ecology and Systematics* **14**, 335–357.

Springer, B., Boettger, E.C., Krischner, P. & Wallace, R.J. Jr (1995a) Phylogeny of the *Mycobacterium chelonae*-like organism based on partial sequencing of the 16S rRNA gene and proposal of *Mycobacterium mucogenicum* sp. nov. *International Journal of Systematic Bacteriology* **45**, 262–267.

Springer, M.S. & Kirsch, J.A.W. (1989) Rates of single-copy DNA evolution in phalangeriform marsupials. *Molecular Biology and Evolution* **6**, 331–341.

Springer, M.S. & Kirsch, J.A.W. (1991) DNA hybridization, the compression effect, and the radiation of diprotodontian marsupials. *Systematic Zoology* **40**, 131–151.

Springer, M. & Krajewski, C. (1989a) DNA hybridization in animal taxonomy: a critique from first principles. *Quarterly Review of Biology* **64**, 291–318.

Springer, M.S. & Krajewski, C. (1989b) Additive distances, rate variation, and the perfect-fit theorem. *Systematic Zoology* **38**, 371–375.

Springer, M.S., Kirsch, J.A.W., Aplin, K. & Flannery, T. (1990) DNA hybridization, cladistics, and the phylogeny of phalangerid marsupials. *Journal of Molecular Evolution* **30**, 298–311.

Springer, M.S., Davidson, E.H. & Britten, R.J. (1992) Calculation of sequence divergence from the thermal stability of DNA heteroduplexes. *Journal of Molecular Evolution* **34**, 379–382.

Springer, M.S., Hooar, L.J. & Kirsch, J.A.W. (1995b) Phylogeny, molecules versus morphology, and rates of character evolution among fruit bats (Chiroptera: Megachiroptera). *Australian Journal of Zoology* **43**, 557–582.

Stackebrandt, E. & Goebel, B.M. (1994) Taxonomic note: a place for DNA–DNA reassociation and 16S rRNA sequence analysis in the present species definition in bacteriology. *International Journal of Systematic Bacteriology* **44**, 846–849.

Templeton, A.R. (1985) The phylogeny of the hominoid primates: a statistical analysis of the DNA–DNA hybridization data. *Molecular Biology and Evolution* **2**, 420–433.

Templeton, A.R. (1986) Further comments on the statistical analysis of DNA–DNA hybridization data. *Molecular Biology and Evolution* **3**, 290–295.

Thorpe, J.P. (1982) The molecular clock hypothesis: biochemical evolution, genetic differentiation and systematics. *Annual Review of Ecology and Systematics* **13**, 139–168.

Vawter, L. & Brown, W.M. (1986) Nuclear and mitochondrial DNA comparisons reveal extreme rate variation in the molecular clock. *Science* **234**, 194–196.

Verneau, O., Renaud, F. & Catzeflis, F.M. (1991) DNA reassociation kinetics and genome complexity of a fish (*Psetta maxima*: Teleostei) and its gut parasite (*Bothriocephalus gregarius*: Cestoda). *Biochemical Physiology* **99B**, 883–886.

Werman, S.D., Springer, M. S. & Britten, R. J. (1996) Nucleic acids I: DNA–DNA hybridization. In: *Molecular Systematics*, 2nd edn (eds D.M. Hillis, C. Moritz & B.K. Mable), pp. 169–204. Sinauer Associates, Sunderland, Massachusetts.

Wheeler, W.C. (1990) Nucleic acid sequence phylogeny and random outgroups. *Cladistics* **6**, 363–367.

Winkler, D.W. & Sheldon, F.H. (1993) Evolution of nest construction in swallows (Hirundinidae): a molecular phylogenetic perspective. *Proceedings of the National Academy of Sciences USA* **90**, 5705–5707.

Zink, R.M. & Weller, S.J. (1991) [Review of] 'Molecular systematics', by D. M. Hillis & C. Moritz]. *Auk* **108**, 452–455.

# CHAPTER 6

# DNA Fingerprinting using Minisatellite Probes

ROYSTON E. CARTER

## 6.1 Introduction

Polymorphic systems based on many biochemical systems such as blood groups, isozymes and restriction fragment-length polymorphisms (RFLPs) have been exploited by forensic scientists to match suspects to crimes (Dodd 1985; Sensabaugh & Crim 1986). In 1985 a new molecular biology technique was developed (Jeffreys *et al.* 1985a) with the ability to detect individual-specific patterns and therefore permit an unequivocal match of a forensic sample from the scene of crime with a suspect. This technique was called DNA fingerprinting.

In its original definition, DNA fingerprinting is the detection of an individual-specific hybridization pattern using various classes of highly polymorphic repetitive DNA sequences known as minisatellites. There are now, however, other techniques based on polymerase chain reaction (PCR) amplification, which detect different classes of highly polymorphic sequences (e.g. microsatellites), that also provide a high degree of individual discrimination. These alternative techniques are often called DNA profiling as they usually do not provide the same discriminatory power as DNA fingerprinting. This chapter will concentrate on DNA fingerprinting methods based on hybridization with minisatellite probes, as PCR-based methods are described in detail in Chapters 3, 9 and 10.

Detection of the minisatellites, usually by Southern blot hybridization techniques, results in patterns of bands on autoradiographs that resemble supermarket bar codes. The bands (alleles) detected at these genetic loci demonstrate extreme allelic length variation, such that the pattern can be individual-specific. A further feature of the bands is that they are generally inherited in a simple Mendelian manner (half of the bands are inherited from each parent on average), and so DNA fingerprints can be used for analysis of close familial relationships.

Provided that suitable probes are available, a similar degree of individual recognition is possible in other species as well. However, for most ecological studies, as well as human paternity testing, the simple inheritance of the bands is utilized for analyses of close relatives and parentage analysis.

## 6.2 Genome organization and minisatellite structure

The genome of eukaryotic organisms is often described as being composed of three classes of DNA, each defined according to its relative abundance. Single copy sequences, those that occur once per haploid genome, include most genes together with their 3′ and 5′ untranslated regions and introns. Moderately repetitive sequences include certain multicopy gene families such as ribosomal RNA (rRNA) genes that occur up to several thousand times per haploid genome. Highly repetitive sequences are usually noncoding DNA. There are two distinct types of repetitive DNA, interspersed repeats and tandemly repetitive sequences. Interspersed repeats consist of individual units of specific sequences distributed throughout the genome. There are two general types differentiated by the size of the repeating unit; short interspersed repeats (SINES) that are less than 500 bp, and those with units longer than 500 bp known as long interspersed repeats (LINES). Tandemly repetitive sequences comprise shorter units, typically ranging from 9 to 250 bp, repeated hundreds or even thousands of times. These are often referred to as satellite DNA, as they commonly have distinct nucleotide composition which confers specific buoyancy such that they can be isolated from genomic DNA on caesium chloride gradients. Each class of sequences is ubiquitous in the genome of all eukaryotes, but repetitive sequences are rare in most prokaryotes. In eukaryotes, repetitive sequences can constitute from 1% to 60% of the genome, with copies of individual sequences ranging from 1 to $10^6$ per genome. The genomes of many amphibians and, in particular, flowering plants are often extremely abundant in repetitive DNA. Related species may differ greatly in their relative abundance. Repetitive DNAs tend to be associated with heterochromatin such as centromeres or telomeres, but many are widely dispersed through the genome, interspersed with single-copy DNA (euchromatin).

Minisatellite DNA, also sometimes called variable number of tandem repeats (VNTRs) (Nakamura *et al.* 1987), is a subtype of repetitive DNA consisting of small to moderate-sized repeated units. These loci generally consist of reiterated sequences that share the property of highly polymorphic length. They have been discovered in the genomes of man and other organisms, and fall into several categories; for example, sequence analysis has shown that 33.6 and 33.15 and many other related probes all contain consensus sequence, since referred to as the 'core', an almost invariant GGGCAGGAXG, that shares similarity to an *Escherichia coli* recombination hot-spot called χ (Jeffreys *et al.* 1985a). Other classes include satellites with unevenly distributed point mutations, e.g. *Taq*I polymorphisms of human satellite III (Fowler *et al.* 1987, 1988), proline-rich exonic repeats (Lyons *et al.*

1988) and regions of the major histocompatibility complex (MHC) (Benoist *et al.* 1983). All of these loci consist of arrays of relatively short (occasionally up to 50 bp) units repeated and arranged in tandem. Heterozygosity estimates of 90–99% for some of these loci make them among the most variable of genetic markers. Microsatellites, also called simple sequence repeats (SSRs) (Tautz 1989), are a different class of polymorphic DNA, consisting of relatively simple mono-, di-, tri- or tetranucleotide repeats of the type [AT]$n$ or [GATC]$n$. They are usually detected using various PCR-based techniques. Details are presented in Chapters 3, 9 and 10.

### 6.2.1 Overview of DNA fingerprinting

The simultaneous detection of multiple hypervariable regions (HVRs), the basis for DNA fingerprinting, was first demonstrated by Alec Jeffreys and coworkers (Jeffreys *et al.* 1985a). Briefly, they used a 33 bp repeated sequence isolated from the 3′ end of the first intron of the human myoglobin gene as a probe at low stringency on Southern blots of human DNA, and detected many different bands, some of which were highly polymorphic. This led them to isolate these related loci from a genomic library (Jeffreys *et al.* 1985a), including two probes named 33.6 and 33.15, which are still among the best and most widely used multilocus fingerprinting probes in humans and other organisms. Using these probes they demonstrated the extreme discriminatory power of DNA fingerprinting and showed that relationships between closely related individuals could be resolved (Jeffreys *et al.* 1985b,c). Since then similar DNA fingerprints have been prepared from many different species for many purposes.

### 6.2.2 Multilocus and single locus DNA fingerprinting

Multilocus DNA fingerprinting is the simultaneous detection of multiple highly polymorphic loci from independent but related (by sequence homology) loci using a multilocus probe (MLP), usually by Southern blot hybridization at low stringency. Single locus fingerprinting is the detection of alleles at a specific minisatellite locus using a single locus probe (SLP), usually by Southern blot hybridization at high stringency.

Many different minisatellite loci scattered throughout the genome can be partitioned into a variety of families related to each other by homology of their monomeric sequence. Consequently, when used as probes under suitable conditions (usually moderate to low stringency), cloned minisatellites can be hybridized to Southern-blotted DNA to detect simultaneously the alleles from several related loci (Jeffreys *et al.* 1985a). The alleles at each locus are inherited in a simple Mendelian manner, and act as codominant characters. Because several loci are detected simultaneously, and the majority of alleles that are detected are rare, the pattern of hybridizing bands is usually very complex. In most cases, it will be specific for the individual from which the DNA was extracted (except for monozygotic twins or clonal populations). Even though closely related individuals (e.g. siblings) share some alleles that they have coinherited from their common parents, they are very unlikely to have inherited exactly the same combination of parental alleles. The profile of an individual is generally stable both in the soma and the germline so that any tissue can be used to give a representative DNA fingerprint. Over evolutionary time, however, minisatellites have a high mutation rate, and this is what causes the high allelic diversity seen in DNA fingerprints. The ability of some cloned minisatellites to cross-hybridize to analogous sequences and detect MLP-like patterns in other organisms was independently demonstrated by several groups during 1987: for example in cattle (Vassart *et al.* 1987), house sparrows (Wetton *et al.* 1987) and other bird species (Burke & Bruford 1987), dogs and cats (Jeffreys & Morton 1987), and mice (Jeffreys *et al.* 1987). DNA fingerprinting studies in most vertebrate groups have since been completed, and cross-hybridization of some probes to invertebrates (Amichot *et al.* 1989; Blanchetot 1991a,b; Carvalho *et al.* 1991; Hauser *et al.* 1992) and plants has also been reported (Dallas 1988; Rogstad *et al.* 1988; Nybom & Schaal 1990a,b).

If a cloned minisatellite is instead hybridized at high stringency, it may detect only the alleles representing its specific locus, which is thus called a single locus probe (SLP). The pattern will normally consist of two different bands (assuming the individual is heterozygous, both alleles are detected intact and that is does not hybridize to related loci). An alternative term, DNA profiling, has been suggested for SLP analysis because the pattern is not individual-specific. Individual single SLP profiles provide less discriminating power, but are much simpler, and so are very much easier to interpret. If, however, several different SLPs are hybridized sequentially to the same DNAs, then a composite DNA fingerprint can be produced which can approach the individual specificity of MLP DNA fingerprints. The disadvantage of SLPs is that, because each probe is locus-specific, several independent probes must be isolated and cloned for fully informative analyses. In most instances SLPs are highly species-specific and therefore new probes must be cloned for each species being studied, thus for some studies of animal species use of SLPs may be impractical.

### 6.2.3 DNA fingerprinting probes

A great many different HVRs have been isolated and are being used as informative fingerprinting probes to detect either single or multiple loci. Most are

GC-rich, but a few AT-rich examples exist. This bias may result from the fact that existing (GC-rich) probes are often used to isolate new sequences from libraries. Some of these, particularly those that detect multilocus patterns, are able to detect analogous patterns in a wide variety of animal and sometimes plant species. Single locus probes are, however, largely species or family specific. Most effort is now focused on attempts to isolate SLPs, in particular, those that are closely linked to important loci such as disease loci in man, and those determining economically important traits in domestic animals or in wild animals for use in ecological studies.

Synthetic oligonucleotides based on the consensus sequence of pre-existing probes, or with repetitive simple-sequences such as $(GATA)_n$ have been shown to detect MLP fingerprint-like patterns in Southern-blotted DNA (Epplen 1988; Schafer *et al.* 1988; Ledwith *et al.* 1990; Menzel *et al.* 1990; Gupta *et al.* 1994; Bhat *et al.* 1995). Under high stringency some simple oligonucletides have detected polymorphic single loci (Ali & Wallace 1988; Celia May, personal observation).

## 6.3 DNA fingerprinting procedures

The interpretation of DNA fingerprints, in particular MLP patterns, requires extremely high band resolution and definition so that bands (alleles) of various sizes that may vary from others by only a small difference in mobility can be distinguished without ambiguity. This is accomplished by attaining high autoradiographic signal and optimal resolution from many closely spaced DNA fragments, which might differ in their intensity because of variation in their homology for the probe. Consequently, each step in the formation of a DNA fingerprint has been modified from the original Southern blotting/DNA hybridization procedures to optimize a required feature of the DNA fingerprint. The procedures described below are optimized for multilocus DNA fingerprinting but can be used for SLP or other RFLP analyses.

### 6.3.1 DNA extraction

High quality, high molecular weight genomic DNA is essential for successful DNA fingerprint analysis. For most studies phenol/chloroform extraction procedures adapted from Hermann & Frischauf (1987) have been used (a protocol for this procedure is provided in Protocols, Chapter 2). However, many other methods of extracting high molecular weight DNA from various biological materials have been described in the literature (see Chapter 2). Some of these are ideal for extraction from some types of sample (e.g. plants), but many of the procedures prove inadequate for fingerprinting.

On occasion it is necessary to extract DNA from solid tissue obtained from biopsies or corpses, or from plant species. In these instances tissues are ground to a fine powder in liquid nitrogen using mortar and pestle and then approximately 10 µg of the powdered tissue is suspended per 1 mL of extraction buffer. These samples can then be processed essentially as described above.

### 6.3.2  Practical considerations for restriction digestions

To visualize the size differences between different minisatellite alleles it is necessary first to isolate them from their surrounding DNA. This is achieved by digesting with a restriction enzyme that is able to cleave the flanking DNA into relatively small fragments, but which leaves the minisatellite DNA intact. Completeness of enzyme digestion is essential for successful DNA fingerprinting analyses, because even a small degree of partial digestion can result in faint bands being present which might be scored mistakenly as independent alleles. In most instances restriction enzymes with 4 bp recognition sequences (*Alu*I, *Hin*fI, *Taq*I, *Rsa*I, *Sau*3A and *Hae*III or their common isoschizomers) are used.

Assays to determine the concentration of undigested DNA are unreliable, particularly when the DNA is viscous. So usually the DNA is digested, then quantitatively and qualitatively assayed. To ensure complete digestion an excess of enzyme is used, often for an extended period, depending upon the enzyme characteristics. However, enzyme concentration must never be permitted to be so great as to promote star activity (see Chapter 2). After the restriction reaction is complete, the DNA is assayed for adequate digestion and DNA concentration.

### 6.3.3  Assays for DNA quantity, quality and the completeness of enzyme digestion

It is necessary to quantify precisely the amount of DNA to be loaded onto the gel in order to standardize the amount of data available from all the DNA fingerprints. Otherwise, lanes with large amounts of DNA may show bands that are too faint to be detected in other samples. It is also important to determine that the DNA is not degraded, and that it has been digested completely with the enzyme.

The characteristic fluorometric properties of a DNA/Hoechst 33258 (bis-benzimidazole) complex have been exploited in the development of an accurate and sensitive assay for DNA concentration (Labarca & Paigen 1980), even in the presence of RNA protein or phenol, all components known to interfere with other assays. We routinely used a dedicated fluorometer (TKO 100;

Hoefer and Amersham Pharmacia Biotech, Little Chalfont-UK, 96-well format fluorometers are now available) for the quantification of fully digested DNA. This assay works extremely well, it is very sensitive (<0.1 μg per assay) and so consumes very little sample, it has a large linear range (over two orders of magnitude), and is highly reproducible and simple, and is quick to perform.

Assay of ethidium-bromide-stained DNA (Maniatis *et al.* 1982; Berger 1987) to quantify either digested or undigested genomic DNA using agarose minigels is difficult if the concentration of the samples varies greatly (range in excess of 2- to 5-fold). However, when performed after the digested samples have been diluted to constant concentration this method is useful for confirming that reading or dilution errors have not been introduced. It is also excellent as a qualitative assay to identify poorly digested or degraded samples.

### 6.3.4  Dilution of samples to a standard concentration

As a further aid to reproducibility, after digestion is complete all samples are diluted to the same concentration with 2× bromophenol blue (BPB) loading dye. For most of the species we have analysed, preliminary experiments indicated that 3–6 μg of DNA in 40 μL (0.075–0.15 μg mL$^{-1}$) was optimal. This amount of DNA was sufficient for strong hybridization signals, but insufficient to overload the gel and cause smearing of the bands.

### 6.3.5  Electrophoretic check of restricted DNA

Typically, an aliquot (3 μL) of the diluted DNA sample is electrophoresed through a 0.8% agarose gel at 80 V until the bromophenol blue marker dye has migrated to within 1 cm of the end of the gel (approximately 45 min). The smear of ethidium bromide-stained DNA is viewed under ultraviolet (UV) light on a transilluminator at 366 nm, and a Polaroid photo is taken. This is then inspected for evidence of complete digestion, lack of degradation and to assess similar DNA concentration among the samples.

### 6.3.6  Internal molecular weight marker

The use of external molecular weight standards run in lanes adjacent to normal samples aids in the estimation of the molecular weight of the resolved bands. However, they are wasteful of gel space and are unable to take into account any uneven mobility among the samples analysed on the gel. Aberrant mobility can be caused by a number of factors, especially heterogeneity in the gel matrix or variation in the ionic composition of the individual DNA samples, as well as mechanical distortion of the gel during blotting. Although diluting all of the samples with buffer to equalize the ionic concentration results in significant improvements in overall quality, it alone is insufficient to eliminate all of the intrasample variation. In a further effort to control for this variation, internal molecular weight markers can be used to assess and compensate for the distortion. These generally consist of a uniform and characteristic set of bands, which can easily be assessed for distortion, that are run together with every sample and thus experience all the electrophoretic variation met by the sample being analysed. The marker we developed consists of a collection of very dilute bacteriophage λ restriction fragments that cover a wide range of molecular weights (mixture of uncut λ DNA and λ *Hin*dIII, λ *Eco*RI and λ *Hin*dIII/*Eco*RI fragments diluted in Orange G loading dye to aid loading). After Southern transfer and detection of appropriate DNA fingerprints, the marker bands can be hybridized to labelled λ DNA and detected by autoradiography (see Fig. 6.1). Superimposing the marker autoradiograph over the DNA fingerprint allows variation between lanes to be assessed, as well as precise estimates of the molecular weights of the fragments.

### 6.3.7 Agarose gel electrophoresis for DNA fingerprint analysis

Genomic DNA that has been digested by restriction enzymes comprises a very complex mixture of DNA fragments. The first step in its analysis is to separate the mixture electrophoretically through an agarose gel. For DNA fingerprinting a maxigel format is used, typically 20×25 cm, although longer gels are also common. Any good quality 'general purpose agarose', such as Seakem LE (FMC Bioproducts, Rockland, Maine), at a concentration of 0.6–1.4% is suitable for DNA fingerprinting analyses (fragment range 50.0–0.30 kb) depending upon the species/enzyme/probe combination analysed.

Several precautionary steps are included in the preparation and running of the gels for DNA fingerprinting to try to minimize variation in the electrophoresis. The appropriate volume of buffer is weighed so that any vapour lost during melting can be replaced. The agarose is added and allowed to swell for about 5–10 min before being melted (medium setting in a microwave oven with periodic mixing), as this reduces small lumps or irregularities. The agarose solution is cooled to below boiling and re-weighed and distilled water added to replace the lost vapour. The gel is then cooled further to 55 °C by immersion in a water bath with occasional swirling. The electrophoresis tray is prepared and levelled as even a small difference in gel thickness can be noticeable, and the molten agarose is poured and allowed to set for at least 20 min before the sample comb is removed. All of the DNA samples (diluted in running buffer) and any molecular weight markers including the internal marker are partially denatured (65 °C for 10 min then quenched on ice) to dis-

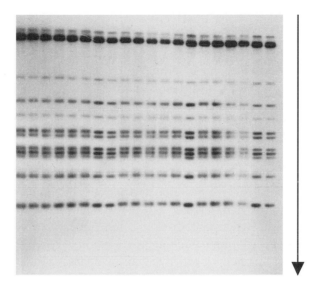

**Fig. 6.1** The figure shows λ markers detected by radiolabelled DNA probe after DNA finger-printing was complete. Note small differences in fragment mobility that were not readily apparent from the original DNA fingerprint, presumably because of irregularities in the gel matrix or small variations in the ionic strength of the DNA samples. Arrow indicates direction of travel.

sociate any noncontiguous DNA fragments that may have annealed. DNA samples are loaded and allowed to equilibrate with the running buffer, and electrophoresis is performed slowly to reduce gel artefacts (typically 40 V for 2 3 days).

### 6.3.8 Sample loading

To reduce the importance of any distortions in DNA fingerprints produced in the above steps, it is common practice always to compare, for example, adults and their putative offspring in adjacent lanes. Whenever possible, the putative parents are run in adjacent tracks so that small mobility differences between their fingerprint bands can more easily be detected, and therefore, any similarities in the offspring can be assigned to one or other parent. When larger cohorts are studied (typically five or more offspring), the parental samples are run in duplicate at both sides of the family group to facilitate accurate band matching.

### 6.3.9 DNA transfer

The direct hybridization of a probe to DNA fragments retained in a gel is possible, and has been used for some DNA fingerprinting studies (Epplen & Zischler 1990). However, it is more usual to transfer them from the fragile gel to a solid support. This is accomplished by some form of blotting process to transfer the DNA fragments from the gel to the membrane while faithfully retaining their relative positions. Capillary blotting, vacuum blotting and electroblotting procedures are used. Southern blotting (Southern 1975), or related alkaline blotting modifications are the most common procedures because they are simple and efficient, and require little specialized equipment. Duplex DNA fragments in a gel are denatured *in situ* by incubation in alkali (Wahl *et al.* 1979), then the gel is neutralized in high salt buffer (to maintain single-strandedness) and a blotting membrane is closely juxtaposed to the gel (see Fig. 2.2). The DNA fragments are then eluted and deposited onto the surface of the membrane by capillary action. Once it has been transferred to the membrane the DNA must be immobilized (fixed) so that it will not wash-off during the subsequent hybridization steps. Further details and a protocol are provided in Chapter 2.

### 6.3.10 Prehybridization, hybridization and stringency washing

Prehybridization is a pretreatment of the membrane with 'blocking agents' designed to suppress nonspecific hybridization of the probe to the membrane. Various recipes are used, e.g. Denhardt's solution (Denhardt 1966), nonfat milk proteins called Blotto (Johnson *et al.* 1984), or high concentrations of sodium dodecyl sulphate (SDS) (Church & Gilbert 1984). These reagents are required to achieve a good signal:noise ratio. Hybridization of the single-stranded probe to its target is followed by posthybridization washing to remove unbound probe from the membrane. This is the usual stage at which the stringency is controlled. However, for multilocus DNA fingerprinting studies we found that regulating the stringency during the hybridization step proved more reproducible (i.e. it is easier to prevent unwanted, poorly matched hybrids from forming than it is reproducibly to dissociate them afterwards without affecting the desired hybrids). For SLPs, prehybridization usually requires the addition of homologous competitor DNA (denatured and sheared for 20 min at 100 °C, and at a concentration of 5–10 $\mu$g mL$^{-1}$) to suppress hybridization to other minisatellites or repetitive DNA sequences. Washing of SLPs usually requires high stringency (0.1× SSC, 0.1% SDS at 65 °C). Detection of SLPs in relatively poor quality DNA (poor digest or partially degraded) can be achieved with less than 1 $\mu$g of DNA in many instances. Further details are provided in Chapter 2.

### 6.3.11  Deprobing and rehybridization

Many different minisatellite probes are currently available (including multi-locus probes and species-specific SLPs). It is therefore desirable to be able to probe individual Southern blot membranes repeatedly with a variety of probes. To do so it is necessary to be able to remove a previously hybridized probe from the membrane without removing the immobilized target DNA. The method currently preferred is to wash nylon membranes in two changes of 0.4 mol NaOH solution at 42–65 °C for approximately 1 h, and then to neutralize them in two changes of 0.2 mol Tris pH 7.5, 0.1 × SSC, 0.1% SDS for 1 h. Probe removal can be monitored using a Geiger tube or by autoradiography.

### 6.3.12  Hybridization probes

DNA fingerprinting probes can be prepared from double- (dsDNA) or single-stranded DNA (ssDNA), RNA (Riboprobes) or synthetic oligonucleotides. They are usually radiolabelled with $^{32}$P for autoradiographic detection. However, nonradioactive alternatives, such as chromogenic and chemiluminescent probes, can also be used in a similar fashion. Commercial probe preparation kits are available and prove cost effective for all of the labelling sytems described below.

Double-stranded DNA probes made by random priming (Feinberg & Vogelstein 1984) can be labelled to $> 10^9$ c.p.m. µg$^{-1}$, and have been used by several groups (Burke & Bruford 1987) for routine DNA fingerprint analyses. These probes are easy to prepare. Sometimes whole, double-stranded vectors containing the sequences of interest are used, at other times the plasmid insert is first gel purified. Random-primed probes in particular are ideal for routine preparation of SLPs, which generally do not require the very high specific activity of MLPs.

Irrespective of how they are manufactured, double-stranded DNA probes must be heat denatured into single strands at 100 °C then quenched on ice immediately prior to use. Unfortunately, the two complementary strands of the probe are able to re-anneal to each other and therefore will be unavailable for hybridization to the target. This is usually a disadvantage. However, under suitable conditions partial re-annealing can allow the formation of structures called hyperpolymers, in which several probe molecules can contribute to the hybridization signal and can result in an amplification of it. Reagents such as dextran sulphate (Wahl et al. 1979) and polyethylene glycol (Renz & Kurz 1984) help to promote hyperpolymer formation and for this reason are often included in hybridization solutions for use with DNA probes.

Single-stranded probes of either RNA or ssDNA can offer significant advantages over double-stranded probes for DNA fingerprinting applications. The absence of a competing second strand in the hybridization can result in improved kinetics. The length of the probe can be predetermined by the choice of restriction enzyme used to linearize the transcription vector. This property can be utilized to prevent cross-hybridization of the probe to the flanking sequences of the target. Single-stranded RNA probes (often called Riboprobes) offer improved hybrid stability over DNA probes and, therefore, the opportunity to use more stringent hybridization conditions, often resulting in lower background. In addition they can routinely be produced at very high specific activity.

Single-stranded DNA probes require cloning of the probe sequence into one of a series of recombinant single-stranded cloning vectors such as bacteriophage M13 derivatives (Hu & Messing 1982; Messing & Vieira 1982; Messing 1983) or phagemids (Zagursky & Baumeister 1987), from which ssDNA can be readily prepared (see Chapter 2). The first DNA fingerprint probes were prepared (Jeffreys *et al.* 1985a) as ssDNA; however, more convenient procedures such as random priming and riboprobes are now generally preferred.

Synthetic oligonucleotides can be labelled as hybridization probes. Increasingly such probes are being used in DNA fingerprinting studies since it was shown that many simple, tandemly repetitive sequences (microsatellites) detect fingerprint-like patterns. Usually the oligonucleotides are end-labelled with [$\gamma$-$^{32}$P] ATP catalysed by $T_4$ polynucleotide kinase (van de Sande *et al.* 1973); however, they may also be concatenated and labelled in a specific priming reaction (May & Wetton 1991). Once labelled they may be hybridized to genomic DNA on membranes or for within-gel hybridizations.

Riboprobes are produced by transcription of DNA sequences cloned downstream of specific transcription promoters by specific DNA-dependent RNA polymerases such as T3, T7 and SP6 promoters/polymerases (Cox *et al.* 1984; Melton *et al.* 1984; Tabor & Richardson 1985; Kreig & Melton 1984, 1987; Little & Jackson 1987). They are each very specific for their own promoter sequence. *In vitro* transcription initiated from these promoters results in large quantities of probe, which is often of very high specific activity ($>5 \times 10^9$). Riboprobe derivatives of the minisatellite probes used for DNA fingerprinting have been developed (Carter *et al.* 1989), and they have proven essential for the analysis of many species, including sparrowhawks and aphids, from which useful DNA fingerprints could otherwise not be produced.

### 6.3.13  Removal of unincorporated nucleotides and assay of radioactive incorporation

To allow assays of radioactive incorporation to be performed and to reduce nonspecific hybridization, unincorporated radionucleotides are usually removed from the newly synthesized probe using spin-column chromatography (Sambrook *et al.* 1989) or, preferably, by coprecipitation with glycogen (Tracy 1981).

The incorporation of radioactivity into a riboprobe is determined as a quality control measure, and to determine the ideal probe concentration needed for the hybridization. One microlitre of the reaction mix and $1\,\mu L$ of either the column eluate or resuspended pellet are counted in $5\,mL$ of an aqueous scintillation fluid (e.g. Ecoscint). After adjusting these counts for sample volume, the percentage incorporation is calculated. A volume equivalent to $120\,000\,c.p.m.\,mL^{-1}$ is used for hybridization.

## 6.4  Reading and interpretation of DNA fingerprints

The reading of DNA fingerprints is a most important step in the procedure. Often, however, this process is difficult as there may be regions of the DNA fingerprints in which the bands are closely spaced, some are faint and, particularly in the lower molecular weight region of the gel, many may be diffuse. To improve the reproducibility of the scoring of bands, simple rules help to standardize the decision-making process; for example, bands where the centres are less than 0.5–1.0 mm apart (depending upon the quality of the autoradiogram being analysed) and where estimated intensities differ by less than threefold (approximately, by eye) are regarded as being the same band. Sometimes, however, these criteria will be insufficient to distinguish correctly between situations where two truly identical alleles are being compared, and when two different alleles have comigrated to the same position. However, any errors will be conservative, i.e. they will serve to make individuals appear more similar rather than to suggest differences that do not exist. This is important because the weight given to evidence for differences proves more important in the interpretation process than evidence for similarities.

In practice, a transparent ruler or similar device placed onto the DNA fingerprint to help align identical bands is very useful. This is gradually moved down the fingerprint and each 'new' band is identified and numbered. Its distribution among the various individuals is recorded into a scoring matrix.

In most species high molecular weight bands (>5–10 kb), if they exist, are intense and sharp, and therefore relatively easy to score. However, it should be noted that in this region of the fingerprint even quite significant differences in actual band molecular weight will result in relatively small differences in mobility. Conversely, in the lower molecular weight range of the gel, bands become obviously more diffuse and hence difficult to position accurately. In this region apparent small differences in mobility might be the result of distortion from gel inconsistencies.

Faint bands seen on short exposures are usually scored from a longer exposure autoradiograph (bands not readily seen in an overnight exposure are not reliable, and are ignored). Broad high molecular weight bands often seen in the overnight and the long exposure can often be identified as doublets or triplets using another long exposure with no-screens (4–14 days).

### 6.4.1 The statistics of band-sharing and discrimination of familial relationships

Various statistical treatments of the band-sharing data can be used to help to interpret the relationships of the individuals being analysed. Probably the simplest is the band-sharing coefficient ($D$; equivalent to $F$ or $s$ of some investigators), used to describe the similarity between two fingerprints as the proportion of bands shared between them:

$$D = 2 \cdot N_{ab}/(N_a + N_b),$$

where $N_{ab}$ is the number of bands of approximately similar intensity and mobility in individuals a and b, respectively, and $N_a$ and $N_b$ are the numbers of bands in a and b that could have been scored in the other individual had they been present.

Band-sharing is expected to increase as relatedness increases (see Wetton et al. 1992) as a result of common descent; for example, unrelated individuals are unlikely to share many bands (low value of $D$), unless alleles are very frequent in the population. Full-sibs on the other hand, will co-inherit many of the same alleles from both parents and will have correspondingly high band sharing. Analysis of the experimental data would be expected to show that each class of first-degree relatives would have a very similar mean band-sharing, although full-sibs would have a higher variance to their band-sharing measurement because, on average, all offspring will inherit half of their bands from each parent. Individual offspring could, however, inherit different subsets of bands from their parents, and thus appear relatively dissimilar to each other.

In addition to the simple analyses of the similarity coefficient described above, several more sophisticated analyses are possible based on population genetic principles (Brookfield 1989; Lynch 1988, 1990; worked examples provided and reviewed in Parkin & Wetton 1991; Bruford *et al.* 1992; Austin *et al.* 1993).

### 6.4.2 Segregation analyses

Statistical analyses of DNA fingerprinting patterns assume Mendelian segregation of the DNA fingerprint bands. Segregation analyses are necessary to ensure that bands are transmitted equally from parents to offspring without any sex bias, and that groups of bands do not segregate as apparent linked haplotypes (see Wetton *et al.* 1992 for counter examples). These analyses are best performed on large family cohorts of at least 10 siblings, and the data are most conveniently analysed by methods based on LOD score analysis (Cavalli-Sforza & Bodmer 1971) using simple computer programs (see Parkin & Wetton 1991; Bruford *et al.* 1992, for descriptions). Ideally, captive-bred family cohorts are available for these analyses. However, in many cases such rigorous analyses are not possible. It is common for many species studied in ecological research that both parents and sufficient offspring may not be available for comparison; for example, paternal samples are often not available, or species may never produce enough offspring. In these instances less satisfactory alternative analyses must be substituted. Simple segregation analyses (Dallas 1988; Meng *et al.* 1989; Austin *et al.* 1993), based on predicted binomial distribution, can also be performed. A model for predicting segregation when only one parent is available has been developed (see Brookfield *et al.* 1993, with convenient tables provided). If such analyses are not possible, care must be exercised when interpreting DNA fingerprint data.

### 6.5 Single locus probes

A method of 'probe-walking', whereby a gridded chromatid (Saito & Stark 1986) library is re-scanned with newly isolated SLPs, was developed (Armour *et al.* 1990) and is described in detail by Hernotte *et al.* (1991) and Bruford *et al.* 1992). Positive clones are screened for usefulness by hybridization to Southern blots of genomic DNA, including panels of unrelated individuals and large family cohorts. The patterns are examined for high heterozygosity (range of different sized bands seen in unrelated individuals). Evidence for sex-linkage of loci present on the X (or Z) chromosome is provided when all heterogametic (XY or WZ) individuals appear hemizygous (one band only), or if present on the Y (or W) chromosome the homogametic sex (e.g. XX) will lack bands altogether. Occasionally individuals may apparently possess only single bands, and this may indicate null alleles (bands may be too small or faint for detection on the Southern blot) or true homozygotes. Segregation analyses should be used to determine whether the apparent homozygotes transmit alleles to their offspring.

If only a few probes are needed for proposed analyses and existing $\lambda$ or even plasmid libraries are available for the species under study, it is probably best to try to isolate SLPs from a small aliquot ($\sim 10^5$ clones) (Taggart & Ferguson 1990). If, however, many probes are needed it is better to invest the time and effort (realistically 6–12 months) in developing an arrayed charomid library. Once appropriate clones are identified, they should be subcloned into plasmid or phagemid vectors for ease of preparation, storage and probe manufacture.

## 6.6  Applications of DNA fingerprinting

### 6.6.1  Analysis of family relationships

The first published application of DNA fingerprinting for the elucidation of complex familial relationships (Jeffreys et al. 1985b), was a particularly intractable immigration dispute concerning a boy wishing to be reunited with his mother. This was particularly difficult because his paternity was uncertain (so there was no paternal sample available for comparison), and the case was further complicated because the claimed mother was suspected of being the aunt. To complete this analysis a paternal fingerprint was reconstructed from that of the claimant's 'brother' and two 'sisters', by subtracting from these all of those bands that could have been maternally derived. It was then found that all of the claimant's bands not present in the mother were present in the reconstructed paternal fingerprint. Statistical modelling of these data gave a high probability that the woman was the boy's mother.

The same technology continues to benefit numerous studies of plants and animals, supplying precise knowledge of individual identity and/or of their familial relationships. One of the most active areas of ecological research is the study of individual variation in reproductive success. Ecologists have speculated on the patterns of cuckoldry that might exist among different species, depending upon the prevailing ecological conditions. Ever since DNA fingerprinting was first demonstrated to be capable of identifying the presence of extra-pair offspring (EPO) in the nests of wild birds (Burke & Bruford 1987; Wetton et al. 1987), its application has become widespread in these studies. To illustrate the usefulness of DNA fingerprinting several key studies are highlighted below.

DNA fingerprint analyses of an established marked population of house sparrows (*Passer domesticus*) have demonstrated that neither the proportion nor the distribution of EPOs varies greatly between years (Wetton et al. 1992; unpublished data), and also that certain males are more prone to cuckoldry than are others (Wetton & Parkin 1991; Whetton *et al.* 1992). The widely held hypothesis that cuckolded males would be young and inexperienced individuals has been demonstrated in purple martins (*Progne subis*) (Morton *et al.* 1990), and supported by observations of indigo bunting (*Passerina cyanea*) in North Carolina (Westneat 1990). Other observations have demonstrated that EPOs occur most frequently in high density populations (Gibbs *et al.* 1990). It has been shown in the barn swallow (*Hirundo rustica*) that so-called 'preferred males' (those with long tail ornaments, see Moller 1988) are cuckolded more often than 'average quality' males (Smith *et al.* 1991). And analyses of the pied flycatcher (*Ficedula hypoleuca*) have shown that even though the polygynous males may leave the nest unattended for long periods of time, they are not cuckolded more often than monogamous males (Lifjeld *et al.* 1991). DNA fingerprinting has been demonstrated to be capable of correctly identifying cuckolding males (Wetton *et al.* 1987; Gibbs *et al.* 1990; Westneat 1990; J.H. Wetton and T. Burke, personal communication), which has been important for testing the prediction that cuckolding males are usually near-neighbours or floaters. DNA fingerprinting was essential to elucidate the detailed variable mating system of the dunnock (*Prunella modularis*) (Burke *et al.* 1989; Davies *et al.* 1992), and similarly the analyses of the European bee-eater (*Merops merops*) (C. Jones and J. Krebs, personal communication) and of the stripe-backed wren (*Campylorhynchus nuchalis*) (Rabenold *et al.* 1990), show that distant relatives (e.g. grandparents) sometimes cooperate with breeders to raise families, and the high philopatry of males results in a tendency for brothers to compete for paternity (Rabenold *et al.* 1990). Intraspecific nest parasitism and possible quasi-parasitism in the zebra finch (*Taeniopygia guttata*) (Birkhead *et al.* 1990), and the 'adoption' of cygnets in the mute swan (*Cygnus olor*) (possible that the mother of the adopted offspring was the daughter of the alloparental male; Meng & Parkin 1991) have been studied. Similar DNA fingerprint analyses have been used to analyse many diverse animal species including: paternity in shrews (Tegelstrom *et al.* 1991); old world monkeys (Weiss *et al.* 1988) and marmosets (Dixson *et al.* 1989; E. Signer, personal communication). Combinatory analyses have been made using several different molecular approaches, such as the analysis of pod structure and paternity in pilot whales (*Globicephala melas*) using both microsatellite (Schlotterer *et al.* 1991) and minisatellite probes (W. Amos, personal communication), and a combination of MLPs and SLPs was used to study wild mice (J. Fang and R. Carter, unpublished data). These studies demonstrate the power of DNA fingerprinting to elucidate complex familial relationships in wild populations, but also demonstrate the need for reliable observational data.

DNA fingerprinting analyses have also been applied to studies to deduce relationships more distant than first or second degree; for example, the demographic study of individual and intrapopulation variation in the African lion

(*Panthera leo*) (Gilbert *et al.* 1991). However, it is recognized that these can be fraught with problems such as possible unrecognized relationships or population subdivision (see Lynch 1988, 1990; Brookfield 1991, for discussion). Generally, fingerprint profiles evolve too rapidly to allow studies at the population level. However, using a combination of molecular techniques Hoelzel & Dover (1991) have studied populations of the killer whale (*Orcinus orca*). Mitochondrial DNA sequence variation indicated that populations were sympatric, while gross minisatellite data suggested that inbreeding occurred within the population, which was consistent with behavioural data. Divergence and the relationship between small isolated populations of the Californian Channel Island fox (*Urocyon littoralis*) (Gilbert *et al.* 1990) and those of naked mole rats (Reeve *et al.* 1990) have been made possible by comparison of fingerprint patterns within and between populations because of the occurrence of numbers of fixed alleles. In the same way, strains of laboratory mice, domestic chickens (Kuhnlein *et al.* 1989) and *Tilapia* spp. (Naish *et al.* 1995, and personal observation) can be distinguished. A study of honey bees (*Apis mellifera*) (Blanchetot 1991b) has used DNA fingerprinting to deduce the probable number of patrilines involved in the development of a colony. DNA fingerprinting also has some utility in the study of clonal species; for example, the analysis of clonal fish species (Turner *et al.* 1990), the estimation of differentiation among isolates of aphid clones (Carvalho *et al.* 1991), and the study of the introduction history of the freshwater snail (*Potamopyrgus antipodarum*) (Hauser *et al.* 1992) have benefited from DNA fingerprint analyses. Other applications of DNA fingerprinting to the study of invertebrates include the discrimination between individuals and strains of house fly (*Musca domestica*) (Amichot *et al.* 1989; Blanchetot 1991a) and the malarial parasite (*Plasmodium falciparum*) (Rogstad *et al.* 1989).

DNA fingerprinting is finding great utility in animal conservation biology. Many zoos and other animal collections are cooperating in captive breeding projects, where DNA fingerprinting is used to assess the diversity in the captive stock and to ascertain parentage in colonies or groups. In this context it has been used to suggest possible pairings which may best maintain or represent the variation present in the population so that the deleterious effects of inbreeding can be minimized; examples of this include Rothschild's mynah (*Leucopsar rochildi*) (D. Ashworth, personal communication), speckled pink pigeon (*Besoenas mayeri*) (M. Bruford, personal communication) and Waldrapp ibis (*Geronticus eremita*) (E. Signer, personal communication). The application of paternity testing to birds of prey that are claimed to be captively bred but suspected of being stolen or having been taken illegally from the wild has also been investigated (Parkin *et al.* 1988; J.H. Wetton *et al.*, personal communication, and personal observations).

By comparison with animal studies, the hybridization-based DNA fingerprinting analyses in plants are less common. Most reported applications are for strain verification, e.g. cultivars of rice (*Oryza* spp.) (Dallas 1988) as well as various genera of the Rosaceae (Nybom 1990), and blackberries and rasberries (Nybom & Schaal 1990a). DNA fingerprinting has also been utilized in parentage studies (Nybom & Schaal 1990b), and for the study of plant population distribution studies (Nybom 1990). For plant studies PCR-based fingerprinting techniques, such as microsatellite analysis, randomly amplified polymorphic DNA (RAPD) analysis and amplified length polymorphism (AFLP) analysis have been preferred. These techniques are described in detail in Chapters 3, 9 and 10.

## References

Ali, S. & Wallace, R.B. (1988) Intrinsic polymorphism of variable number tandem repeat loci in the human genome. *Nucleic Acids Research* **16**, 8487–8496.

Amichot, M., Fournier, D. & Berge, J.-B. (1989) Nucleotide sequence of an element detecting highly polymorphic regions in insect genomes. *Nucleic Acids Research* **17**, 5863.

Armour, J.A.L., Povey, S., Jeremiah, S. & Jeffreys, A.J. (1990) Systematic cloning of human minisatellites from ordered array charomid libraries. *Genomics* **8**, 501–512.

Austin, J.J., Carter, R.E. & Parkin, D.T. (1993) Genetic evidence for extra-pair fertilisations in socially monogamous Short-tailed Shearwaters, *Puffinus tenuirostris* (Procellariiformes: Procellariidae), using DNA fingerprinting. *Australian Journal of Zoology* **41**, 1–11.

Benoist, C.O., Mathis, D.J., Kanter, M.R., Williams, V.E. & McDevitt, H.O. (1983) The murine I-alpha chains, E-alpha and A-alpha show a suprising degree of sequence homology. *Proceedings of the National Academy of Sciences USA* **80**, 534–538.

Berger, S.L. (1987) Quantifying 32-P labelled and unlabelled nucleic acids. *Methods in Enzymology* **152**, 49–54.

Birkhead, T.R., Burke, T., Zann, R., Hunter, F.M. & Krupa, A.P. (1990) Extra-pair paternity and intraspecific brood parasitism in wild zebra finches *Taeniopygia guttata*, revealed by DNA fingerprinting. *Behavioural Ecology and Sociobiology* **27**, 315–324.

Blanchetot, A. (1991a) A *Musca domestica* satellite sequence detects individual polymorphic regions in insect genome. *Nucleic Acids Research* **19**, 929–932.

Blanchetot, A. (1991b) Genetic relatedness in honeybees as established by DNA fingerprinting. *Journal of Heredity* **82**, 391–396.

Brookfield, J.F.Y. (1989) Analysis of DNA fingerprinting data in cases of disputed paternity. *IMA Journal of Mathematics Applied in Medicine and Biology* **6**, 111–132.

Brookfield, J.F.Y. (1991) The statistical interpretation of hypervariable DNAs. In: *Molecular Techniques in Taxonomy* (ed. G.M. Hewitt), pp. 159–169. Springer-Verlag, Berlin.

Brookfield, J.F.Y., Carter, R.E., Mair, G.C. & Skibinski, D.O.F. (1993) A case study of the interpretation of linkage data using DNA fingerprinting probes. *Molecular Ecology* **2**, 209–218.

Bruford, M.W., Hannotte, O., Brookfield, J.F.Y. & Burke, T. (1992) Single-locus and multilocus DNA fingerprinting. In: *Molecular Genetic Analysis of Populations. A Practical Approach* (ed. A.R. Hoelzel), pp. 225–269. IRL Press, Oxford.

Burke, T. & Bruford, M.W. (1987) DNA fingerprinting in birds. *Nature* **327**, 149–152.

Burke, T., Davies, N.B., Bruford, M.W. & Hatchwell, B.J. (1989) Parental care and mating behaviour in polyandrous dunnocks *Prunella modularis* related to paternity by DNA fingerprinting. *Nature* **338**, 249–251.

Carter, R.E., Wetton, J.H. & Parkin, D.T. (1989) Improved genetic fingerprinting using RNA probes. *Nucleic Acids Research* **17**, 5867.

Carvalho, G.R., Maclean, N., Wratten, S.D., Carter, R.E. & Thurston, J.P. (1991) Differentiation of aphid clones using DNA fingerprints from individual aphids. *Proceedings of the Royal Society of London B* **243**, 109–114.

Cavalli-Sforza, L.L. & Bodmer, W.F. (1971) *The Genetics of Human Populations.* W.H. Freeman, San Francisco.

Church, G.M. & Gilbert, W. (1984) Genomic sequencing. *Proceedings of the National Academy of Sciences USA* **81**, 1991–1995.

Cox, K.H., DeLeon, D.V., Angerer, L.M. & Angerer, R.C. (1984) Detection of mRNAs in sea urchin embryos by *in situ* hybridization using asymmetric RNA probes. *Developmental Biology* **101**, 485–502.

Dallas, J.F. (1988) Detection of DNA 'fingerprints' of cultivated rice by hybridization with a human minisatellite DNA probe. *Proceedings of the National Academy of Sciences USA* **85**, 6831–6835.

Davies, N.B., Hatchwell, B.J., Robson, T. & Burke, T. (1992) Paternity and parental effort in dunnocks *Prunella modularis*: how good are chick feeding rules? *Animal Behaviour* **43**, 729–745.

Denhardt, D.T. (1966) A membrane-filter technique for the detection of complementary DNA. *Biochemical and Biophysics Research Communications* **23**, 641–646.

Dixson, A.F., Hastie, N., Patal, I. & Jeffreys, A.J. (1989) DNA 'fingerprinting' of captive family groups of common marmosets (*Callithrix jacchus*). *Folia Primatologia* **51**, 52–55.

Dodd, B.E. (1985) DNA fingerprinting in matters of family and crime. *Nature* **318**, 506–507.

Epplen, J.T. (1988) On simple repeated GATA/GACA sequences: a critical reappraisal. *Journal of Heredity* **79**, 409–417.

Feinberg, A.P. & Vogelstein, B. (1984) A technique for radiolabelling restriction endonuclease fragments to high specific activity. *Analytical Biochemistry* **137**, 266–267.

Fowler, C., Drinkwater, R., Skinner, J. & Burgoyne, L. (1988) Human satellite-III DNA: an example of a macrosatellite polymorphism. *Human Genetics* **79**, 265–272.

Fowler, J.C.S., Burgoyne, L.A., Scott, A.C. & Harding, H.W.J. (1987) Repetitive deoxyribonucleic acid (DNA) and human genome variation—a concise review relevant to forensic biology. *Journal of Forensic Sciences* **33**, 1111–1126.

Gibbs, H.L., Weatherhead, P.J., Boag, P.T., White, B.N., Tabak, L.M. & Hoysak, D.J. (1990) Realized reproductive success of polygynous red-winged blackbirds revealed by DNA markers. *Science* **250**, 1394–1397.

Gilbert, D.A., Lehman, N., O'Brien, S.J. & Wayne, R.K. (1990) Genetic fingerprinting reflects population differentiation in the Californian Channel Island fox. *Nature* **344**, 764–767.

Gilbert, D.A., Packer, C., Pusey, A.E., Stephens, J.C. & O'Brien, S.J. (1991) Analytical DNA fingerprinting in lions: Parentage, genetic diversity, and kinship. *Journal of Heredity* **82**, 378–386.

Hanotte, O., Burke, T., Armour, J.A.L. & Jeffreys, A.J. (1991) Hypervariable minisatellite DNA sequences in the Indian peafowl Pavo Cristatus. *Genomics* **9**, 587–597.

Hauser, L., Carvalho, G.R., Hughes, R.N. & Carter, R.E. (1992) Clonal structure of the introduced freshwater snail, *Potamopyrgus antipodarum* (Prosobranchia: Hydrobiidae), as revealed by DNA fingerprinting. *Proceedings of the Royal Society of London B* **249**, 19–25.

Hermann, B.G. & Frischauf, A.-M. (1987) Isolation of genomic DNA. *Methods in Enzymology* **152**, 180–183.

Hoelzel, A.R. & Dover, G.A. (1991) Genetic differentiation between sympatric Killer whale populations. *Heredity* **66**, 191–195.

Hu, N.-T. & Messing, J. (1982) The making of strand-specific M13 probes. *Gene* **17**, 271–277.

Jeffreys, A.J. & Morton, D.B. (1987) DNA fingerprints of dogs and cats. *Animal Genetics* **18**, 1–15.

Jeffreys, A.J., Wilson, V. & Thein, S.L. (1985a) Hypervariable 'minisatellite' regions in human DNA. *Nature* **314**, 67–73.

Jeffreys, A.J., Brookfield, J.F.Y. & Semeonoff, R. (1985b) Positive identification of an immigration test-case using DNA fingerprints. *Nature* **317**, 818–819.

Jeffreys, A.J., Wilson, V. & Thein, S.L. (1985c) Individual specific 'fingerprints' of human DNA. *Nature* **316**, 76–79.

Jeffreys, A.J., Wilson, V., Kelly, R., Taylor, B.A. & Bulfield, G. (1987) Mouse DNA 'fingerprints': analysis of chromosome localization and germ-line stability of hypervariable loci in recombinant inbred strains. *Nucleic Acids Research* **15**, 2832–2836.

Johnson, D.A., Gautsch, J.W., Sportsman, J.R. & Elder, J.H. (1984) Improved technique utilizing nonfat dry milk for analysis of proteins and nucleic acids transferred to nitrocellulose. *Gene Analysis Techniques* **1**, 3–8.

Kreig, P.A. & Melton, D.A. (1984) Functional messenger RNAs are produced *in vitro* transcription of cloned cDNAs. *Nucleic Acids Research* **12**, 7057–7070.

Kreig, P.A. & Melton, D.A. (1987) *In vitro* RNA synthesis with SP6 RNA polymerase. *Methods in Enzymology* **155**, 397–415.

Kuhnlein, U., Dawe, Y., Zadworny, D. & Gavora, J.S. (1989) DNA fingerprinting: a tool for determining genetic distances between strains of poultry. *Theoretical and Applied Genetics* **77**, 669–672.

Labarca, C. & Paigen, K. (1980) A simple, rapid, and sensitive DNA assay procedure. *Analytical Biochemistry* **102**, 344–352.

Ledwith, B.J., Manam, S., Nichols, W.W. & Bradley, M.O. (1990) Preparation of synthetic tandem-repetitive probes for DNA fingerprinting. *Biotechniques* **9**, 149–153.

Lifjeld, J.T., Slagsvold, T. & Lampe, H.M. (1991) Low frequency of extra-pair paternity in pied flycatchers revealed by DNA fingerprinting. *Behavioural Ecology and Sociobiology* **29**, 95–101.

Little, P.F.R. & Jackson, I.J. (1987) Application of plasmids containing promoters specific for phage-encoded RNA polymerases. In: *DNA Cloning* (ed. D.M. Glover), pp. 1–18. IRL Press, Oxford.

Lynch, M. (1988) Estimation of relatedness by DNA fingerprinting. *Molecular Biology and Evolution* **5**, 584–599.

Lynch, M. (1990) The similarity index and DNA fingerprinting. *Molecular Biology and Evolution* **7**, 478–484.

Lyons, K.M., Stein, J.H. & Smithies, O. (1988) Length polymorphism in human proline-rich protein genes generated by intragenic unequal crossing over. *Genetics* **120**, 267–278.

Maniatis, T., Fritsch, E.F. & Sambrook, J. (1982) *Molecular Cloning. A Laboratory Manual.* Cold Spring Harbor, New York.

May, C.A. & Wetton, J.H. (1991) DNA fingerprinting by specific priming of concatenated oligonucleotides. *Nucleic Acids Research* **19**, 4557.

Melton, D.A., Kreig, P.A., Rebagliati, M.R., Maniatis, T., Zinn, K. & Green, M.R. (1984) Efficient *in vitro* synthesis of biologically active RNA and RNA hybridization probes from plasmids containing a bacteriophage SP6 promoter. *Nucleic Acids Research* **12**, 7035–7056.

Meng, A. & Parkin, D.T. (1991) Alloparental behaviour in mute swans *Cygnus olor* detected by DNA fingerprinting. *Wildfowl Supplement* **1**, 310–318.

Meng, A., Carter, R.E. & Parkin, D.T. (1989) The variability of DNA fingerprints in three species of swan. *Heredity* **64**, 73–80.

Menzel, A., Lagoda, P.J.L. & Issinger, O.-G. (1990) Hypervariable individual-specific DNA band patterns revealed by a 22mer promoter-specific oligonucleotide probe containing an SP1 site. *Nucleic Acids Research* **18**, 4287.

Messing, J. (1983) New M13 vectors for cloning. *Methods in Enzymology* **101**, 20–78.

Messing, J. & Vieira, J. (1982) A new pair of M13 vectors for selecting either strand of double-digest restriction fragments. *Gene* **19**, 269–276.

Moller, A.P. (1988) Female choice selects for male sexual tail ornaments in the monogamous swallow. *Nature* **332**, 640–642.

Morton, E.S., Forman, L. & Braun, M. (1990) Extrapair fertilizations and the evolution of colonial breeding in purple martins. *Auk* **107**, 275–283.

Naish, K.A., Warren, M., Bardakci, F., Skibinski, D.O., Carvalho, G.R. & Mair, G.C. (1995) Multilocus DNA fingerprinting and RAPD reveal similar genetic relationships between strains of *Orechromis niloticus* (Pisces: Cichlidae). *Molecular Ecology* **4**, 271–274.

Nakamura, Y., Leppert, M., O'Connell, P. *et al.* (1987) Variable number tandem repeat (VNTR) markers for human gene mapping. *Science* **235**, 1616–1622.

Nybom, H. (1990) Genetic variation in ornamental apple trees and their seedlings (Malus, Rosaceae) revealed by DNA 'fingerprinting' with the M13 repeat probe. *Hereditas* **113**, 17–28.

Nybom, H. & Schaal, B. (1990a) DNA 'fingerprints' reveal genotypic distributions in natural populations of blackberries and raspberries (Rubus, Rosaceae). *American Journal of Botany* **77**, 883–888.

Nybom, H. & Schaal, B.A. (1990b) DNA 'fingerprints' applied to paternity in apples (*Malus x domestica*). *Theoretical and Applied Genetics* **79**, 763–768.

Parkin, D.T. & Wetton, J.H. (1991) DNA fingerprinting. In: *NATO ASI Series* (eds G.M. Hewitt, A.W.B. Johnston & J.P.W. Young). pp. 145–157 Springer-Verlag, Berlin.

Parkin, D.T., Hutchinson, I. & Wetton, J.H. (1988) *Genetic Fingerprinting and its Role in Bird Research and Law Enforcement*. RSPB Conservation Review. RSPB, Sandy, Beds.

Rabenold, P.P., Rabenold, K.N., Piper, W.H., Haydock, J. & Zack, S.W. (1990) Shared paternity revealed by genetic analysis in cooperatively breeding tropical wrens. *Nature* **348**, 538–540.

Reeve, H.K., Westneat, D.F., Noon, W.A., Sherman, P.W. & Aquadro, C.F. (1990) DNA 'fingerprinting' reveals high levels of inbreeding in colonies of the eusocial naked mole rat. *Proceedings of the National Academy of Sciences USA* **87**, 2496–2500.

Renz, M. & Kurz, C. (1984) A colorimetric method for DNA hybridization. *Nucleic Acids Research* **13**, 3435.

Rogstad, S.H., Patton, J.C. & Schaal, B.A. (1988) M13 repeat probe detects DNA minisatellite-like sequences in gymnosperms and angiosperms. *Proceedings of the National Academy of Sciences USA* **85**, 9176–9178.

Rogstad, S.H., Herwaldt, B.L., Schlesinger, P.H. & Krogstad, D.J. (1989) The M13 repeat probe detects RFLP's between two strains of protozoan malaria parasite *Plasmodium falciparum*. *Nucleic Acids Research* **17**, 3610.

Saito, I. & Stark, G.S. (1986) Charomids: Cosmid vectors for efficient cloning and mapping of large or small restriction fragments. *Proceedings of the National Academy of Sciences USA* **83**, 8664–8668.

Sambrook, J., Fritsch, E.F. & Maniatis, T. (1989) *Molecular Cloning: a Laboratory Manual*, 2nd edn. Cold Spring Harbor Laboratory Press, Cold Spring Harbor.

Schafer, R., Zischler, H. & Epplen, J.T. (1988) (CAC) 5, a very informative oligonucleotide probe for DNA fingerprinting. *Nucleic Acids Research* **16**, 5196.

Schlotterer, C., Amos, W. & Tautz, D. (1991) Conservation of polymorphic simple sequence loci in cetacean species. *Nature* **354**, 63–65.

Sensabaugh, G.F. & Crim, D. (1986) Forensic biology—is recombinant DNA technology in its future? *Journal of Forensic Sciences* **31**, 393–396.

Smith, H.G., Montgomerie, R., Poldmaa, T., White, B.N. & Boag, P.T. (1991) DNA fingerprinting reveals relation between tail ornaments and cuckoldry in Barn Swallows, *Hirundo rustica*. *Behavioural Ecology* **2**, 90–98.

Southern, E.M. (1975) Detection of specific sequences among DNA fragments separated by gel electrophoresis. *Journal of Molecular Biology* **98**, 503–517.

Tabor, S. & Richardson, C.C. (1985) A bacterial T7 RNA polymerase/promoter system for controlled exclusive expression of specific genes. *Proceedings of the National Academy of Sciences USA* **82**, 1074–1078.

Taggart, J.B. & Ferguson, A. (1990) Hypervariable minisatellite DNA single locus probes for the Atlantic salmon, *Salmo salar* L. *Journal of Fish Biology* **37**, 991–993.

Tautz, D. (1989) Hypervariability of simple sequences as a general source for polymorphic DNA markers. *Nucleic Acids Research* **17**, 6463–6471.

Tegelstrom, H., Searle, J., Brookfield, J.F.Y. & Mercer, S. (1991) Multiple paternity in wild common shrews (*Sorex araneus*) is confirmed by DNA-fingerprinting. *Heredity* **66**, 373–380.

Tracy, S. (1981) Improved rapid methodology for the isolation of nucleic acids from agarose gels. *Preparative Biochemistry* **11**, 251–268.

Turner, B.J., Elder, J.F.J., Laughlin, T.F. & Davis, W.P. (1990) Genetic variation in clonal vertebrates detected by simple-sequence DNA fingerprinting. *Proceedings of the National Academy of Sciences USA* **87**, 5653–5657.

van de Sande, J.H., Kleppe, K. & Khorana, H.G. (1973) Reversal of bacteriophage T4 induced polynucleotide kinase action. *Biochemistry* **12**, 5050–5055.

Vassart, G., Georges, M., Monsieur, R., Brocas, H., Lequarre, A.S. & Christophe, D. (1987) A sequence of M13 phage detects hypervariable minisatellites in human and animal DNA. *Science* **235**, 683–684.

Weiss, M.L., Wilson, V., Chan, C., Turner, T. & Jeffreys, A.J. (1988) Application of DNA fingerprinting probes to Old World monkeys. *American Journal of Primatology* **16**, 73–79.

Westneat, D.F. (1990) Genetic parentage in the indigo bunting: a study using DNA fingerprinting. *Behavioural Ecology and Sociobiology* **27**, 67–76.

Wetton, J.H. & Parkin, D.T. (1991) An association between fertility and cuckoldry in the House Sparrow (*Passer domesticus*). *Proceedings of the Royal Society of London B* **245**, 227–233.

Wetton, J.H., Carter, R.E., Parkin, D.T. & Walters, D. (1987) Demographic study of wild house sparrow population by DNA fingerprinting. *Nature* **327**, 147–149.

Wetton, J.H., Parkin, D.T. & Carter, R.E. (1992) The use of genetic markers for parentage analysis in *Passer domesticus* (House Sparrows). *Heredity* **69**, 243–254.

Zagursky, R. & Baumeister, K. (1987) Construction and use of pBR322 plasmids that yield single stranded DNA for sequencing. *Methods in Enzymology* **155**, 139–155.

# CHAPTER 7

# Mitochondrial DNA

ETTORE RANDI

## 7.1 Introduction

The new molecular genetic methods, particularly polymerase chain reaction (PCR) amplification and DNA sequencing, which were implemented over the last two decades, have been extensively applied to the study of population genetics and phylogenetics of living organisms. The genome of mitochondria (mtDNA) has been the workhorse of this molecular revolution. Mitochondrial DNA is a small and simple haploid molecule, which apparently does not recombine. It is clonally and usually maternally inherited, although cases of paternal leakage and inheritance have been described. Each unique mtDNA sequence is a haplotype, which is transmitted intact through generations and can be reliably used to reconstruct genealogies in populations and phylogenies among species. Mitochondrial genes evolve faster than nuclear genes, on average, and accumulate genetic variability within and between populations in short evolutionary timespans. However, conserved positions of protein-coding mitochondrial genes and translated amino acid sequences can retain phylogenetic signals up to several tens of million years (Ma), thus allowing phylogenetic inference at deeper divergence times. Analyses of mtDNA, either using restriction fragment-length polymorphisms (RFLP) or nucleotide sequencing, have been instrumental in defining conceptual issues and empirical descriptions of population variability, intraspecific phylogeography, historical biogeography, hybridization, gene flow and species boundaries, patterns and rates of molecular evolution, conservation and phylogenetic biology (Avise 1994).

The aim of this chapter is to summarize the information on the genetics and molecular evolution of mtDNA that is relevant for research in molecular ecology and evolution. First, a general description is included of mtDNA structure, function and inheritance. Second, rates and patterns of molecular evolution of mitochondrial genes, which are controlled by different selective constraints, are described. Finally, there is a review of the array of laboratory, statistical and computational tools which can be used for population genetic and phylogenetic analyses of mtDNA.

## 7.2 Mitochondria and mitochondrial DNA

The genetic material of eukaryotic cells is compartmentalized in the nucleus and in two different cytoplasmic organelles: the mitochondria of eukaryotes, and the plastids, limited to the photosynthetic plants. Mitochondrial DNA is double-stranded, usually circular, much smaller and simpler than the nuclear genome, and it replicates independently. Complementary mtDNA strands are called the heavy H-strand, with high G+T content, and the light L-strand. Mitochondrial DNA genomes range in size from 13.8 kb, in the nematode worm *Caenorhabditis elegans*, up to 2500 kb, in the musk melon *Cucumis melo*. Some lower invertebrates have linear mtDNAs (Nosek *et al.* 1998).

The mitochondrial genome has variable structure, organization and gene content in plant and animal species. The mtDNAs in fungi are circular and of different sizes in Oomycetes (36–78 kb), Ascomycetes (17–115 kb) and Basidiomycetes (33–176 kb). Structural organization of these mtDNAs is variable because of the presence of introns, inverted repeats and of AT- or GC-rich insertions, which separate conserved single-copy sequences. Rearrangements of gene order are widespread among fungi. However, the rates of nucleotide substitutions in fungal mtDNA genes are lower than in nuclear genes (Clark-Walker 1992).

The mitochondrial genomes of higher plants are large (100–2500 kb), usually circular, and highly dynamic as a result of frequent recombination, which generates populations of molecules with different size and organization. Coding sequences are highly conserved, they have introns and are flanked by highly variable noncoding sequences. Therefore, rearrangements are common events in the evolution of plant mtDNAs. However, single-copy mtDNA sequences of plants have relatively low rates of base substitution. Mitochondrial mutations may have important phenotypical consequences in plants, including cytoplasmic male sterility, sensitivity to fungal toxins and abnormal morphology of leaves (Hanson & Folkerts 1992).

The gene content of mtDNA of Metazoa is remarkably conserved (Fig. 7.1). All mtDNAs have genes coding for enzymatic proteins of the respiratory chain: apocytochrome *b* (CYB), four to eight subunits of nicotinamide adenine dinucleotide hydroxide dehydrogenase (NADH-1 to 7, plus NADH-4L), up to three subunits of cytochrome *c* oxidase (COX-1 to 3), up to three subunits of adinoriutriphosphate synthetase (ATP-6, 8 and 9). Moreover, mtDNAs have genes coding for components of the protein-synthesizing system: two ribosomal RNAs (the small and large subunits 12S and 16S rRNA), and 20–28 genes for transfer RNAs (tRNA). The other proteins necessary for the function, replication, transcription and translation of mtDNA are encoded by nuclear DNA and imported from the cytoplasm. The main noncoding sequence is the control-region (named the D-loop in vertebrates and AT-rich region in invertebrates),

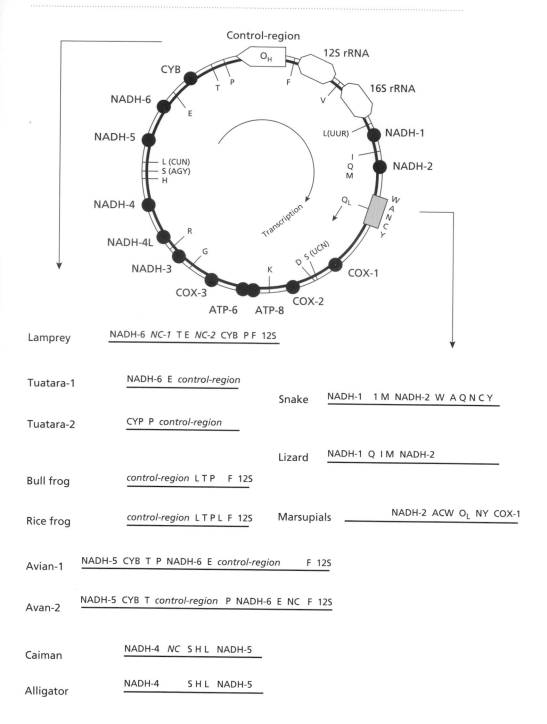

which regulates the replication and transcription of the entire mitochondrial genome (Wolstenholme & Jeon 1992).

### 7.2.1 Evolution of gene order in vertebrate mtDNAs

Mitochondrial gene order is mostly conserved within, but varies between, different phyla or orders (Fig. 7.1; see Table 7.1 for references). All the described mtDNA rearrangements involve transpositions of some tRNAs or the control-region (Fig. 7.1). The mechanisms of mtDNA reorganization are unknown, although it has been suggested that intermolecular recombination (Lunt & Hyman 1997), variable location of replication origins (Macey *et al.* 1997) and tandem duplications as a result of slipped-strand mispairing (Moritz *et al.* 1987), could generate duplicate segments, which are later partially deleted during the evolutionary process (Mindell *et al.* 1998).

Mitochondrial DNA rearrangements could provide molecular markers for phylogenetic analyses (Boore *et al.* 1995). However, while most of the described rearrangements of vertebrate mtDNAs identify monophyletic groups (Macey *et al.* 1997), the recently described parallel transpositions in birds (Mindell *et al.* 1998) warn against generalizations. Finally, knowlege of mtDNA organization is necessary to design PCR primers for population and phylogenetic studies.

### 7.2.2 Origins of mitochondria and interactions with the nuclear genome

Mitochondria originated through the integration of an anaerobic bacterium into a eukaryotic cell (Gray 1992). Symbiotic organellar and nuclear genomes coevolved by selective elimination of mitochondrial redundant genes and gene transfer to the nucleus, and replication of mtDNA became dependent on nuclear-encoded enzymes. Nuclear–mitochondrial coevolution is still ongoing (Zevering *et al.* 1991; Kilpatrick & Rand 1995), and transpositions of mtDNA genes to the nucleus ('numts'; Lopez *et al.* 1994) have been documented in plants and animals (Zhang & Hewitt 1996). Numts are considered 'fossil

---

**Fig. 7.1** (*Opposite.*) Circular map of the most frequent gene order in the vertebrate mtDNA (top). $O_H$ and $O_L$ are the origins of H- and L-strand replication. The arrows in the control region box and at $O_L$ indicate the directions of H- and L-strand replication. Direction of transcription is indicated by the arrow. Abbreviations for genes are given in section 7.3.2. Transfer RNAs are indicated by their standard one-letter amino acid code. Linear maps indicate: (1) rearrangements which involve the control region (left side); (2) rearrangements which involve the IQM and WANCY tRNA clusters (right side). NC indicates the presence of duplicated noncoding regions in the lamprey, avian-2 and caiman mtDNAs (references are in Macey *et al.* 1997; Mindell *et al.* 1998).

**Table 7.1** List of completely sequenced eukaryotic mitochondrial DNA genomes, with publication references and European Molecular Biology Laboratory gene bank accession numbers. The National Center for Biotechnology Information (NCBI) now offers a collection of complete mitochondrial and plastid genome sequences at URL: www.ncbi.mlm.mik.gov/IMGifs/Genomes/organelles.html

| Taxonomic group and species name | Common name | Reference | Accession number |
| --- | --- | --- | --- |
| **Plants** | | | |
| *Reclinomonas americana* | | Lang *et al.* (1997) *Nature* **387**, 493–497 | AF007261 |
| *Chlamydomonas eugametos* | | Denovan-Wright *et al.* (1998) *Plant Mol. Biol.* **36**, 285–295 | AF008237 |
| *Neurospora crassa* | | Narang *et al.* (1984) *Cell* **38**, 441–453 | K03295 |
| *Allomyces macrogynus* | | Paquin & Lang (1996) *J. Mol. Biol.* **255**, 688–701 | U41288 |
| *Podospora anserina* | | Cummings *et al.* (1990) *Curr. Genet.* **1**, 375–402 | X55026 |
| **Protozoa** | | | |
| *Paramecium aurelia* | | Pritchard *et al.* (1990) *Nucleic Acid Res.* **18**, 173–180 | X15917 |
| *Acanthamoeba castellanii* | | Burger *et al.* (1995) *J. Mol. Biol.* **245**, 522–537 | U12386 |
| **Metazoa/invertebrates/cnidarians** | | | |
| *Metridium senile* | Sea anemone | Wolstenholme (1992) In: *Mitochondrial Genomes* (Wolstenholme & Jeon eds) | AF000023 |
| **Nematodes** | | | |
| *Onchocerca volvulus* | Nematode | Keddie *et al.* (1998) *Mol. Biochem. Parasitol.* **95**, 111–127 | AF015193 |
| *Caenorhabditis elegans* | Soil nematode | Okimoto *et al.* (1992) *Genetics* **130**, 471–498 | X54252 |
| *Ascaris suum* | Pig gut nematode | Okimoto *et al.* (1992) *Genetics* **130**, 471–498 | X52453 |
| **Anellids** | | | |
| *Lumbricus terrestris* | Earthworm | Boore & Brown (1995) *Genetics* **141**, 305–319 | U24570 |
| **Molluscs** | | | |
| *Katharina tunicata* | Black chiton | Boore & Brown (1994) *Genetics* **138**, 423–443 | U09810 |
| *Albinaria coerulea* | Land snail | Hatzoglou *et al.* (1995) *Genetics* **140**, 1353–1366 | X8339 |

| | | | |
|---|---|---|---|
| *Cepaea nemoralis* | Snail | Terret *et al.* (1994) *The Nautilus* **108**, 79–84 | U23045 |
| *Mytilus edulis* | Blue mussel | Hoffmann *et al.* (1992) *Genetics* **131**, 397–411 | M83756-M83762 |
| *Loligo bleekeri* | Squid | Tomita *et al.* (1998) *Biochem. Biophys. Acta* **1399**, 78–82 | AB009838 |
| **Arthropods** | | | |
| *Anopheles gambiae* | African malaria mosquito | Beard *et al.* (1993) *Insect Mol. Biol.* **2**, 103–124 | L20934 |
| *Locusta migratoria* | Grasshopper | Flook *et al.* (1995) *J. Mol. Evol.* **41**, 928–941 | X80245 |
| *Apis mellifera* | Honeybee | Crozier & Crozier (1993) *Genetics* **133**, 97–117 | L06178 |
| *Drosophila yakuba* | Fruit fly | Clary & Wolstenholme (1985) *J. Mol. Evol.* **22**, 252–271 | X03240 |
| *D. melanogaster* | Fruit fly | Garesse (1988) *Genetics* **118**, 649–663 | Y00610 |
| *Artemia franciscana* | Artemia | Valverde *et al.* (1994) *J. Mol. Evol.* **39**, 400–408 | X69067 |
| **Hemichordates** | | | |
| *Balanoglossus carnosus* | | Castresana *et al.* (1998; *Proc. Natl Acad. Sci. USA* **97**, 3703–3707 | AF051097 |
| **Echinoderms** | | | |
| *Asterina pectinifera* | Starfish | Asakawa *et al.* (1995) *Genetics* **140**, 1047–1060 | D16387 |
| *Paracentrotus lividus* | Sea urchin | Cantatore *et al.* (1989) *J. Biol. Chem.* **264**, 10965–10975 | J04815 |
| *Strongylocentrotus purpuratus* | Sea urchin | Jacobs *et al.* (1988) *J. Mol. Biol.* **202**, 185–217 | X12631 |
| *Arbacia lixula* | Sea urchin | de Giorgi *et al.* (1996) *Mol. Phylogenet. Evol.* **5**, 323–332 | X80936 |
| *Florometra serratissima* | Crinoid | Scouras & Smith (1998) unpublished information | AF049132 |
| **Vertebrates/fish** | | | |
| *Petromyzon marinus* | Sea lamprey | Lee & Kocher (1995) *Genetics* **139**, 873–888 | U11880 |
| *Branchiostoma floridae* | Lancelet | Naylor & Brown (1998) *Syst. Biol.* **47**, 61–76 | AF035164-AF035176 |
| *Myxine glutinosa* | Hagfish | Rasmussen *et al.* (1998) *J. Mol. Evol.* **46**, 382–388 | Y15182-Y15192 |
| *Polypterus ornatipinnis* | Bichir | Noack *et al.* (1996) *Genetics* **144**, 1165–1180 | U62532 |
| *Scyliorhinus canicula* | Dogfish | Barriel *et al.* (1998) *Genetics* **150**, 331–344 | Y16067 |
| *Protopterus dolloi* | African lungfish | Zardoya & Meyer (1996) *Genetics* **142**, 1249–1263 | L42813 |
| *Latimeria chalumnae* | Coelacanth | Zardoya & Meyer (1997) *Genetics* **146**, 995–1010 | U82228 |

*Continued on p. 142*

**Table 7.1** *Continued*

| Taxonomic group and species name | Common name | Reference | Accession number |
|---|---|---|---|
| *Mustelus manazo* | Shark | Cao et al. (1998) Mol. Biol. Evol. **15**, 1637–1646 | AB015962 |
| *Oncorhynchus mykiss* | Rainbow trout | Zardoya et al. (1995) J. Mol. Evol. **41**, 942–951 | L29771 |
| *Cyprinus carpio* | Carp | Chang et al. (1994) J. Mol. Evol. **38**, 138–155 | X61010 |
| *Carassius auratus* | Carassius | Murakami et al. (1998) Zool. Sci. **15**, 335–337 | AB006953 |
| *Crossostoma lacustre* | Loach | Tzeng et al. (1992) Nucl. Acid Res. **20**, 4853–4858 | M91245 |
| *Gadus morhua* | Atlantic cod | Johansen et al. (1990) Nucl. Acid Res. **18**, 38–39 | X17660 |
| *Salmo salar* | Atlantic salmon | Hurst et al. (1998) unpublished information | U12145 |
| **Amphibians** | | | |
| *Xenopus laevis* | Clawed toad | Roe et al. (1985) J. Biol. Chem. **260**, 9729–9774 | X02890 |
| **Reptiles** | | | |
| *Dinodon semicarinatus* | Snake | Kumazawa et al. (1998) Genetics **15**, 313–329 | AB008539 |
| *Alligator mississippiensis* | Alligator | Janke & Arnason (1997) Mol. Biol. Evol. **14**, 1266–1272 | Y13113 |
| **Birds** | | | |
| *Rhea americana* | Greater rhea | Härlid et al. (1998) J. Mol. Evol. **46**, 669–679 | Y16884 |
| *Struthio camelus* | Ostrich | Härlid et al. (1997) Mol. Biol. Evol. **14**, 754–761 | Y12025 |
| *Gallus gallus* | Domestic chicken | Desjardins & Morais (1990) J. Mol. Biol. **212**, 599–634 | X52392 |
| **Mammals** | | | |
| *Homo sapiens* | Human (chimaeric sequence) | Anderson et al. (1981) Nature **290**, 457–465 | V00662 |
| *Homo sapiens* | Human (Caucasian) | Arnason et al. (1996) J. Mol. Evol. **42**, 145–152 | X93334 |
| *Pan troglodytes* | Common chimpanzee | Horai et al. (1995) Proc. Natl Acad. Sci. USA **92**, 532–536 | D38113 |
| *Pan paniscus* | Pigmy chimpanzee | Horai et al. (1995) Proc. Natl Acad. Sci. USA **92**, 532–536 | D38116 |

| Species | Common name | Reference | Accession |
|---|---|---|---|
| *Gorilla gorilla* | Western lowland gorilla | Xu & Arnason (1996) *Mol. Biol. Evol.* **3**, 691–698 | D93347 |
| *Pongo pygmaeus* | Bornean orang-utan | Horai et al. (1995) *Proc. Natl Acad. Sci. USA* **92**, 532–536 | D38115 |
| *Pongo abelii* | Sumatran orang-utan | Xu & Arnason (1996) *J. Mol. Evol.* **43**, 431–437 | X97707 |
| *Hylobates lar* | White-handed gibbon | Arnason et al. (1996) *Hereditas* **124**, 185–189 | X99256 |
| *Equus caballus* | Horse | Xu & Arnason (1994) *Gene* **148**, 357–362 | X79547 |
| *Equus asinus* | Donkey | Xu et al. (1996) *J. Mol. Evol.* **43**, 438–446 | X97337 |
| *Ceratotherium simum* | White rhinoceros | Xu & Arnason (1997) *Mol. Phylogenet. Evol.* **7**, 189–194 | Y07726 |
| *Rhinoceros unicornis* | Indian rhinoceros | Xu et al. (1996) *Mol. Biol. Evol.* **13**, 1167–1173 | X97336 |
| *Phoca vitulina* | Harbor seal | Arnason & Johnsson (1992) *J. Mol. Evol.* **34**, 493–505 | X63726 |
| *Halichoerus grypus* | Grey seal | Arnason et al. (1993) *J. Mol. Evol.* **37**, 323–330 | X72004 |
| *Felis catus* | Domestic cat | Lopez et al. (1996) *Genomics* **33**, 229–246 | U20753 |
| *Canis familiaris* | Domestic dog | Kim et al. (1998) *Mol. Phylogenet. Evol.* **10**, 210–220 | U96639 |
| *Balaenoptera physalus* | Fin whale | Arnason et al. (1991) *J. Mol. Evol.* **33**, 556–568 | X61145 |
| *Balaenoptera musculus* | Blue whale | Arnason & Gullberg (1993) *J. Mol. Evol.* **37**, 312–322 | X72204 |
| *Hippopotamus amphibius* | Hippopotamus | Ursing & Arnason (1998) *Proc. R. Soc. Lond. B* **265**, 2251–2255 | AJ010957 |
| *Bos taurus* | Cow | Anderson et al. (1982) *J. Mol. Biol.* **156**, 683–717 | J01394 |
| *Glis glis* | Fat dormouse | Reyes et al. (1998) *Mol. Biol. Evol.* **15**, 499–505 | AJ001562 |
| *Oryctolagus cuniculus* | Domestic rabbit | Gissi et al. (1998) *Genomics* **49**, 161–169 | AJ001588 |
| *Cavia porcellus* | Guinea pig | D'Erchia et al. (1996) *Nature* **381**, 597–599 | AJ222767 |
| *Rattus norvegicus* | Rat | Gadaleta et al. (1989) *J. Mol. Evol.* **28**, 497–516 | X14848 |
| *Mus musculus* | Mouse | Bibb et al. (1981) *Cell* **26**, 167–180 | J01420 |
| *Mus domesticus* | Mouse | Loveland et al. (1990) *Cell* **60**, 971–980 | L07095 |
| *Dasypus novemcinctus* | Long-nosed armadillo | Arnason et al. (1997) *Mol. Biol. Evol.* **14**, 762–768 | Y11832 |
| *Erinaceus europaeus* | Hedgehog | Krettek et al. (1995) *J. Mol. Evol.* **41**, 952–957 | X88898 |
| *Artibeus jamaicensis* | Fruit bat | Pumo et al. (1998) *J. Mol. Evol.* **47**, 709–717 | AF061340 |
| *Didelphis virginiana* | Opossum | Janke et al. (1994) *Genetics* **137**, 243–256 | Z29573 |
| *Ornithorhynchus anatinus* | Platypus | Janke et al. (1996) *J. Mol. Evol.* **42**, 153–159 | X83427 |
| *Macropus robustus* | Wallaroo | Janke & Arnason (1997) *Proc. Natl Acad. Sci. USA* **94**, 1276–1281 | Y10524 |

molecules' evolving slower than their original mitochondrial counterparts as a result of the lower substitution rates of genes encapsulated in the nuclear 'environment'. Therefore, numts could retain ancestral pretranslocation states of the genes and could be identified as outgroup sequences in a phylogenetic tree (Arctander 1995; but see also Lopez *et al.* 1997). Detection of recent numts can be problematic, and laboratory precautions are recommended to avoid their PCR amplification (Sorenson & Quinn 1998).

### 7.2.3 Inheritance and population genetics of organellar genomes

Mitochondria in somatic cells comprise populations of about 10 DNA molecules, and each cell has $10^2$–$10^5$ mitochondria, depending on tissue and physiological state. The inheritance of mtDNA is prevalently maternal: sperm contribute only about 50 molecules to the eggs, which contain up to $10^5$ copies of maternal mtDNA. Rare cases of paternal transmission have been reported, particularly in interspecific crosses, but paternal mtDNA is rapidly degraded in the oocyte cytoplasm in intraspecific fertilization (Lightowlers *et al.* 1997). On the contrary, marine and freshwater mussels have two distinct mtDNA types inherited by paternal and maternal transmission (Hoeh *et al.* 1996). Although intra- or inter-mtDNA recombination is possible (Lunt & Hyman 1997), normally the mitochondrial genome is maternally transmitted as a single haploid supergene.

High rates of mutation should continually generate new mtDNA haplotypes by point mutation and insertion/deletions (indels), thus at any time mtDNA populations can be genetically heterogeneous (intra-individual heteroplasmy). If not selected, polymorphic mitochondrial genomes might replicate and segregate randomly among daughter organelles and cells. However, the mtDNA population size of each mitochondrion is lower than 10, and thus heteroplasmy should be lost rapidly by random drift. Moreover, mtDNA populations pass through a series of bottlenecks during oogenesis, and heteroplasmy should be severely reduced before fertilization (Lightowlers *et al.* 1997). Indeed, Hauswirth & Laipis (1982) showed that two mtDNA types cotransmitted within maternal cow lineages switched to one of the two alternative homoplastic states within two generations. However, stable heteroplasmy has been discovered particularly in the control region and elsewhere in the genome (Zevering *et al.* 1991; Gemmell *et al.* 1996; Casane *et al.* 1997).

## 7.3 Evolution of mitochondrial DNA

Mitochondrial DNA sequences evolve, on average, 5–10 times faster than nuclear genes (Wilson *et al.* 1985). Closely related mtDNA sequences differ mainly in transitions (Ti) rather than transversions (Tv), and values of Ti:Tv

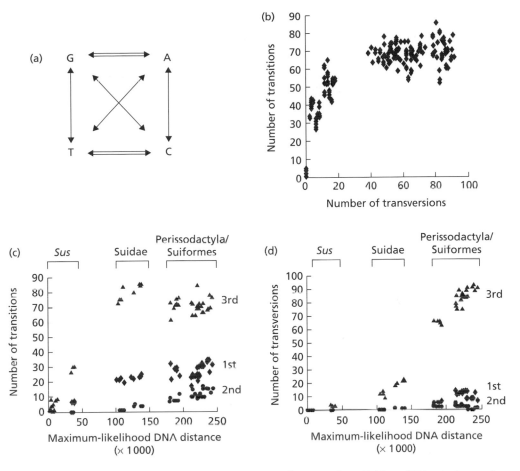

**Fig. 7.2** Saturation plots of transitions (Ti) and transversions (Tv) in mtDNA cytochrome *b* genes of Suiformes. (a) Ti and Tv substitutions. (b) Pairwise relations of the numbers of Ti vs. Tv substitutions, showing that saturation of Ti is strong after the accumulation of 10–20 Tv. A plateau is reached at about 70 Ti. (c) Plotting of the number of Ti versus the estimated pairwise maximum-likelihood DNA distances. Transitions at the third codon positions accumulate faster at small genetic distances, but saturate quickly in pairwise comparisons within family Suidae. On the contrary, Ti substitutions at the first and second codon positions accumulate slowly and are less saturated also in comparisons with the outgroups. (d) Transversions accumulate very slowly at the first and second codon positions, and faster and linearly at the third codon positions in pairwise comparisons among Suidae and the outgroups.

ratios can be 10–20 or higher in intraspecific comparisons. However, Tv accumulate with time, and Ti:Tv values become lower than 1, indicating mutational saturation and increasing occurrence of multiple Ti at the same sites (Irwin *et al.* 1991; Fig. 7.2).

The average overall mtDNA divergence rate has been estimated about 1–2% Ma$^{-1}$ in vertebrates (Wilson *et al.* 1985), but relative and absolute rates can vary widely among lineages and genes (Mindell & Thacker 1996). Third positions of codons, synonymous sites and the hypervariable parts of the control region can evolve 10–20 times faster than average rates for mtDNA coding sequences. Transitional hot spots in the control region can evolve 100 times faster than any other sites, making it hard to calibrate evolutionary rates (Wills 1995; Loewe & Scherer 1997). On average mtDNA will saturate in about 10–20 Ma, although the different genes and substitution classes show different dynamics to saturation (Fig. 7.2). However, concatenated sequences from protein-coding mitochondrial genes have provided phylogenetic resolution up to about 250 Ma, particularly if the inferred amino acid sequences are used (Janke & Arnason 1997).

Although current methods in population genetics and phylogenetics are pervasively inspired by the neutral theory of molecular evolution, it is worth remembering that neutrality of mtDNA substitutions cannot be assumed (Ballard & Kreitman 1995), and whenever possible it should be tested using computer programs such as DnaSP and ARLEQUIN (Table 7.2).

### 7.3.1 Evolution of the control region

Replication of mtDNA is unidirectional and asymmetric (Fig. 7.1). Synthesis of the H-strand starts at the $O_H$ within the control-region, originates a triplex structure with the parent H-strand displaced (the D-loop) and often ends at termination-associated sequences (TASs) in domain I (Fig. 7.3). Most of the daughter H-strands are abortive and they are continually degraded. Sometimes H-strand replication continues until it reaches the $O_L$ (Fig. 7.1), which is exposed as a single-stranded template and activates the synthesis of the L-strand (Clayton 1992).

The control region is usually the most variable part of mtDNA, but site mutability and structural rearrangements are not distributed randomly across the entire region, and affect particular hypervariable sites and domains (Yang 1996). Hypervariable sequences are extremely useful in high resolution population analyses (Stanley *et al.* 1996; Baker & Marshall 1997; Good *et al.* 1997; Fry & Zink 1998), although multiple substitutions can generate homoplasy and affect the analysis of genetic diversity and genealogical relationships at the population level (Wills 1995). Alignments of the entire control region have provided phylogenetic information among teleost fish (Faber & Stepien 1997), birds (Kidd & Friesen 1998; Randi & Lucchini 1998), whales (Arnason *et al.* 1993), and ungulate mammals (Douzery & Randi 1997).

**Table 7.2** List of computer programs which can be used for phylogenetic and population genetic analyses. The www sites should be visited to download the updated program versions and the relevant references

| Computer program | Authors (year) | URL and address |
|---|---|---|
| DnaSP (v.3.14) | Rozas & Rozas (1999) | http://www.bio.ub.es/~julio/DnaSP.html |
| ARLEQUIN (v.2.0) | Schneider et al. (1999) | http://anthropologie.unige.ch/arlequin |
| MFOLD | Gilbert (1992) | ftp://ftp.bio.indiana.edu/molbio |
| LOOPDLOOP | Gilbert (1992) | ftp://ftp.bio.indiana.edu/molbio |
| OLIGO | MedProbe A.S. | http://www.medprobe.com |
| OMIGA | MedProbe A.S. | http://www.medprobe.com |
| MacVector | Oxford Molecular Group | http://www.oxmol.com |
| REAP | McElroy et al. (1984) | email: mcelrdm@wkuvx1.wku.edu |
| NUCLEODIV | Ho Singer (1998) | ftp://darwin.eeb.uconn.edu/pub/ |
| PHYLIP (v.3.5c) | Felsenstein (1995) | http://evolution.genetics.washington.edu/phylip.html |
| PAUP (v.4.0) | Swofford (1998) | http://www.sinauer.com |
| HENNIG86 | Farris (1988) | email:arnold.g.kluge@um.cc.umich.edu |
| CLUSTAL X | Thompson et al. (1997) | http://www-igbmc.u-strasbg.fr/ |
| MALIGN (v.2.7) | Wheeler & Gladstein (1998) | ftp://ftp.amnh.org/pub/molecular/ |
| GENEDOC (v.2.0) | Nicholas & Nicholas (1997) | http://www.cris.com/~ketchup/genedoc.html |
| Se-Al (v.1.0) | Rambaut (1996) | http://evolve.zoo.ox.ac.uk/Se-Al/Se-Al.html |
| SITES (v.1.1) | Hey (1997) | http://heylab.rutgers.edu/ |

Continued on p. 148

**Table 7.2** *Continued*

| Computer program | Authors (year) | URL and address |
|---|---|---|
| FLUCTUATE (v.1.1) | Kuhner *et al.* (1996) | http://evolution.genetics.washington.edu/pub/lamarc |
| POPGENE (v.1.21) | Yeh *et al.* (1997) | http://www.ualberta.ca/~fyeh |
| GENEPOP (3.1d) | Raymond & Rousset (1999) | http://www.cefe.cnrs-mop.fr/ |
| FSTAT (v.1.2) | Goudet (1994) | email: jerome.goudet@izea.unil.ch |
| HEAPBIG | Palumbi (1998) | http://www.oeb.harvard.edu/cceg/information |
| MACLADE (v.3) | Maddison & Maddison (1992) | http://www.sinauer.com |
| MIGRATE (v.0.7) | Beerli & Felsenstein (1999) | http://evolution.genetics.washington.edu/lamarc.html |
| Bi-De | Rambaut *et al.* (1994) | http://evolve.zps.ox.uk/ |
| End-Epi | Rambaut *et al.* (1994) | http://evolve.zps.ox.uk/ |
| LAMARC | Felsenstein *et al.* (1999) | http://evolution.genetics.washington.edu/lamark.html |
| GENETREE | Griffith (1998) | http://www.maths.monash.edu.au/~mbahlo/mpg/gtree. html |
| MISMATCH (v.2c) | Rogers (1995) | ftp://ftp.anthro.utah.edu |
| ANALYSE | Barton & Baird (1998) | http://helios.bto.ed.ac.uk/evolgen/Mac/Analyse/ |
| MOLPHY (v.2.2) | Adachi & Hasegawa (1994) | ftp://ftp.sunmh.ism.ac.jp/pub/molphy* |
| PAML (v.2.0a) | Yang (1999) | http://abacus.gene.ucl.ac.uk/pub/paml/ |
| PUZZLE (v.4.0) | Strimmer & von Haeseler (1996) | http://www.zi.biologie.uni-muenchen.de/~strimmer/puzzle.html |
| FastDNAML (v.1.0) | Olsen *et al.* (1992) | ftp://ftp.bio.indiana.edu/molbio |
| GAML | Lewis (1998) | http://biology.unm.edu/~lewisp/gaml.html |

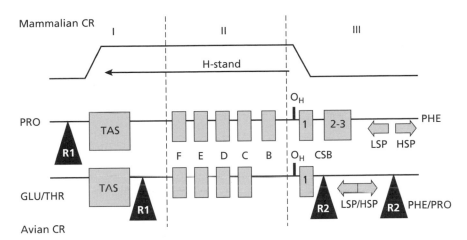

**Fig. 7.3** Organization of the mammalian and avian mtDNA control regions. Domains I, II (with conserved sequence blocks B-F) and III are mapped. Synthesis of the H-strand starts at $O_H$ in domain III and stops at termination-associated sequences (TAS) in domain I. CSB-1, -2 and -3 are sites of functional importance in controlling H-strand replication. LSP and HSP are the promoters of H- and L-strand transcription. Indels and VNTRs have been mapped in different species, usually around the TAS region (R1) and the CSBs region (R2).

## 7.3.2 Evolution of the protein-coding genes

Transcription of the mtDNA is controlled by the H- and L-strand promoters (HSP and LSP) within the control region (Fig. 7.2), and produces a single poly-cistronic RNA that is processed and spliced into the individual mRNAs (Clayton 1992). In vertebrate mtDNA all genes are encoded by the H-strand, except NADH-6 and eight tRNAs. The mitochondrial genetic code is simpler, more degenerate and less constrained than the nuclear code. Consequently, substitutions at the third codon positions are synonymous and do not change the amino acid sequences, while all substitutions at the second and most first codon positions are nonsynonymous and result in amino acid replacements (Moritz *et al.* 1987).

Constraints to amino acid changes determine faster accumulation of syn-onymous than nonsynonymous substitutions. Therefore, the three codon positions mutate at different rates (Fig. 7.2). Synonymous sites evolve about 8–10 times faster than second positions and 2–4 times faster than first positions in vertebrates (Kocher & Carleton 1997), and can be used for estimating genetic diversity within populations (Bermingham *et al.* 1997) and relation-ships among closely related species and genera (Nunn & Cracraft 1996). In contrast, nonsynonymous substitutions and substitutions in the inferred amino acid sequences can provide more conserved phylogenetic information to be used for inferring patterns of deeper evolutionary divergence (Ledje &

Arnason 1996). Evolutionary rates and patterns of substitution in protein-coding genes are constrained also by selective forces acting to maintain the biochemical functions of polypeptides and their structural interactions with the inner mitochondrial membrane. In consequence, mtDNA genes evolve at different rates and have different phylogenetic performances (Zardoya & Meyer 1996).

The *CYB* gene has been extensively used both for phylogenetic and population genetic studies (Kocher *et al.* 1989; Irwin *et al.* 1991; Friesen *et al.* 1996; Moore & DeFilippis 1997 among the many others), although shortcomings and limitations have been noted (Meyer 1994). Structural models of the CYB protein identify transmembrane, inner and outer domains (Howell 1989), which evolve at different rates (Griffith 1997).

### 7.3.3 Structure and evolution of RNA genes

The products of ribosomal genes in mtDNA are two noncoding single-stranded RNAs which bind to proteins and form a functional ribosome. The appropriate RNA–protein interactions are determined by the secondary structure, which comprises double-stranded stems and single-stranded loops and that is maintained by natural selection. Also the secondary cloverleaf structures of tRNAs are strictly controlled by natural selection (Kumazawa & Nishida 1993). Stems are stabilized by basepairing and compensatory mutations, and loops have conserved motifs responsible for association to ribosomal subunits, binding to proteins and tRNAs. These functional constraints determine the highly variable substitution rates among domains and sites of RNA genes (Springer *et al.* 1995).

Aligning tRNAs is straightforward, and can be done by eye. On the contrary, the alignment of rRNAs is complicated, and computer alignments should be improved manually by fitting the sequences to a model of secondary structure (Hickson *et al.* 1996). Variable substitution rates and compensatory substitutions call for differential weighting in parsimony analyses and require evolutionary models which account for correlated (nonindependent) mutations (Rzhetsky 1995). Although mitochondrial tRNAs and rRNAs evolve faster than their nuclear counterparts, they evolve slowly relative to the other mtDNA genes, and have been used to study deep phylogenetic relationships (Kumazawa & Nishida 1993; Mindell *et al.* 1997; Ortí & Meyer 1997). RNA secondary structures can be reconstructed, drawn and edited using MFOLD and LOOPDLOOP (Table 7.2).

### 7.4 Molecular tools for nucleotide sequence analyses

Updated laboratory protocols have been recently published (Sambrook *et al.* 1989; Hoelzel 1992; Ferraris & Palumbi 1996; Hillis *et al.* 1996; Chapters 2 and

3), and will be not extensively discussed here, except for aspects which are important in mtDNA analyses.

### 7.4.2 Primer design and PCR optimization for the amplification of mtDNA sequences

Efficient PCR primers should be highly specific for the target sequence, should not anneal with other regions of the mitochondrial or nuclear DNA, and not dimerize or form internal hairpins. These and other criteria are implemented in computer programs (e.g. OLIGO, OMIGA and MacVector; Table 7.2). However, the performances of each primer pair should be carefully optimized in PCR experiments (Chapter 3).

Homologous primers can be designed using published mtDNA sequences accessed through the genetic data banks (Stoesser *et al.* 1998,1999) and scientific literature (Table 7.1). Specialized mtDNA databases are also available (mitBASE: human mtDNA sequences; MmtDB: Metazoan mtDNAs; see Attimomelli *et al.* 1998a,b). Alternatively, heterologous primers should be designed using mtDNA conserved regions, such as conserved blocks in the rRNAs (Sullivan *et al.* 1995), conserved anticodon stem and loops of tRNAs flanking the target protein-coding genes (Kumazawa & Nishida 1993), and the conserved sequence blocks of the central domain of the control region (Kocher *et al.* 1989). Compilations of 'universal' or 'versatile' primers, which can be used succesfully to amplify mtDNA sequences in a variety of closely or distantly related taxa, have been published by Kocher *et al.* (1989), Simon *et al.* (1994) and Palumbi (1996).

### 7.4.3 Electrophoresis and visualization of mtDNA fragments and sequences

Purified mtDNA or PCR fragments can be digested with restriction endonucleases, bacterial enzymes which cut the DNA at specific sequences (see the database REBASE at EMBL site: www.ebi.ac.uk). Fragment size can vary among individuals as a result of the presence/absence of particular restriction sites, indels or variable number of tandem repeats (VNTR), and can be estimated through agarose gel electrophoresis of the digested samples run in parallel with DNA standards of known molecular size (Fig. 7.4). Artefacts can arise from fragmented target DNA, incomplete digestions and biparental mtDNA inheritance.

Agarose gels are used to check for quantity and quality of DNA extractions, control the results of PCR amplifications, analyse RFLP and to purify PCR products in low-temperature-melting agarose gel (Jones 1998). Nondenaturing acrylamide gels are used to improve the resolution of fragments of less than a few thousand nucleotides. Denaturing acrylamide gels are

```
      1                          EcoRI                    BamHI                                              60
0   GATCATTTCACTA   GAATTC   GTACTATT   GGATCC   TACTGCGTCGCGTTCCTGTATCTTTCAG
1   GATCATTTCACTA   GAATTC   GTACTATT   GGATCC   TACTGCGTCGCGTTCCTGTATCTTTCAG
2   GATCATTTCACTA   GAATTC   GTACTATT   GGATCC   TACTGCGTCGCGTTCCTGTATCTTTCAG
3   GATCATTTCACTA   GAATTC   GTACTATT   GGATCC   TACTGCGTCGCGTTCCTGTATCTTTCAG

4   GATCATTTCACTA   GAATCC   GTACTATT   GGATCC   TACTGCGTCGCGTTCCTGTATCTTTCAG
5   GATCATTTCACTA   GAATCC   GTACTATT   GGATCC   TACTGCGTCGCGTTCCTGTATCTTTCAG

6   GATCATTTCACTA   GAATTC   GTACTATT   GGATCT   TACTGCGTCGCGTTCCTGTATCTTTCAG
7   GATCATTTCACTA   GAATTC   GTACTATT   GGATCT   TACTGCGTCGCGTTCCTGTATCTTTCAG

8   GATCATTTCACTA   GAGTTC   GTACTATT   GGGTCC   TACTGCGTCGCGTTCCTGTATCTTTCAG
9   GATCATTTCACTA   GAGTTC   GTACTATT   GGGTCC   TACTGCGTCGCGTTCCTGTATCTTTCAG
```

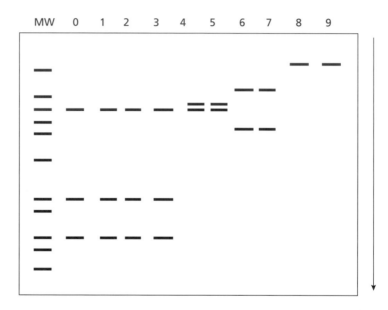

**Fig. 7.4** An alignment of 10 fictitous sequences with two polymorphic restriction sites (top) and a draft of the expected agarose gel (bottom). In these sequences there are four different haplotypes defined by the presence of two *Eco*RI and *Bam*HI restriction sites (individuals 0–3), only one *Eco*RI restriction site (4–5), only one *Bam*HI restriction site (6–7), and no restriction sites (8–9). The recognition sequences are in bold in individuals 0–3. The restriction sites have been lost in individuals 4–9 as a result of mutations (in bold) within the recognition sequences. The expected agarose gel restriction fragment-length polymorphism (RFLP) pattern, and a molecular weight ladder (MW), are shown.

used in single-strand conformation polymorphism (SSCP) detection, denaturing gradient gel electrophoresis (DGGE) and other techniques for screening point mutations (Pfeiffer 1996), and for nucleotide sequencing (Hillis *et al.* 1996).

DNA fragments in agarose and acrylamide gels are stained using EtBr or silver-staining procedures. Oligonucleotide or cloned DNA probes can be labelled nonisotopically, by incorporating biotinylated nucleotides or by immunoenzymatic methods, using very efficient commercial kits. Highly purified mtDNA fragments are end-labelled using isotopic methods.

Nucleotide sequencing using Sanger's dideoxy-termination method is widely used and implemented in commercial kits. Manual sequencing involves the use of isotopically labelled primers or terminators and the products of sequencing reactions are separated in denaturing acrylamide gels. Automatic sequencers produce high throughput of very reliable sequence data and avoid using unstable and hazardous radioactive labelling.

### 7.4.4  Comparison of the various methods: applications and limits

Although nucleotide sequencing provides more resolution and details, RFLP offers a fast method to analyse large sample numbers, particularly in population and ecological genetic projects. When fresh or well preserved tissues are available, mtDNA can be obtained by inexpensive alkaline lysis and restriction fragments can be visualized by EtBr, silver staining or isotopic end-labelling. The PCR offers unlimited possibilities to amplify specific fragments long enough to be digested with restriction enzymes, and which can be directly visualized in electrophoretic gels. Analysis using PCR-RFLP is versatile, fast and inexpensive. In molecular ecology it is advisable to obtain a small number of preliminary sequences in samples encompassing most of the presumed genetic variability, produce restriction site maps of the sequences and then analyse the entire sample set using informative restriction enzymes. Particularly if generated using 4-base cutters, RFLPs can reveal huge amounts of mtDNA variability within and between populations, and can be used to estimate haplotype and nucleotide diversity, individual relatedness, mating systems and geographical variation. The identification of diagnostic restriction sites in parental allopatric populations allows the study of hybridization, maternal introgression and gene flow.

Restriction fragment-length polymorphisms can underestimate sequence diversity because of the nonrandom distribution of restriction sites in the genome. Multiple substitutions within the recognition sequence can produce convergent site losses and bias estimates of sequence diversity. The homology of fragments is deduced from their identical electrophoretic mobility, but small size differences might go undetected and different fragments can comigrate. Fragment homology can be better assessed by mapping the cleavage sites (a laboratory protocol is detailed by Dowling *et al.* 1996). The true mitochondrial origins of RFLPs obtained by PCR fragments should be controlled to avoid nonhomologous comparisons with numt sequences.

Nucleotide sequencing of mtDNA fragments amplified by PCR is a very versatile, albeit more expensive, technique, which can be adapted to achieve insightful results in a variety of research projects in population genetics, molecular ecology, conservation biology, phylogenetics, molecular taxonomy, biogeography and comparative biology. Almost any kind of fresh or degraded biological sample can be used as a source of DNA for PCR. Because mtDNA is effectively one linked gene, in many research projects it is advisable to complement mtDNA data with genetic information on nuclear genes such as polymorphic allozyme loci and microsatellites in population genetic studies, or sequences of autosomal or other sex-linked genes in phylogenetics.

## 7.5  Mitochondrial DNA variability within and among populations

### 7.5.1  Estimating sequence variability from RFLP

When entire mtDNA or PCR fragments are digested by restriction enzymes the different haplotypes can be identified by gel electrophoresis and their number ($k$) and frequency ($x_i$) are directly computed (Fig. 7.4). These measures are dependent on sample size, and population variability is more appropriately estimated by the haplotypic diversity ($H$), equivalent to average heterozygosity; mean number of restriction site differences ($v$); and nucleotide diversity ($\pi$; Nei 1987). Analyses with RFLP at the population level can be performed using REAP (Table 7.2), the AMOVA procedure included in ARLEQUIN, and NUCLEODIV (Table 7.2), which compute estimates of nucleotide and haplotype diversity and test for heterogeneity of allele frequences in different groups using randomization and other procedures.

Mitochondrial DNA RFLPs have been used in many studies of population genetics (Andersen *et al.* 1998), differentiation of human populations (Excoffier & Langaney 1989), genetic divergence among subspecies and sibling species (Zink & Dittman 1993), hybridization and gene flow (Hare & Avise 1996), comparative phyogeography (Avise *et al.* 1987; Zink 1996), conservation genetics (Cunningham & Moritz 1998) and calibration of mtDNA sequence divergence (Shields & Wilson 1987). Mitochondrial DNA RFLP analyses are also being used for population genetic and phylogeographic studies in plants, especially in trees (recent results have been revised by Newton *et al.* 1999).

From a phylogenetic perspective the utility of RFLP is limited to the study of relationships among closely related taxa (Avise 1994).The program RESTSITE (Nei & Miller 1990; program available from the authors) can be used to compute distance matrices from restriction sites for phylogenetic trees.

Programs in PHYLIP (Table 7.2), and parsimony programs such as PAUP and HENNIG86 (Table 7.2), which allow a discrete classification of presence (1) or absence (0) of homologous restriction sites, can be used to analyse RFLP data in a cladistic framework.

### 7.5.2  Aligning nucleotide sequences

A correct alignment is necessary to establish homologies among nucleotides and codons before proceeding further (Wheeler 1995). Aligning protein-coding genes and the short tRNAs is usually simple and can often be undertaken visually, also using start and stop codons, number and clusters of conserved amino acid domains, and models of secondary structures as landmarks for correct visual alignments. Aligning rRNAs and control regions is more problematic and preliminary alignments of linear sequences should be fitted to a model of secondary structure.

Computer alignments can be obtained using CLUSTAL X and MALIGN, while GENEDOC and Se-Al can be used for editing those alignments (see Table 7.2).

### 7.5.3  Analysis of sequence variability within and between populations

Descriptive DNA sequence statistics, such as tables of nucleotide and amino acid compositions, codon usage, nucleotide pair frequencies, indel frequencies, patterns of variability analysed in sliding windows and polymorphic sites, can be computed with MEGA (Kumar et al. 1993), SITES (Table 7.2), ARLEQUIN and DnaSP.

Observed values of population diversity are provided by the number of haplotypes ($k$) and their frequency ($x_i$), and by the average number ($s$) and percent ($p_n$) of polymorphic (segregating) sites in $n$ sequences. These values are dependent on sample size ($n$), and population variability is more appropriately estimated by haplotype diversity ($H$) and nucleotide diversity ($\pi$; Nei 1987). In randomly mating populations, $H$ and $\pi$ are estimates of heterozygosity at the gene and nucleotide levels. If the population is in mutation/drift equilibrium, it is possible to estimate the parameter

$$\Theta = 2N_{ef}\mu$$

(where $N_{ef}$ is the female effective population size, and $\mu$ is the mutation rate per nucleotide per generation), a measure of population DNA polymorphism. In nonequilibrium conditions (historical fluctuation of population size), $\Theta$ is more appropriately estimated using the coalescent theory (Kuhner et al. 1998;

program FLUCTUATE; Table 7.2). If the mutation rate is known, it is possible to use $\Theta$ and estimate $N_{ef}$, and vice versa.

The average number of nucleotide substitutions between two populations $x$ and $y$ is:

$$d_{xy} = \sum_{ij} x_i y_j d_{ji},$$

where $d_{ji}$=nucleotide substitutions between haplotype i from $x$ and j from $y$. The net genetic divergence among populations:

$$d_A = d_{xy} - (\pi_x \times + \pi_y)/2$$

is related to the time $T$ since divergence of the two populations by: $d_A = 2\,\mu T$ (with constant substitution rate $\mu$ per site per year; Nei 1987). Estimates of DNA polymorphism with their variances, divergence between populations and tests of neutrality of nucleotide substitutions are computed using POPGENE (Table 7.2), DnaSP, SITES and ARLEQUIN.

Nucleotide differentiation between populations is computed using the $F$-statistics (Wright 1951), namely by estimating the proportion of total variance which is distributed among populations,

$$F_{ST} = 1 - (Hw/Hb),$$

where Hw and Hb are the average number of pair-wise differences (polymorphic DNA sites) within and between populations, respectively (Hudson *et al.* 1992). In a finite island model of population structure, $F_{ST}$ equilibrates more rapidly for mitochondrial than for nuclear genes (Birky *et al.* 1989). Mitochondrial genes will show stronger geographical subdivision than nuclear genes, if the breeding sex ratio is 1:1 and dispersal favours females. The effective population size of mtDNA is 1/4 that of the nuclear genes ($N_f = 1/4N_e$) if the sex ratio is $N_m/N_f = 1$, and therefore mtDNA variability is sensitive to demographic bottlenecks and random drift in small and isolated populations (Avise *et al.* 1984). $F_{ST}$ can be estimated by different computational methods, which make different assumptions of the effect of sampling variances on the estimates, as Weir's $\Theta$ (Weir & Cockerham 1984), Nei's $G_{ST}$ (Nei 1987) and Hudson's $F_{ST}$ (Hudson *et al.* 1992). Computer programs GENEPOP and FSTAT estimate $F$-statistics using Weir and Cockerham's $\Theta$, while HEAPBIG implements Hudson's $F_{ST}$. Both approaches are implemented in ARLEQUIN (Table 7.2).

### 7.5.4 Reconstructing genetic population structure: isolation and gene flow

Parsimony, maximum likelihood or distance methods (in MEGA, PAUP and

PHYLIP) can be used at the population level to infer genealogical relationships among haplotypes within and among populations. Results are represented as unrooted (network) or rooted (at the most divergent sequence or using appropriate outgroups) phylogenetic trees which can be used, in parsimony or coalescent frameworks, to derive information on genetic population structure and dynamics (Avise *et al.* 1987; Avise 1994; Slatkin 1994a,b; Harvey *et al.* 1996).

Recently isolated populations might share the same ancestral polymorphisms. Avise *et al.* (1984) suggested that the expected time to mtDNA monomorphism is $1/4N_{ef}$ in isolated populations of stable size, and the expected time to reciprocal monophyly is about $3–4N_{ef}$ (Neigel & Avise 1986). However, the behaviour of mtDNA haplotypes in real populations can be complex because social and geographical structure and effective population size can affect the period of lineage sorting and determine the retention of mtDNA polymorphisms for a long time (Hoelzer *et al.* 1998).

Alternatively, geographical isolation might be incomplete and populations are connected by ongoing migration and gene flow. Gene flow is estimated as $N_m$, the effective population size times the fraction of migrating individuals, using different models (Slatkin 1994a). In Wright's $F_{ST}$-based models $N_m$ is estimated as $[(1/F_{ST})-1]/2$ (for haploid organisms or mtDNA) and $F_{ST}$ is estimated by $\Theta$. Estimates of $F_{ST}$-based mitochondrial gene flow can be computed using ARLEQUIN and DnaSP. Cladistic methods are based on reconstructed phylogenies of DNA sequences on which are mapped (as a multistate character) the geographical sample locations (Slatkin & Maddison 1989). Cladistic estimates of gene flow can be obtained first by estimating the maximum parsimony tree with PAUP, then by counting the minimum number of migrations in the tree with MACLADE (Maddison & Maddison 1992; Table 7.2), and finally by converting *s* into $N_m$ using table 1 in Slatkin & Maddison (1989). When populations are not in equilibrium it could be appropriate to estimate gene flow using the coalescent program MIGRATE (Table 7.2).

Pairwise genetic distances can increase with increasing geographical distances among continuously distributed populations, according to the isolation-by-distance model, which predicts that gene frequencies might diverge among populations with low dispersal rates and low local effective population size. Slatkin & Maddison (1990) showed that the regression of $\log(M)$, a pairwise estimate of $F_{ST}$, on log(distance) is expected to be linear in case of isolation-by-distance, and can be used to estimate migration rates between adjacent demes (Slatkin 1994b). This regression can be computed using GENEPOP and FSTAT.

The coalescent theory (Hudson 1990) uses haplotype genealogies to estimate population genetic parameters, such as the time to the most recent common ancestor (MRCA), effective population size, long-term population growth, historical bottlenecks, gene flow, selection, rates of lineage speciation and extinction (Harvey *et al.* 1996). Computer programs Bi-De and End-Epi, SITES, LAMARC, GENETREE and MISMATCH are listed in Table 7.2.

Real populations might not meet the theoretical assumptions of population models. Discrepancies among direct short-term and long-term genetic estimates of $N_{ef}$ have been documented by Avise *et al.* (1988). Long-term fluctuating population size and historical bottlenecks have been hypothesized to account for lower than expected mtDNA sequence diversity and $N_{ef}$ values (O'Brien *et al.* 1996). The northern hemisphere global climatic changes related to glacial–interglacial cycles might have resulted in repeated bottlenecks and have affected the distribution of terrestrial and marine organisms. In many cases, populations may not have re-established genetic equilibrium. Under these conditions it is difficult to distinguish between current and historical gene flow. High resolution markers (such as mtDNA control region sequences and microsatellites) should be used to describe population structure (Andersen *et al.* 1998; Paetkau *et al.* 1998), and provide indirect estimates of gene flow that can be compared with direct estimates of dispersal through banding or radiotelemetry experiments.

### 7.5.5 Hybridization and introgression of mtDNA

Genetically distinct populations and parapatric species may meet and hybridize in areas of secondary contact (Barton & Hewitt 1989). Many hybrid zones are stable and narrow relative to the distribution range and dispersal rates of parental populations, and are maintained by a balance between dispersal and selection against hybrids (tension zones; Barton & Hewitt 1989). The genetic structure of hybridizing populations is characterized by clinal gradients in allele frequencies, which can be coincident and change steeply at the centre of the hybrid zone. The S-shaped form of clines describes introgression of alleles, which can be unconstrained or limited by natural selection. Quantitative analyses of clines can therefore allow estimates of the intensity of the genetic factors (gene flow, selection, recombination), which affect the interactions among hybridizing genomes.

Genetic clines are reconstructed and analysed using diagnostic loci with fixed allele differences among allopatric parental populations (using program ANALYSE; Table 7.2). While nuclear genes usually show similar clines, mtDNA often shows extensive asymmetrical introgression; for example, mtDNA of the house mouse (*Mus musculus domesticus*) has introgressed 750 km into Scandinavian populations of *M. m. musculus*, while the nuclear gene clines are only 50 km wide (Gyllensten & Wilson 1987). Introgression of nuclear genes can be limited by direct selection or by chromosomal linkage, while mtDNA behaves more neutrally. However, non-neutral interactions among mitochondrial and nuclear genomes in hybrids have been reported by Kilpatrick & Rand (1995), which raised cautions about inferring population genetic parameters in hybrid zones assuming mtDNA neutrality. The use of mitochondrial DNA in conjunction with nuclear genes allows the study of cytonuclear disequilibria (Arnold 1993) and investigation of reproductive interactions among populations, such as mating asymmetry and female choice (Patton & Smith 1993), and sex-biased dispersal and gene flow (Tegelström & Gelter 1990). Unexpected cases of hybridization have been discovered among Canidae and Felidae using mtDNA and nuclear gene markers (O'Brien *et al.* 1996; Wayne 1996).

### 7.5.6 Phylogeography and molecular biogeography

Historical biogeography aims to reconstruct origins, dispersal, patterns of endemism, vicariance and eventual extinction of populations and communities within a phylogenetic framework. Mitochondrial and chloroplast DNAs are particularly suited to reconstruct the historical patterns of geographical distribution of monophyletic lineages and identify the manifold determinants of biodiversity. When the structure of haplotype trees overlaps with the geographical distributions of sampled populations, genetic diversity has a geographical structure that is called intraspecific phylogeography (Avise *et al.* 1987). Deep phylogenetic discontinuities can be ascribed to deep geographical isolation because of past population fragmentation and evolution in allopatry, and often indicates the existence of biogeographical barriers to dispersal and gene flow. Deep gaps can be concordantly shown by different genes and phenotypic characters, and then constitute a strong base for taxonomic inference (Avise & Ball 1990). Unrelated taxa may have concordant phylogeography, which helps in delineating the boundaries of biogeographical provinces and areas of endemism (Avise 1996). Phylogeographic structure has been detected in terrestrial, freshwater, coastal and marine species (Avise 1996), suggesting the existence of persistent barriers to dispersal even if obvious physical barriers appear to be absent. Deep genetic gaps and phylogeographic patterns have been documented also in oceanic species with great potential for passive or active dispersal (see papers by Baker & Palumbi, Bowen & Avise, Allendorf & Waples and Graves published in Avise & Hamrick 1996). Colonization sequences in oceanic islands have been elucidated by mtDNA phylogenies (Shaw 1996). Using a molecular clock, dates of colonization have been compared with the geological origins of islands (Thorpe *et al.* 1994).

   Palaeoclimatic changes during upper Pliocene and Pleistocene glacial–interglacial cycles in the last 2.5 Ma have shaped the patterns of current biodiversity (Klicka & Zink 1997; Avise & Walker 1998). Models of geographical isolation–expansion of populations related to glacial cycles have been developed to explain their current distributions, differentiation and speciation (Hewitt 1996). Comparative phylogeography has been used to locate refuge areas and postglacial dispersal routes in the Northen Hemisphere (Zink 1996; Jaarola *et al.* 1997) and in tropical areas (Joseph *et al.* 1995; Roy 1997). Unravelling deep geographical structure and subdivision of conspecific populations has relevant implications for ecological genetics, taxonomy and conservation biology (Avise & Hamrick 1996).

### 7.5.7  Using mitochondrial DNA sequences in phylogenetic biology

Molecular phylogenies represent hypotheses of evolutionary relationships among genes and organisms which can be used to map the evolutionary patterns of morphological, behavioural or life-history traits (Brooks & McLennan 1991; Eggleton & Vane-Wright 1994). When a molecular clock is applied, molecular phylogenies can be used to define temporal frameworks, which correlate with other abiotic and biotic information (Thorpe *et al.* 1994; Radtkey 1996; Shaw 1996; Bermingham *et al.* 1997; Omland 1997). Correct phylogenies are necessary to evaluate the role of genealogical relations or current ecological factors in adaptation and to describe the patterns of biodiversity at different taxonomic ranks (Edwards & Naeem 1993; Lanyon 1994; Richman 1996). Understanding the distribution of genetic diversity in a phylogenetic context is often necessary to implement appropriate conservation strategies (Moritz 1996; O'Brien *et al.* 1996).

Phylogenetic relationships among haplotypes are reconstructed by distance, parsimony and maximum likelihood methods (Swofford *et al.* 1996). Distance matrices are computed by MEGA, PHYLIP, PAUP and other phylogenetic programs. Parsimony methods are implemented in PAUP, HENNIG86 and PHYLIP. Maximum likelihood trees are computed by PAUP, MOLPHY, PAML, PUZZLE, FastDNAML, GAML and DNAML in PHYLIP (Table 7.2).

There are at least three major problems that must be solved to ensure the correct use of mtDNA sequences in phylogenetic biology:

1  sample size — the number of sampled sequences/taxa must be representative of the biodiversity of the studied taxonomic groups;

2  sequence length, which is strictly dependent on the rate of molecular evolution of the studied genes and the levels (and time) of divergence of the taxonomic groups of interest; and

3  choice of the appropriate substitution model and its computational implementation.

Mitochondrial DNA has been extensively used in phylogenetic biology. A mitochondrial phylogeny is a single-gene tree, which might not represent the species phylogeny, because of possible horizontal gene transfer through hybridization and deep coalescence of allelic lineages (Maddison 1997). However, the effective population size of mtDNA is 1/4 that of nuclear genes, hence mitochondrial haplotypes have shorter coalescence time, polymorphisms will be sorted quickly among species and there is a high probability that a mtDNA tree represents the species tree (Moore 1995).

Substitution rates are variable among evolutionary lineages, genes and gene regions in the same mtDNA genome. Therefore, notwithstanding that individual genes are linked in a single mitochondrial supergene and share an identical evolutionary history, discordant phylogenies are often produced because the resolution level can vary according to the rate of evolution of the studied genes (Mindell & Thacker 1996). The increasing number of entire mtDNA genomes that are becoming available (Table 7.1) allows researchers to reconstruct phylogenies using the concatenated sequences of all the protein-coding genes. The optimistic suggestion that longer sequences will produce reliable phylogenies among many taxa also using simple and 'assumption-free' models as unweighted parsimony, has been contradicted by empirical findings (Sullivan & Swofford 1997). Understanding the patterns of molecular evolution and using the appropriate phylogenetic models is also inescapable for handling entire mtDNAs.

## 7.6 Summary

Since its discovery in 1965, studies of organization and sequence evolution of mtDNA have had a deep impact on the development of modern comparative biology, phylogenetics, taxonomy, population and conservation genetics of plant and animal species. The simplicity of mtDNA organization, which is usually clonally and maternally inherited, make this molecule a unique marker for phylogenetic and population genetic studies. With the exception of plants, the mitochondrial genes evolve faster than nuclear genes, although conserved domains can retain phylogenetic signals at deeper divergence times. The discovery of PCR and the recent improvement in automatic nucleotide sequencing, have fostered an explosive production of mtDNA studies covering almost the entire range of living organisms and applications in the fields of population and evolutionary biology. This chapter has summarized the information on the genetics and evolution of mtDNA, and the main laboratory and computational tools that are relevant for research in molecular ecology and evolution.

## References

Andersen, L.W., Born, E.W., Gjertz, I., Wiig, O., Holm, L.-E. & Bendixen, C. (1998) Population structure and gene flow of the Atlantic walrus (*Odobenus rosmarus rosmarus*) in the eastern Atlantic Arctic based on mitochondrial DNA and microsatellite variation. *Molecular Ecology* **7**, 1323–1336.

Arctander, P. (1995) Comparison of a mitochondrial gene and a corresponding nuclear pseudogene. *Proceedings of the Royal Society of London B* **262**, 13–19.

Arnason, U., Gullberg, A. & Widegren, B. (1993) Cetacean mitochondrial DNA control region: Sequences of all extant baleen whales and two sperm whale species. *Molecular Biology and Evolution* **10**, 960–970.

Arnold, J. (1993) Cytonuclear disequilibria in hybrid zones. *Annual Review of Ecology and Systematics* **24**, 521–554.

Avise, J.C. (1994) *Molecular Markers, Natural History and Evolution*. Chapman & Hall, New York.

Attimonelli, M., Calò, D., de Montalvo, J.M. *et al.* (1998a) The Mit BASE human detaset structure. *Nucleic Acid Research* **26**, 116–119.

Attimonelli, M., Calò, D., de Montalvo, T.M. *et al.* (1998b) Update of MmtDB: a Metazoan mitochonotial DNA variants database *Nucleic Acid Research* **26**, 120–125.

Avise, J.C. (1996) Toward a regional conservation genetics perspective: Phylogeography of faunas in the southeastern United States. In: *Conservation Genetics* (eds J.C. Avise & J.L. Hamrick), pp. 431–470. Chapman & Hall, New York.

Avise, J.C. & Ball, R.M. (1990) Principles of genealogical concordance in species concept and biological taxonomy. *Oxford Survey of Evolutionary Biology* **7**, 45–67.

Avise, J.C. & Hamrick, J.L. (eds) (1996) *Conservation Genetics*. Chapman & Hall, New York.

Avise, J.C. & Walker, D.E. (1998) Pleistocene phylogeographic effects on avian populations and the speciations process. *Proceedings of the Royal Society of London B* **265**, 457–463.

Avise, J.C., Neigel, J.E. & Arnold, J. (1984) Demographic influences of mitochondrial DNA lineage survivorship in animal populations. *Journal of Molecular Evolution* **20**, 99–105.

Avise, J.C., Arnold, J., Ball, R.M. *et al.* (1987) Intraspecific phylogeography: The mitochondrial DNA bridge between population genetics and systematics. *Annual Review of Ecology and Systematics* **8**, 489–522.

Avise, J.C., Ball, R.M. & Arnold, J. (1988) Current versus historical population sizes in vertebrate species with high gene flow: a comparison based on mitochondrial DNA lineages and inbreeding theory for neutral mutations. *Molecular Biology and Evolution* **5**, 331–344.

Baker, A.J. & Marshall, H.D. (1997) Mitochondrial control region sequences as tools for understanding the evolution of avian taxa. In: *Avian Molecular Evolution and Systematics* (ed. D.P. Mindell), pp. 49–80. Academic Press, New York.

Ballard, J.W.O. & Kreitman, M. (1995) Is mitochondrial DNA a strictly neutral marker? *Trends in Ecology and Evolution* **10**, 485–488.

Barton, N.H. & Hewitt, G.M. (1989) Adaptation, speciation and hybrid zones. *Nature* **341**, 497–503.

Bermingham, E., McCafferty, S.S. & Martin, A.P. (1997) Fish biogeography and molecular clocks: Perspectives from the Panamanian isthmus. In: *Molecular Systematics of Fishes* (eds T.D. Kocher & C.A. Stepien), pp. 113–128. Academic Press, New York.

Birky, C.W. Jr, Fuerst, P. & Maruyama, T. (1989) Organelle gene diversity under migration, mutation and drift: equilibrium expectations, approach to equilibrium, effects of heteroplasmic cells, and comparison to nuclear genes. *Genetics* **121**, 613–627.

Boore, J.L., Collins, T.M., Stanton, D., Daehler, L.L. & Brown, W.M. (1995) Deducing the pattern of arthropod phylogeny from mitochondrial DNA rearrangements. *Nature* **376**, 163–165.

Brooks, D.R. & McLennan, D.A. (1991) *Phylogeny, Ecology, and Behavior: A Research Program in Comparative Biology*. University of Chicago Press, Chicago.

Casane, D., Dennebouy, N., de Rochambeau, H., Mounolou, J.C. & Monnerot, M. (1997) Nonneutral evolution of tandem repeats in the mitochondrial DNA control region of Lagomorphs. *Molecular Biology and Evolution* **14**, 779–789.

Clark-Walker, G.D. (1992) Evolution of mitochondrial genomes in Fungi. In: *Mitochondrial*

*Genomes. International Review of Cytology*, vol. 141 (eds D.R. Wolstenholme & K.W. Jeon), pp. 89–127. Academic Press, San Diego.

Clayton, D.A. (1992) Transcription and replication of animal mitochondrial DNAs. In: *Mitochondrial Genomes. International Review of Cytology*, vol. 141 (eds D.R. Wolstenholme & K.W. Jeon), pp. 217–323. Academic Press, San Diego.

Cunningham, M. & Moritz, C. (1998) Genetic effects of forest fragmentation on a rainforest restricted lizard (Scincidae: *Gnypetoscincus queenslandie*). *Biological Conservation* **83**, 19–30.

Douzery, E. & Randi, E. (1997) The mitochondrial control region of Cervidae: Evolutionary patterns and phylogenetic contents. *Molecular Biology and Evolution* **14**, 1154–1166.

Dowling, T.E., Moritz, C., Palmer, J.D. & Rieseberg, L.H. (1996) Nucleic acids III: analysis of fragments and restriction sites. In: *Molecular Systematics* (eds D.M. Hillis, C. Moritz & B.K. Mable), pp. 302–314. Sinauer Associates, Sunderland, Massachusetts.

Edwards, S.V. & Naeem, S. (1993) The phylogenetic component of cooperative breeding in perching birds. *American Naturalist* **141**, 754–789.

Eggleton, P. & Vane-Wright, R. (eds) (1994) *Phylogenetics and Ecology*. Linnean Society Symposium Series no. 17. Academic Press, London.

Excoffier, L. & Langaney, A. (1989) Origin and differentiation of human mitochondrial DNA. *American Journal of Human Genetics* **44**, 73–85.

Faber, J.E. & Stepien, C.A. (1997) The utility of mitochondrial DNA control region sequences for analyzing phylogenetic relationships among populations, species, and genera of the Percidae. In: *Molecular Systematics of Fishes* (eds T.D. Kocher & C.A. Stepien), pp. 129–143. Academic Press, San Diego.

Ferraris, J.D. & Palumbi, S.R. (eds) (1996) *Molecular Zoology*. Wiley-Liss, New York.

Frantzen, M.A.J., Silk, J.B., Ferguson, J.W.H., Wayne, R.K. & Kohn, M.H. (1998) Empirical evaluation of preservation methods for faecal DNA. *Molecular Ecology* **7**, 1423–1428.

Friesen, V.L., Montevecchi, W.A., Baker, A.J., Barrets, R.T. & Davidson, W.S. (1996) Population differentiation and evolution in the common guillemot *Uria aalge*. *Molecular Ecology* **5**, 793–805.

Fry, A.J. & Zink, R.M. (1998) Geographic analysis of nucleotide diversity and song sparrow (Aves: Emberizidae) population history. *Molecular Ecology* **7**, 1303–1313.

Gemmell, N.J., Western, P.S., Watson, J.M. & Marshall-Graves, J.A. (1996) Evolution of the mammalian mitochondrial control region—Comparisons of control region sequences between monotreme and therian mammals. *Molecular Biology and Evolution* **13**, 798–808.

Gerloff, U., Schlotterer, C., Rassmann, K. *et al.* (1995) Amplification of hypervariable simple sequence repeats (microsatellites) from excremental DNA of wild living Bonobos (*Pan paniscus*). *Molecular Ecology* **4**, 515–518.

Good, S.V., Williams, D.F., Ralls, K. & Fleischer, R.C. (1997) Population structure of *Dipodomys ingens* (Heteromydae): The role of spatial heterogeneity in maintaining genetic diversity. *Evolution* **51**, 1296–1310.

Gray, M.W. (1992) The endosymbiont hypothesis revisited. In: *Mitochondrial Genomes*, vol. 141 (eds D.R. Wolstenholme & K.W. Jeon), pp. 233–357. Academic Press, San Diego.

Griffith, C.S. (1997) Correlation of functional domains and rates of nucleotide substitution in cytochrome *b*. *Molecular Phylogenetics and Evolution* **7**, 352–365.

Gyllensten, U.B. & Wilson, A.C. (1987) Interspecific mitochondrial DNA transfer and the colonization of Scandinavia by mice. *Genetics Researches* **49**, 25–29.

Hanson, M.R. & Folkerts, O. (1992) Structure and function of the higher plant mitochondrial genome. In: *Mitochondrial Genomes. International Review of Cytology*, vol. 141 (eds D.R. Wolstenholme & K.W. Jeon), pp. 129–172. Academic Press, San Diego.

Hare, M.P. & Avise, J.C. (1996) Molecular genetic analysis of a stepped multilocus cline in the american oyster (*Crassostrea virginica*). *Evolution* **50**, 2305–2315.

Harvey, P.H., Leigh Brown, A.J., Maynard Smith, J. & Nee, S. (1996) *New Uses for New Phylogenies*. Oxford University Press, Oxford.

Hauswirth, W.W. & Laipis, P.J. (1982) Mitochondrial DNA polymorphism in a maternal lineage of Holstein cows. *Proceedings of the National Academy of Sciences USA* **79**, 4686–4690.

Hewitt, G.M. (1996) Some genetic consequences of ice ages, and their role in divergence and speciation. *Biological Journal of the Linnean Society* **58**, 247–279.

Hickson, R.E., Simon, C., Cooper, A., Spicer, G.S., Sullivan, J. & Penny, D. (1996) Conserved sequence motifs, alignment and secondary structure for the third domain of animal 12S RNA. *Molecular Biology and Evolution* **13**, 150–169.

Hillis, D.M., Moritz, C. & Mable, B.K. (eds) (1996) *Molecular Systematics.* Sinauer Associates, Sunderland, Massachusetts.

Hoeh, W.R., Stewart, D.T., Sutherland, B.W. & Zouros, E. (1996) Multiple origins of gender-associated mitochondrial DNA lineages in bivalves (Mollusca: Bivalvia). *Evolution* **50**, 2276–2286.

Hoelzel, A.R. (ed.) (1992) *Molecular Genetic Analysis of Populations: a Practical Approach.* IRL Press at Oxford University Press, New York.

Hoelzer, G.A., Wallman, J. & Melnick, D.J. (1998) The effects of social structure, geographical structure, and population size on the evolution of mitochondrial DNA: II. Molecular clocks and the lineage sorting period. *Journal of Molecular Evolution* **47**, 21–31.

Höss, M. & Pääbo, S. (1993) DNA extraction from Pleistocene bones by a silica-based purification method. *Nucleic Acid Research* **21**, 3913–3914.

Howell, N. (1989) Evolutionary conservation of protein regions in the protonmotive cytochrome *b* and their possible roles in redox catalysis. *Journal of Molecular Evolution* **29**, 157–169.

Hudson, R.R. (1990) Gene genealogies and the coalescent process. *Oxford Survey of Evolution Biology* **7**, 1–44.

Hudson, R.R., Boos, D.D. & Kaplan, N.L. (1992) A statistical test for detecting geographic subdivision. *Molecular Biology and Evolution* **9**, 138–151.

Irwin, D.M., Kocher, T.D. & Wilson, A.C. (1991) Evolution of the cytochrome *b* gene of mammals. *Journal of Molecular Evolution* **32**, 128–144.

Jaarola, M., Tegelström, H. & Fredga, K. (1997) A contact zone with noncoincident clines for sex-specific markers in the field vole (*Microtus agrestis*). *Evolution* **51**, 241–249.

Janke, A. & Arnason, U. (1997) The complete mitochondrial genome of *Alligator mississippiensis* and the separation between recent archosauria (birds and crocodiles). *Molecular Biology and Evolution* **14**, 1266–1272.

Jones, P. (1998) *Gel Electrophoresis of Nucleic Acids.* BIOS by John Wiley & Sons, Chichester, UK.

Joseph, L., Moritz, C. & Hugall, A. (1995) Molecular support for vicariance as a source of diversity in rainforest. *Proceedings of the Royal Society of London B* **260**, 117–182.

Kidd, M.G. & Friesen, V.L. (1998) Analysis of mechanisms of microevolutionary change in *Cepphus* guillemots using patterns of control region variation. *Evolution* **52**, 1158–1168.

Kilpatrick, S.T. & Rand, D.M. (1995) Conditional hitchhiking of mitochondrial DNA: Frequency shifts of *Drosophila melanogaster* mtDNA variants depend on nuclear genetic background. *Genetics* **141**, 1113–1124.

Klicka, J. & Zink, R.M. (1997) The importance of recent ice ages in speciation: a failed paradigm. *Science* **277**, 1666–1669.

Kocher, T.D. & Carleton, K.L. (1997) Base substitution in fish mitochondrial DNA: Patterns and rates. In: *Molecular Systematics of Fishes* (eds T.D. Kocker & C.A. Stepien), pp. 13–24. Academic Press, San Diego.

Kocher, T.D., Thomas, W.K., Meyer, A. *et al.* (1989) Dynamics of mitochondrial DNA evolution in animals: amplification and sequencing with conserved primers. *Proceedings of the National Academy of Sciences USA* **86**, 6196–6200.

Kuhner, M.K., Yamato, J. & Felsenstein, J. (1998) Maximum likelihood estimation of population growth rates based on the coalescent. *Genetics* **149**, 429–434.

Kumar, S., Tamura, K. & Nei, M. (1993) MEGA: *Molecular Evolutionary Genetic Analysis*, v. 1.01. The Pennsylvania State University, University Park, PA.

Kumazawa, Y. & Nishida, M. (1993) Sequence evolution of mitochondrial tRNA genes and deep-branch animal phylogenetics. *Journal of Molecular Evolution.* **37**, 380–398.

Lanyon, S.M. (1994) Polyphyly of the blackbird genus *Agelaius* and the importance of assumptions of monophyly in comparative studies. *Evolution* **48**, 679–693.

Ledje, C. & Arnason, U. (1996) Phylogenetic analyses of complete cytochrome *b* genes of the order Carnivora with particular emphasis on the caniforms. *Journal of Molecular Evolution* **42**, 135–144.

Lightowlers, R.N., Chinnery, P.F., Turnbull, D.M. & Howell, N. (1997) Mammalian mitochondrial genetics: Heredity, heteroplasmy and disease. *Trends in Genetics* **13**, 450–455.

Loewe, L. & Scherer, S. (1997) Mitochondrial Eve: The plot thickens. *Trends in Ecology and Evolution* **12**, 422–423.

Longmire, J.L., Lewis, A.K., Brown, N.C. *et al.* (1988) Isolation and molecular characterization of a highly polymorphic centromeric tandem repeat in the family Falconidae. *Genomics* **2**, 14–24.

Lopez, J.V., Yuhki, N., Masuda, R., Modi, W. & O'Brien, S.J. (1994) Numt, a recent transfer and tandem amplification of mitochondrial DNA to the nuclear genome of the domestic cat. *Journal of Molecular Evolution* **39**, 174–191.

Lopez, J.V., Culver, M., Claiborne Stephens, J., Johnson, W.E. & O'Brien, S.J. (1997) Rates of nuclear and cytoplasmic mitochondrial DNA sequence divergence in mammals. *Molecular Biology and Evolution* **14**, 277–286.

Lunt, D.H. & Hyman, B.C. (1997) Animal mitochondrial DNA recombination. *Nature* **387**, 247.

Macey, J.R., Larson, A., Ananjeva, N.B., Fang, Z. & Papenfuss, T.J. (1997) Two novel gene orders and the role of light-strand replication in rearrangement of the vertebrate mitochondrial genome. *Molecular Biology and Evolution* **14**, 91–104.

Maddison, W.P. (1997) Gene trees in species trees. *Systematic Biology* **46**, 523–536.

Maddison, W.P. & Maddison, D.R. (1992) *MacClade. Analysis of phylogeny and character evolution*. Sinauer Associates, Sunderland, MA.

Meyer, A. (1994) Shortcomings of the cytochrome *b* gene as a molecular marker. *Trends in Ecology and Evolution* **9**, 278–280.

Mindell, D.P. & Thacker, C.E. (1996) Rates of molecular evolution: Phylogenetic issues and applications. *Annual Review of Ecology and Systematics* **27**, 279–303.

Mindell, D.P., Sorenson, M.D., Huddleston, C.J. *et al.* (1997) Phylogenetic relationships among and within select avian orders based on mitochondrial DNA. In: *Avian Molecular Evolution and Systematics* (ed. D.P. Mindell), pp. 213–247. Academic Press, New York.

Mindell, D.P., Sorenson, M.D. & Dimcheff, D.E. (1998) Multiple independent origins of mitochondrial gene order in birds. *Proceedings of the National Academy of Sciences USA* **95**, 10693–10697.

Moore, W.S. (1995) Inferring phylogenies from mtDNA variation: Mitochondrial-gene trees versus nuclear-gene trees. *Evolution* **49**, 718–726.

Moore, W.S. & DeFilippis, V.R. (1997) The window of taxonomic resolution for phylogenies based on mitochondrial cytochrome *b*. In: *Avian Molecular Evolution and Systematics* (ed. D.P. Mindell), pp. 83–119. Academic Press, New York.

Moritz, C. (1996) Uses of molecular phylogenies for conservation. In: *New Uses for New Phylogenies* (eds P.H. Harvey, A.J. Leigh Brown, J. Maynard Smith & S. Nee), pp. 203–214. Oxford University Press, New York.

Moritz, C., Dowling, T.E. & Brown, W.M. (1987) Evolution of animal mitochondrial DNA: Relevance for population biology and systematics. *Annual Review of Ecology and Systematics* **18**, 269–292.

Nei, M. (1987) *Molecular Evolutionary Genetics*. Columbia University Press, New York.

Nei, M. & Miller, J.C. (1990) A simple method for estimating average number of nucleotide substitutions within and between populations from restriction data. *Genetics* **125**, 873–879.

Neigel, J.E. & Avise, J.C. (1986) Phylogenetic relationships of mitochondrial DNA under various demographic models of speciation. In: *Evolutionary Processes and Theory* (eds E. Nevo & S. Karlin), pp. 515–534. Academic Press, New York.

Newton, A.C., Allnut, T.R., Gillies, A.C.M., Lowe, A.J. & Ennos, R.A. (1999) Molecular phylogeography, intraspecific variation and the conservation of tree species. *Trends in Ecology and Evolution* **14**, 140–145.

Nosek, J., Tomaska, L., Fukuhara, H., Suyama, Y. & Kovac, L. (1998) Linear mitochondrial genomes: 30 years down the line. *Trends in Genetics* **14**, 184–188.

Nunn, G.B. & Cracraft, J. (1996) Phylogenetic relationships among the major lineages of the birds-of-paradise (Paradisaeidae) using mitochondrial DNA gene sequences. *Molecular Phylogenetics and Evolution* **5**, 445–459.

O'Brien, S.J., Martenson, J.S., Miththapala, S. *et al.* (1996) Conservation genetics of the Felidae. In: *Conservation Genetics* (eds J.C. Avise & J.L. Hamrick), pp. 50–74. Chapman & Hall, New York.

Omland, K.E. (1997) Correlated rates of molecular and morphological evolution. *Evolution* **51**, 1381–1393.

Ortì, G. & Meyer, A. (1997) The radiation of characiform fishes and the limits of resolution of mitochondrial ribosomal DNA sequences. *Systematic Biology* **46**, 75–100.

Paetkau, D., Shields, G.F. & Strobeck, C. (1998) Gene flow between insular, coastal and interior populations of brown bears in Alaska. *Molecular Ecology* **7**, 1283–1292.

Palumbi, S.R. (1996) Nucleic acids II: The polymerase chain reaction. In: *Molecular Systematics* (eds D.M. Hillis, C. Moritz & B.K. Mable), pp. 205–247. Sinauer Associates, Sunderland, Massachusetts.

Patton, J.L. & Smith, M.F. (1993) Molecular evidence for mating asymmetry and female choice in a pocket gopher *(Thomomys)* hybrid zone. *Molecular Ecology* **2**, 3–8.

Pfeiffer, G.P. (1996) *Technologies for Detection of DNA Damage and Mutations.* Plenum Press, New York.

Radtkey, R.R. (1996) Adaptive radiation of day-geckos (*Phelsuma*) in the Seychelles archipelago: a phylogenetic analysis. *Evolution* **50**, 604–623.

Randi, E. & Lucchini, V. (1998) Organization and evolution of the mitochondrial DNA control region in the avian genus *Alectoris. Journal of Molecular Evolution* **47**, 449–462.

Richman, A.D. (1996) Ecological diversification and community structure in the Old World leaf warblers: a molecular phylogenetic perspective. *Evolution* **50**, 2461–2470.

Roy, M.S. (1997) Recent diversification in African greenbuls (Pycnonotidae: *Andropadus*) supports a montane speciation model. *Proceedings of the Royal Society of London B* **264**, 1337–1342.

Rzhetsky, A. (1995) Estimating substitution rates in ribosomal RNA genes. *Genetics* **141**, 771–783.

Sambrook, E., Fritsch, F. & Maniatis, T. (1989) *Molecular Cloning.* Cold Spring Harbor Press, Cold Spring Harbor, New York.

Shaw, K. (1996) Sequential radiations and patterns of speciation in the Hawaiian cricket genus *Laupala* inferred from DNA sequences. *Evolution* **50**, 237–255.

Shields, G.F. & Wilson, A.C. (1987) Calibration of mitochondrial DNA evolution in geese. *Journal of Molecular Evolution* **24**, 212–217.

Simon, C., Frati, F., Beckenbach, A., Crespi, B., Liu, H. & Flook, P. (1994) Evolution, weighting and phylogenetic utility of mitochondrial gene sequences and a compilation of conserved polymerase chain reaction primers. *Annals of the Entomologic Society of America* **87**, 651–701.

Slatkin, M. (1994a) Gene flow and population structure. In: *Ecological Genetics* (ed. L.A. Real), pp. 3–17. Princeton University Press, Princeton, New Jersey.

Slatkin, M. (1994b) Cladistic analysis of DNA sequences from subdivided populations. In: *Ecological Genetics* (ed. L.A. Real), pp. 18–34. Princeton University Press, Princeton, New Jersey.

Slatkin, M. & Maddison, W.P. (1989) A cladistic measure of gene flow inferred from the phylogenies of alleles. *Genetics* **123**, 603–613.

Slatkin, M. & Maddison, W.P. (1990) Detecting isolation by distance using phylogenies of genes. *Genetics* **126**, 249–260.

Sorenson, M.D. & Quinn, T.W. (1998) Numts: a challenge for avian systematics and population biology. *Auk* **115**, 214–221.

Springer, M.S., Hollar, L.J. & Burk, A. (1995) Compensatory substitutions and the evolution of the mitochondrial 12S rRNA gene in mammals. *Molecular Biology and Evolution* **12**, 1138–1150.

Stanley, H.F., Casey, S., Carnahan, J.M., Goodman, S. Harwood, J., & Wayne, R.K. (1996) Worldwide patterns of mitochondrial DNA differentiation in the harbor seal (*Phoca vitulina*). *Molecular Biology and Evolution* **13**, 368–382.

Stoesser, G., Tuli, M.A., Lopez, R. & Sterk, P. (1999) The EMBL nucleotide sequence database. *Nucleic Acids Research* **27**, 18–24.

Sullivan, J. & Swofford, D.L. (1997) Are Guinea pigs rodents? The importance of adequate models in molecular phylogenetics. *Journal of Mammalian Evolution* **4**, 77–85.

Sullivan, J., Holsinger, K.E. & Simon, C. (1995) Among-site rate variation and phylogenetic analysis of 12S rRNA in sigmodontine rodents. *Molecular Biology and Evolution* **12**, 988–1001.

Swofford, D.L., Olsen, G.J., Waddell, P.J. & Hillis, D.M. (1996) Phylogenetic inference. In: *Molecular Systematics* (eds D.M. Hillis, C. Moritz & B.K. Mable), pp. 407–514. Sinauer Associates, Sunderland, Massachusetts.

Tegelström, H. & Gelter, H.P. (1990) Haldane's rule and sex biased gene flow between two hybridizing flycatcher species (*Ficedula albicollis* and *Ficedula hypoleuca*, Aves: Muscicapidae). *Evolution* **44**, 2012–2021.

Thorpe, R.S., McGregor, D.P., Cumming, A.M. & Jordan W.C. (1994) DNA evolution and colonization sequence of island lizards in relation to geological history: mtDNA rflp, cytochrome B, cytochrome oxidase, 12s rRNA sequence, and nuclear RAPD analysis. *Evolution* **48**, 230–240.

Walsh, P.S., Metzger, D.A. & Higuchi, R. (1991) Chelex 100 as a medium for simple extraction of DNA for PCR based typing from forensic material. *Biotechniques* **10**, 506–513.

Wayne, R.K. (1996) Conservation genetics in the Canidae. In: *Conservation Genetics* (eds J.C. Avise & J.L. Hamrick), pp. 75–118. Chapman & Hall, New York.

Weir, B.S. & Cockerham, C.C. (1984) Estimating F-statistics for the analysis of population structure. *Evolution* **38**, 1358–1370.

Wheeler, W.C. (1995) Sequence alignment, parameter sensitivity, and the phylogenetic analysis of molecular data. *Systematic Biology* **44**, 321–331.

Wills, C. (1995) When did Eve live? An evolutionary detective story. *Evolution* **49**, 593–607.

Wilson, A.C., Cann, R.L., Carr, S.M. *et al.* (1985) Mitochondrial DNA and two perspectives on evolutionary genetics. *Biological Journal of the Linnean Society* **26**, 375–400.

Wolstenholme, D.R. & Jeon, K.W. (eds) (1992) *Mitochondrial Genomes*. International Review of Cytology, vol. 141. Academic Press, San Diego.

Wright, S. (1951) The genetical structure of populations. *Annals of Eugenetics* **15**, 323–354.

Yang, Z. (1996) Among-site rate variation and its impact on phylogenetic analyses. *Trends in Ecology and Evolution* **11**, 367–372.

Zardoya, R. & Meyer, A. (1996) Phylogenetic performance of mitochondrial protein-coding genes in resolving relationships among vertebrates. *Molecular Biology and Evolution* **13**, 933–942.

Zevering, C.E., Moritz, C., Heideman, A. & Sturm, R.A. (1991) Parallel origins of duplications and the formation of pseudogenes in mitochondrial DNA from parthenogenetic lizards (*Heteronotia binoei*, Gekkonidae). *Journal of Molecular Evolution* **33**, 431–441.

Zhang, D.-X. & Hewitt, G.M. (1996) Nuclear integrations: Challenges for mitochondrial DNA markers. *Trends in Ecology and Evolution* **11**, 247–251.

Zink, R.M. (1996) Comparative phylogeography in North American birds. *Evolution* **50**, 308–317.

Zink, R.M. & Dittmann, D.L. (1993) Gene flow, refugia, and evolution of geographic variation in the song sparrow (*Melospiza melodia*). *Evolution* **47**, 717–729.

# CHAPTER 8

# Characterization and Evolution of Major Histocompatibility Complex (MHC) Genes in Non-Model Organisms, with Examples from Birds

SCOTT V. EDWARDS, JOHN NUSSER AND JOE GASPER

## 8.1 Introduction

The major histocompatibility complex (MHC) is a multigene family encoding a diversity of molecules involved in the immune response and self–nonself recognition. Major histocompatibility complex molecules are part of the immunoglobulin superfamily and are structurally related to other molecules of the immune system, such as antibodies and T-cell receptors (Klein & O'Huigin 1993). The function of MHC molecules is to bind protein fragments (peptides) of bacterial and viral pathogens, display them on the cell surface, and present them to T-cells to elicit an appropriate immune response (Germain & Margulies 1993). Major histocompatibility complex genes are the most polymorphic coding regions of the vertebrate genome, with over 200 alleles at some loci (Parham & Ohta 1996). This genetic diversity is thought to be maintained by a complex molecular arms race with rapidly evolving pathogens; its sources include balancing selection (heterozygote advantage or frequency-dependent selection), recombination, point mutation and gene conversion (Edwards & Hedrick 1998). Balancing selection makes sense as a mechanism to elevate MHC diversity: if each MHC receptor binds a particular subset of peptides in its peptide-binding region (PBR), then having two such receptors for a given gene (i.e. being heterozygous), or having a rare PBR type that a pathogen has not yet encountered or evolved adaptations to avoid (rare-allele advantage), would increase fitness. Polymorphic MHC genes are also thought to serve as cues for mate choice and avoidance of inbreeding, forces that elevate MHC diversity via dissassortative mating preferences (Apanius *et al.* 1997). The evolutionary genetics of human MHC genes (known as human leucocyte antigens (HLA)) are the best understood, and the importance of linkage disequilibrium, selection and the evolution of noncoding regions has been quantified and more thoroughly explored than in any other species (Hedrick 1994; Bergström *et al.* 1998; Satta *et al.* 1999). But MHC genes have been cloned and compared in a variety of vertebrates, including sharks (Hashimoto *et al.* 1992; Bartl & Weissman 1994; Bartl *et al.* 1997), bony fish (Dixon *et al.* 1995; van Erp *et al.* 1996; Graser *et al.* 1998), reptiles (Grossberger & Parham 1992; Radtkey *et al.* 1996), amphibians (Sato *et al.* 1993; Shum *et al.* 1993; Sammut *et al.* 1997) and birds (Kroemer *et al.* 1990; Wittzell *et al.* 1994; Edwards *et al.* 1998, 1999). Thus far MHC molecules have not yet been found in cyclostomes (hagfish and lampreys), but phylogenetic analyses suggest distant relationships to heat-shock proteins, such as HSP70, which are found throughout all life forms (Hughes & Nei 1993).

MHC genes can generally be categorized as 'classical' or 'nonclassical'. Classical MHC genes are so called because they exhibit the traits found in the loci that were first characterized genetically in mice: high polymorphism and a strong influence on the immune response to challenge by nonself proteins. Nonclassical MHC genes are, in general, expressed at much lower levels than classical genes, and are much less polymorphic (Arnaiz-Villena *et al.* 1997). However, some nonclassical MHC genes also bind foreign peptides, and may evolve in a conservative manner so as to target equally conserved peptide motifs in a wide range of bacterial and other pathogens (Lindahl *et al.* 1997). For this reason, nonclassical MHC genes may also be of great interest to molecular ecologists.

MHC genes can also be categorized as class I or class II. Class I and II genes differ in their structure, pattern of expression in different tissues and in the origin and fate of the peptides they process (Kaufman *et al.* 1994). In the case of class I molecules, the two chains comprising the peptide-binding region (PBR), called $\alpha1$ and $\alpha2$, are encoded by different exons in a single gene, whereas these chains in class II molecules, $\beta1$ and $\alpha1$, are encoded by separate genes (Fig. 8.1). (The genes encoding the $\beta$ chains of class II molecules are called class II B genes, whereas the genes encoding both the $\alpha1$ and $\alpha2$ chains of the class I molecules are called class IA genes.) In general both chains of class I molecules are polymorphic, whereas only the $\beta$ chain of class II molecules is polymorphic. Class I molecules are expressed on all cell types and stimulate cytotoxic (killer) T-cells, whereas class II molecules are expressed primarily on cells related to the immune system, are produced at particularly high levels in the spleen and stimulate helper T-cells. Whereas class I molecules bind and present foreign peptides that are synthesized within cells, such as viral peptides, class II molecules present peptides that have been phagocytosed by macrophages, such as those from bacteria. All of the above differences among MHC molecules have important evolutionary implications that will not be reviewed here. These differences, however, will dictate the choice of target gene(s) and molecular protocols that will be employed in their isolation.

(a)

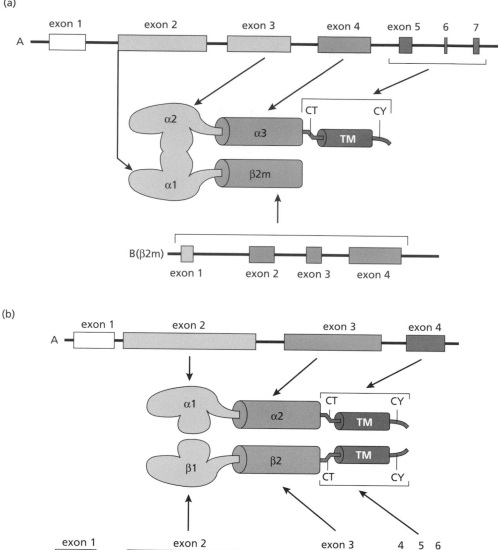

**Fig. 8.1** Diagram of the structure and intron exon organization of MHC class I (a) and II (b) molecules and genes. (Modified from Klein *et al.* 1997 with permission.)

## 8.2 Why study MHC genes?

MHC genes are currently of great interest to ecologists because they are extremely polymorphic and because they are functionally relevant to many classic questions in ecology and evolution (Edwards & Hedrick 1998). The role of MHC genes in immune recognition against parasites makes them of interest in virtually any arena of questions in which parasites play a role—a very large arena in the case of vertebrates (Apanius *et al.* 1997). Studies of fitness, the expression of secondary sexual characters and mate choice can benefit from knowledge of MHC variation (von Schantz *et al.* 1996). MHC genes play an acknowledged and potentially important role in conservation genetics (Edwards & Potts 1996), and their extraordinary diversity can offer novel insights into the population genetic history of populations (Takahata 1993). Finally, the evolution of the locus itself is of intrinsic interest as a model system for multigene family and genome evolution (Nei *et al.* 1997).

The complexity and frequent redundancy of the MHC is fascinating in its own right, but is also at the heart of why it can be practically challenging for molecular ecologists to gain access to MHC genes. It is becoming increasingly clear that understanding of the evolutionary dynamics of the multigene family can greatly facilitate molecular access to the genes for use in ecological and behavioural studies (Fig. 8.2); for example, concerted evolution is a pattern

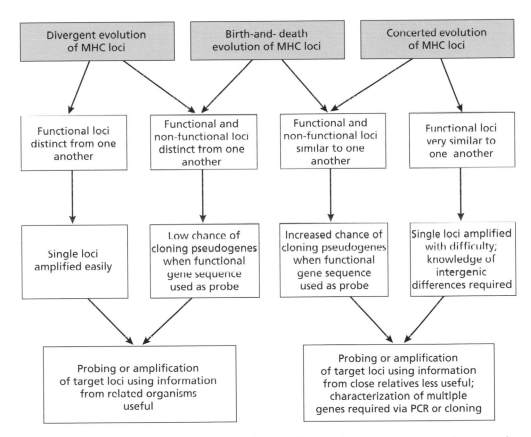

**Fig. 8.2** Influence of long-term evolutionary history of MHC genes on ease and strategy of isolation of individual genes.

exhibited by different genes in a gene family, arrayed on the same or different chromosomes, in which gene members within a species or genome are more closely related to one another than to gene members in other species (Dover 1982; Li 1997). It can be the result of a variety of molecular processes, including interlocus gene conversion and unequal alignment or crossing-over of two arrays of repeated gene sequences. If the MHC genes of a given species are undergoing concerted evolution, as has been suggested for some birds (Edwards *et al.* 1995b), the similarities among genes created by homogenization might specify a particular isolation protocol, or at least signal the need to be aware of the presence of highly related genes in the same genome. By contrast, if, as in mammals, the MHC genes are evolving in a primarily divergent mode (Nei *et al.* 1997), in which different duplicated loci remain evolutionarily independent, then different genes will probably differ by a number of substitutions, hence enabling easier access to single loci. The goal of this chapter is to review means for isolating MHC genes for ecological and evolutionary studies. Potts (1996) has also reviewed useful methods for obtaining access to MHC genes. However, it is first useful to confirm that MHC genes are indeed what is desired in a particular study, because, like many specialized classes of molecular markers, their isolation and characterization in a form that is maximally useful can sometimes require considerable effort. Below are briefly outlined some of the reasons why evolutionary biologists might and might not consider studying MHC diversity, either within or between species, and which type of loci might be most appropriate for such studies.

### 8.2.1 Ecological immunology

Studies aiming to correlate fitness, MHC variability and aspects of the humoral or cellular immune responses (Sheldon & Verhulst 1996) will aim to focus on functional loci, as opposed to pseudogenes (but see below). This requirement makes it useful to begin a study by focusing on transcribed and translated genes, i.e. at the level of RNA and protein. Even some MHC genes that are at first glance psuedogenes, such as the primate class II *DRB6* gene, are eventually discovered to be transcribed and functional (Fernandez-Soria *et al.* 1998). The vast majority of MHC molecules from nonmodel organisms have not been directly characterized at the protein level, and even fewer, if any, have been assayed for peptide binding. Although the immunological methods for isolating MHC molecules at the protein level, and even for assaying the diversity of peptides directly bound to them, are probably applicable to a wide range of vertebrates, such studies will frequently be beyond the scope and immediate interests of many ecological projects.

Polymorphism itself can be a good guide to functional relevance in the immune system. In chickens, the genes expressed at the highest levels, and, by inference, the most important functionally, are also the most polymorphic (Kaufman & Salomonsen 1997). For this reason, accessing MHC genes, first by polymerase chain reaction (PCR) amplification of genomic DNA, may well be adequate for embarking on MHC studies in genetically unstudied organisms. However, little is known about variability in MHC pseudogenes (Hess *et al.* 2000), and the low polymorphism of nonclassical genes should not be used as evidence against their immunological importance (Lindahl *et al.* 1997). Many regions of the MHC, including pseudogenes and noncoding markers such as microsatellites, may exhibit strong linkage disequilibrium with polymorphic functional genes (Hedrick 1994), in which case assaying linked regions might be as useful as assaying the MHC genes themselves (e.g. Paterson *et al.* 1998). Unfortunately, although microsatellites frequently occur in introns of MHC genes, particularly intron 2 of class II B genes (Schwaiger *et al.* 1993; Jin *et al.* 1996), many such markers occur relatively far from functional genes (Meager & Potts 1997; Guillaudeux *et al.* 1998), and in low frequency in some verte-brate groups (Primmer *et al.* 1997), and knowledge of the details of such linked regions will be difficult to gain without extensive cloning.

There is some argument for the case that class I genes are of more direct relevance than class II genes to studies of host–parasite coevolution involving agents such as bacteria and viruses, because class I molecules in general will directly bind such endogenously synthesized pathogens; for example, peptides of several well-studied human pathogens are bound by class I molecules (Hill *et al.* 1992; Goulder *et al.* 1997; Lenz & Bevan 1997). Furthermore, class I molecules stimulate cytotoxic T-cells, which can kill infected cells directly, whereas class II molecules exert their influence on clearance of infection less directly via helper T-cells that stimulate the proliferation of antibodies. Thus, variation in class II molecules might be of interest in correlations with anti-body production in nature (Puel *et al.* 1998). However, many detailed studies of resistance to infectious disease in nature find associations of resis-tance and particular haplotypes composed of both class I and II molecules, sug-gesting a role for both molecules in these cases (Hill *et al.* 1991; Thursz *et al.* 1997).

### 8.2.2  Sexual selection and mate choice

Functional MHC loci are again of primary interest for studies of sexual selec-tion and choice of mate, although linkage disequilibrium might make many regions in the MHC useful for genetic typing in this context. Restriction fragment-length polymorphism (RFLP) and Southern blot analysis of a single class I gene was used to type mice in the studies of MHC-based mating prefer-ences and inbreeding avoidance in semi-natural populations (Potts *et al.* 1991; Manning *et al.* 1992). However, these authors used this class I gene as a proxy

MHC will emerge as a major systematic tool as other neutrally evolving gene regions are examined in as much detail.

### 8.2.6 Population genetics

The MHC has served as an important model system for examining some of the most thorny issues in population genetics, namely the interaction of selection and drift, and the effect of linkage to selected sites on variability at neutral sites (Hedrick *et al.* 1991; Paterson 1998; Grimsley-Garrigan & Edwards 1999). Major histocompatibility complex surveys in several South American Indian tribes have revealed an important role for genetic drift, despite evidence for balancing selection (Watkins *et al.* 1990; Black & Hedrick 1997). The rapidly growing database on noncoding regions linked to functional MHC genes (Beck *et al.* 1996; Mizuki *et al.* 1997; Guillaudeux *et al.* 1998) means that the stage is set, at least in human populations, for examination in detail of the effects over large regions of the MHC of selection on a few regions. The population genetics of MHC pseudogenes is virtually unexplored (Grimstey *et al.* 1998; Hess *et al.* 2000), and could potentially yield novel insights into mechanisms generating variability at functional loci. Clearly it is of interest to compare levels of intraspecific variation and geographical subdivision of MHC loci and neutral loci to obtain a better understanding of the forces moulding MHC variability in natural populations (e.g. Boyce *et al.* 1997).

### 8.2.7 Evolution of the immune system

Major histocompatibility complex genes are a critical component of the adaptive (pathogen-specific) immune response, so confirmation of an adaptive immune response in basal vertebrates requires identification of these MHC genes. To study the origin and evolution of the adaptive immune system, several groups have investigated MHC genes in basal groups, such as sharks, with exciting results (Bartl & Weissman 1994). The recent cloning of class I genes in sharks (Bartl *et al.* 1997; Okamura *et al.* 1997) suggests that all molecular components of the adaptive immune response were in place in the ancestors of this lineage, suggesting that the attenuated immune response in sharks is not a result of absence of critical molecules (Bartl *et al.* 1997).

### 8.2.8 Evolution of multigene families

Major histocompatibility complex genes are excellent models for investigations of molecular evolution and multigene family evolution. Different MHC gene families have been found to evolve according to different models of multigene family evolution, each of which has specific phylogenetic consequences (Nei & Hughes 1992; Nei *et al.* 1997). The relative frequency and rela-tionships of functional genes and pseudogenes is of interest here (Geraghty *et al.* 1992), hence an approach that provides an equal chance of isolating both types of MHC genes would be favoured. There are indications that the long-term evolutionary dynamics of MHC genes in nonmammalian vertebrates could differ from those in mammals (Klein *et al.* 1994; Edwards *et al.* 1995b), but more data are needed to support this, particularly if we are to make clear predictions of optimal methods of surveying populations in molecular ecologi-cal studies from long-term patterns of molecular evolution.

### 8.2.9  Reasons for characterizing MHC genes

The preceding discussion suggests that the best reasons to embark on charac-terizations of MHC genes in an evolutionary context are:
1   to obtain markers whose amino acid variability will ultimately influence fitness, mate choice and life-history components;
2   to access a model system for examining the interactions of linkage, selec-tion and drift; and
3   to examine issues related to long-term molecular evolution and evolution of multigene families. There seems less justification in specifically targeting MHC genes to obtain polymorphic markers for parentage studies or to make specific inferences about the demographic history of populations or phylo-genetic issues, although these latter two points are still debated (Edwards & Hedrick 1998). Because cloning MHC genes from divergent lineages can be difficult, it is worth considering these issues prior to beginning laboratory work.

### 8.3  Overview of approaches for isolating MHC genes

For species with well characterized MHCs, such as humans or mice, it is easy to choose a particular gene or region of interest and apply appropriate PCR primers for amplification. Because intron sequences of MHC genes in these species are known, exons can easily be amplified in their entirety without relying on conserved regions of those exons (e.g. She *et al.* 1991). However, for nearly all other vertebrates, including species whose MHC is relatively well studied, such as *Xenopus*, it may take more than designing a pair of primers to acquire the data of interest. The specific approach used to characterize MHC genes will be dictated by the mode of evolution of the MHC multigene family (Fig. 8.2) and the types of questions posed by the particular study (Fig. 8.3). Key goals in many ecological uses of MHC variation in any species will be locus specificity and amplification of entire, as opposed to partial, peptide-binding regions (PBRs). Because MHC genes frequently come in related clusters of genes within which interlocus gene similarity can be quite high (Edwards *et al.*

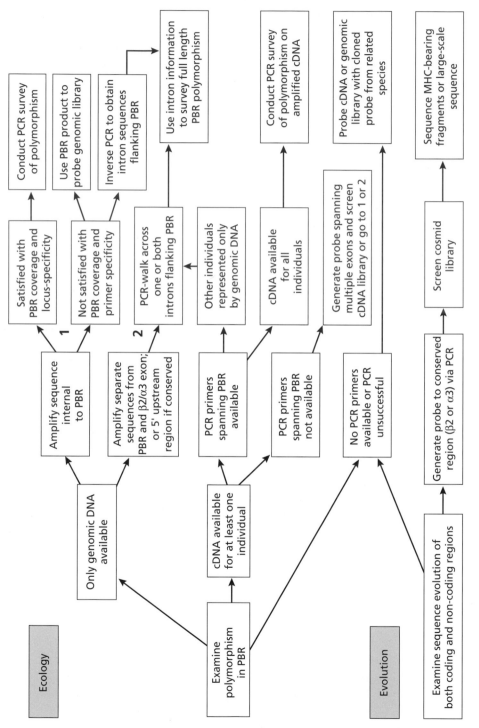

**Fig. 8.3**  Flow diagram for isolation of MHC sequences for various types of molecular ecological and evolutionary questions.

1995b; Wittzel *et al.* 1998), a major difficulty for several fish and bird species has been achieving an assay of variability that can distinguish different loci from one another. Thus, if functional and nonfunctional genes are very closely related to one another because of recent duplications or gene conversion, there is a possibility that longer gene sequences accessed by PCR or library-screening approaches may yield nonfunctional genes even when transcribed sequences are used as probes or as a basis for PCR primer design.

As more and more DNA sequences, particularly genomic sequences, of MHC genes from nonmodel species (e.g. Edwards *et al.* 1998; Westerdahl *et al.* 1998) become available, the task of identifying locus-specific motifs for eventual selective PCR amplification will become easier. It is already clear that placing PCR primers in introns flanking functional exons is more likely, although by no means guaranteed, to achieve locus specificity than placing primers in exons themselves (Gyllensten & Erlich 1989; Gyllensten *et al.* 1990; She *et al.* 1990; Zoorob *et al.* 1993; Edwards *et al.* 1998). Placing primers in introns flanking, say, the polymorphic PBR exon(s), will also permit characterization of the entirety of these coding regions, an important goal if the functional consequences of allelic diversity are of interest. Because partial exon sequences can result in different alleles being scored as identical, incomplete characterization of PBR exons, although technically easier than complete characterization, can result in lower resolution studies (Murray *et al.* 1995; Houlden *et al.* 1996).

For molecular evolutionary studies, it will frequently be the goal to obtain complete sequences of genes, or at minimum, complete coding regions. Phylogenetic analysis of isolated PBRs is troublesome because they are short and because they are subject to evolutionary forces such as recombination and gene conversion that violate the assumptions of standard phylogenetic analysis. Furthermore, it is often the comparison of the evolutionary dynamics of PBRs with other domains that sheds light on broader issues (Garrigan & Edwards 1999; Seddon & Baverstock 2000). Therefore, complete or near-complete sequences of genes are a worthwhile goal. The genetic mapping of MHC genes will not be covered here, despite the fact that rapid progress in this arena has been made recently for several nonmodel species of fish, using haploid embryos or inbred lines (McConnell *et al.* 1998; Málaga-Trillo *et al.* 1998).

### 8.3.1 Isolation of proteins with antibodies

Ultimate proof of expression of MHC molecules as functional proteins has often been achieved by precipitating them from cells or cell lines with monoclonal antibodies recognizing conserved motifs (immunoprecipitation). The precipitated proteins representing multiple loci and alleles can then be separated by one-dimensional isoelectric focusing on polyacrylamide gels. In many cases, antibodies developed to recognize MHC molecules, say, chickens, will cross-react and recognize homologous molecules of distantly related species (Kaufman *et al.* 1990, 1991). This property has not been exploited for many species, although it certainly holds promise for characterizing MHC gene products in nonmodel organisms; for example, Gyllensten *et al.* (1994) used a mouse antibody directed towards a conserved region of human class I MHC molecules to detect a variety of related proteins in a New World monkey, the endangered cotton-top tamarins (*Saguinus oedipus*). Jarvi (Jarvi & Briles 1992; Jarvi *et al.* 1992) and coworkers have defined a functional and polymorphic MHC system in pheasants and cranes using antisera produced by reciprocal immunizations of whole blood among family members. Resolution of proteins using antibodies can sometimes be used to detect recombinant alleles, i.e. alleles that resemble more common alleles in one segment and other alleles in adjacent segments. Such patterns in indigenous South American Indian tribes led to the discovery of novel recombinant alleles in these populations (Belich *et al.* 1992; Watkins *et al.* 1992).

Despite its importance in defining which loci and alleles are functionally expressed in a given individual, serological analysis of MHC proteins is not likely to become widespread in the community of zoologists working on diverse lineages of vertebrates. Serological analysis has relatively limited capacity to detect all polymorphisms, as indicated by comparisons of serologically defined alleles with alleles defined by DNA sequence analysis, and it usually requires the establishment of cell lines. When combined with PCR analyses of cDNA, and expression of such cDNAs in cells, however, serology can very precisely define the set of expressed loci belonging to a particular MHC subfamily (Watkins *et al.* 1990).

### 8.3.2  PCR approaches to cloning MHC genes

By far the fastest way to gain access to MHC gene sequences is by using PCR. Conserved regions in the second and third exons ($\beta$1 and $\beta$2 domains) of class II B genes and in exons 2, 3 and 4 ($\alpha$1–3 domains) of class IA genes have been identified in a variety of vertebrates (Fig. 8.4), and have been used to design primers that successfully amplify intervening segments from genomic DNA or cDNA (Table 8.1). Alternatively, 3′ or 5′ RACE PCR (Rapid Amplification of DNA Ends by PCR; Frohman 1988; Troutt *et al.* 1992) can be used to amplify terminal segments of expressed MHC genes, as has been undertaken for frogs (Sato *et al.* 1993), salamanders (Sammut *et al.* 1997), reptiles (Grossberger & Parham 1992) and birds (Edwards *et al.* 1995b). If sequences from two adjacent exons can be obtained (for example, exons 2 and 3 of class II B genes), then it should be possible to obtain the sequence of the intervening intron 2 using PCR (e.g. Klein *et al.* 1993), although in some species this approach may

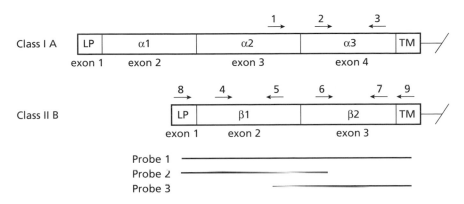

**Fig. 8.4** Diagram showing locations of various primers targeted to conserved regions of MHC class IA and IIB genes in vertebrates. Sequences for some representative primers are given in Table 8.1. The black bars labelled probe 1–3 indicate the size and location of class II B probes generated by PCR for analysis on Southern blots of songbird genomic DNA. Probe 1 was generated with primer forward and reverse primers 8 and 10, respectively, of Edwards *et al.* (1995b). Probe 2 was generated with primer 8 and 62, and probe 3 was generated with primer 61 and 10. The sequence of primer 61 is (5′–3′) GCTTCAGTGGACACGTACTG; primer 62 GCAGGGTAGAAATCCATCAC, designed from sequences in (Edwards *et al.* 1995b). Primers 3F and 3R and 305 and 306 were used to generate PCR products from cosmid clones directly. The sequence of primer 3F is GTGATGGATTTCTACCCTGC; and primer 3R is CAGCTGTAGGTAGCCC, designed from sequences in (Edwards *et al.* 1995b). The sequences of primers 305 and 306 are given in Edwards *et al.* 1995a.

be difficult, either because of high GC content of the MIIC region or because of long introns, as in mammals or frogs (Kobari *et al.* 1995). Conserved priming sites in exon 1 or 5′ upstream regulatory regions have yet to be discovered, but could, in principle, emerge as new genomic sequences are obtained. If the goal is amplification of complete chains of the PBR from genomic DNA, then knowledge of intron 1 in the case of class II B genes, and of introns 1 and 2 in the case of class I A genes, is essential. To obtain such sequences, ligation-anchored PCR approaches (Edwards *et al.* 1995b), library construction (Kasahara *et al.* 1992; Sato *et al.* 1993; Shum *et al.* 1993), or inverse PCR approaches (Kobari *et al.* 1995) can be used.

### 8.3.3 Transcribed genes

To guarantee that effort is focused on transcribed genes, with a greater likelihood of functionality than a genomic sequence, reverse-transcriptase PCR (RT-PCR) and cDNA library approaches can be used. Many of the MHC genes of *Xenopus* and fish (Kasahara *et al.* 1992; Sato *et al.* 1993; Shum *et al.* 1993; Sültmann *et al.* 1993; Sato *et al.* 1996), as well as other immunologically important genes in species such as cyclostomes (Kandil *et al.* 1996), were

**Table 8.1** Degenerate primer pairs designed to amplify segments of MHC class I and II genes from specific groups of vertebrates

| Class | Location | 5′–3′ Primer sequence<br>Forward primer, reverse primer | Reference |
|---|---|---|---|
| **Basal vertebrates (sharks, fish)** | | | |
| Shark — I and II | α3, β2 | TGYTMNGTGACNGRYTTCTAYCC / AGRCTKGKRTGCTCCACNTGRCA | Hashimoto et al. 1990 |
| Zebrafish — II | β2 | TCYMKKGYCVWMTGRYTTCTAYCC / RRGCTGSYGTGMWYCACMWSRCA | Ono et al. 1992 |
| Shark — II | β2 | CCAAGCTTTGYIIIGITWVIGGITTYTAYCC / CCGGATCCARISWRTGIWYIACIYKRCA | Bartl & Weissman 1994 |
| Shark — I | α3 | TCAGGATCCTGYTMNGTGACNGRYTTCTAYCC / GCAGAATTCNRYYTGRWANGTNCCRTC | Okamura et al. 1997 |
| Coelacanth (*Latimeria*) — I | α3 | TGCTGGGYSACSGGCTTCTACCC / AGRCTKGKGTGCTCCACNTGRCA | Betz et al. 1994 |
| **Amphibians** | | | |
| Frog (*Xenopus*) — II | β2 | GRIGAIGTITAYWCITGYCIIGTISAICA† / TGYIIIGYIGWIGGITTYTAYCC† | Sato et al. 1993 |
| Salamander (*Ambystoma*) — I | α3 | | Sammut et al. 1997 |
| Salamander (*Ambystoma*) — I | α2, α3 | TACCTGGAGGGCCCGTGCGTNGAATGG / CTCGATATCCGGNCCRTARAA | Grossberger & Parham 1992 |
| **Reptiles** | | | |
| Sauria (snakes, lizards) — I | α2, α3 | TACCTGGAGGGCCCGTGCGTNGAATGG / CTCGATATCCGGNCCRTARAA | Grossberger & Parham 1992 |
| Geckos — I | α2 | GGGTCTCACACSBTSCARHGGATGT / GTAACTSCGSAGCCACTCCACGC | Radtkey et al. 1996 |

**Archosaurs**

| | | | | |
|---|---|---|---|---|
| Birds | II | β1 | TTCATTAACGGCACGGAG<br>GGCGGAACACCTCGTAGT | Edwards et al. 1998‡ |
| Songbirds | II | β1 | GAAAGCTCGAGTGTCACTTCACGAACGGC<br>GGGTGACAATCCGGTAGTTGTGCCGGCAG | Vincek et al. 1997 |
| Birds | I | α2 | TGGCAGTGGATGTAYGGCTGTGA<br>CCTGCCCAGCTYKCCTTCCCRTA | Westerdahl et al. 1998 |
| Budgerigars | II | β1 | AGGCAMRTCTACAAACGG<br>ACCCCGTAGTTGTKCCG | Edwards et al. 1999‡ |

**Mammals**

| | | | | |
|---|---|---|---|---|
| Eutherians | II | DRβ1 | CTCGGATCCGCATGTGCTACTTCACCAACG<br>GAGCTGCAGGTAGTTGTGTCTGCACAC | Gyllensten et al. 1990 |
| Anthropoid Primates | II | DRβ1 | TGTRACYGGATCGTTCKTGTCCCC<br>CTCACAGGGACYCAGGCCC | Figueroa et al. 1994 |
| Primates | II | DQα | GTGCTGCAGGTGTAAACTTGTACCAC<br>CACGGATCCGGTAGCAGCGGTAGAGTTG | Gyllensten & Ehrlich 1988 |
| Whales | II | DRβ1 | CTCGCCGCTGCATGAAAC<br>CCCCACAGCAGTTTCTTG | Murray & White 1998 |
| Mice (Mus) | II | DQβ1 | CACGGCCCGCGCGGCTCCCGC<br>CGGGCTGACCGCGTCCGTCCGCAG | She et al. 1991 |

† Used in a 3′ RACE PCR in conjunction with an oligo-dT primer.
‡ Known to amplify homologous regions of some birds other than those for which they were originally designed.

originally cloned by generating RT-PCR products from first strand cDNA for use as probes on a cDNA library. Although cDNA libraries can be challenging to set up, the advantage is that they provide a permanent resource for subsequent cloning of many other genes of subsequent interest, and they often yield complete coding sequences, thereby making intervening intron sequences accessible using PCR.

For species such as humans and relatives whose MHC genes are relatively well characterized, RT-PCR amplification and cloning is a rapid way of characterizing alleles at known loci; for example, human class I alleles from the A, B and C loci are typically amplified simultaneously (using primers in regions conserved between all three loci). Analysis of the sequences of cloned PCR products not only identifies the locus of origin of a sequence, but its allelic affinities. This protocol routinely results in PCR-cloning artefacts (Borriello & Krauter 1990; Ennis *et al.* 1990; L'Abbe *et al.* 1992). In well studied species, these artefacts can readily be identified. However, in less well studied organisms this approach can be problematic for making detailed evolutionary inferences, because the number of loci is often unknown and heteroduplex formation can be frequent. Edwards *et al.* (1995a) identified likely recombinant amplicons and *Taq*-induced point mutations in RT-PCR products of class II B genes in birds.

### 8.3.4 Southern blots

Southern blot hybridizations are an excellent way to survey the diversity of MHC and MHC-like sequences in the genome of a genetically unstudied animal. Often a cDNA clone, conventional PCR product, or RT-PCR product is used to detect the diversity of related sequences in the genome (e.g. a segment of exon 2 of class II B genes or exon 3 of class I; Schneider *et al.* 1991; Westerdahl *et al.* 1998). Use of different segments and hybridization conditions can yield complementary sets of data; for example, hybridization of a PBR segment under stringent conditions results in detection of polymorphic bands corresponding to a limited number of loci in outbred populations. By contrast, hybridization with non-PBR exons under mild conditions detects numerous bands, some of which probably correspond to MHC pseudogenes (Fig. 8.5) (Edwards *et al.* 1995a; Bartl *et al.* 1997).

Southern blots are highly recommended prior to cloning or amplification from genomic DNA as a means of assessing genomic complexity of the group of MHC genes one intends to study. Blots yield a rough estimate of the number of related sequences in the genome, and therefore an estimate of the ease with which individual loci will be amplifiable from total genomic DNA. In our experience (see below), bands detected on MHC Southern blots can be weak or strong, the former sometimes indicating pseudogenes. Knowledge of the complexity of the MHC subfamily to be characterized is of enormous

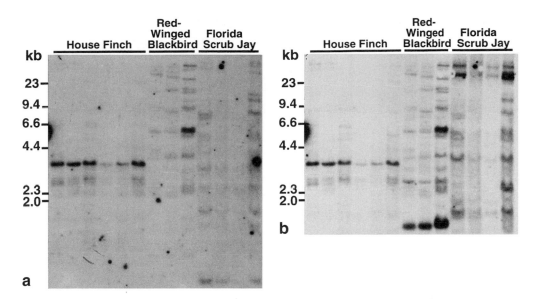

**Fig. 8.5** Autoradiographs of Southern hybridization of probes 2 (a) and 3 (b) to genomic DNA of house finches, red-winged blackbirds, and scrub jays cut with *Pst*I. Probes 2 and 3 are described in materials and methods and in Fig. 8.4. Note that the complexity of the blot is less when probe 2, containing primarily peptide-binding region (PBR) sequences, is used, as compared to when a probe consisting primarily of exon 3 sequences is used.

help in interpreting subsequent genomic cloning, cDNA cloning or PCR approaches.

### 8.3.5 Genomic cloning

Cloning MHC genes in lambda, cosmid or 'bacterial artificial chromosome' (BAC) vectors can be a convenient first step to characterization of complete primary structures for analysis of molecular evolution and ecology. For many molecular ecologists this approach may be 'overkill', but in our experience it is a useful prerequisite to unambiguous determination of complete gene sequences, including introns and flanking regions, thereby paving the way to more confident PCR-typing strategies and more complete understanding of selective pressures acting on MHC genes. It is also an appropriate approach for molecular evolutionary characterization of multiple genes in a gene family and it is in this way that the entire human MHC, and large segments of the chicken MHC (Kaufman *et al.* 1999) have been sequenced. The sCos-1 cosmid cloning system (Evans *et al.* 1989; Longmire *et al.* 1993; Stratagene La Jolla, California) allows one to clone DNA inserts in the 25–40 kb range. Although MHC genes are typically much smaller than this insert size, for some taxa with small MHCs or MHC genes such as chickens, multiple genes may lie on a single cosmid, paving the way for analyses of linkage and larger-scale MHC structure. Furthermore, knowledge of sequences flanking the gene of interest, particularly if they contain polymorphic markers such as Alu sequences or microsatellites (Schonbach *et al.* 1994), can be used to great advantage in later stages of MHC-typing because many such markers show strong linkage disequilibrium with MHC polymorphisms (Schwaiger *et al.* 1993; Jin *et al.* 1996). Cloning MHC genes in BACs, with average insert sizes of 150 kb, is an even more powerful method for characterizing MHC genes at the genomic level, because extended haplotypes of multiple genes are immediately available, allowing easy establishment of linkage far from the MHC gene itself (Graser *et al.* 1998). BAC libraries require very high quality starting DNA and are only available for a few model species, but should prove useful as availability increases. Analysis of large segments of nuclear DNA is likely to become an important tool in systematics generally (Baker *et al.* 1997), and the analysis of MHC gene evolution in natural populations can be greatly facilitated by this approach (Edwards *et al.* 1998).

### 8.3.6  Large-scale sequencing

Large-scale sequencing refers to the determination of complete sequences of cosmid-scale inserts using a combination of traditional shotgun sequencing techniques and novel computer analysis of multiple aligned reads obtained from shotgun subclones (Fig. 8.6). Analysis of multigene families via large-scale sequencing is but one spin-off from the Human Genome Project (Rowen *et al.* 1996). At the time of writing the human MHC has been completely sequenced (MHC sequencing consortium 1999) and a number of efforts are afoot to sequence in their entirety the MHCs of such model species as mouse, chicken and pufferfish (Milne *et al.* 1997; Guillaudeux *et al.* 1998). It is of interest to molecular ecologists that the availability of relatively simple protocols, augmented by kits and automated equipment for rapid isolation and sequencing of large numbers of shotgun subclones, allows such approaches to be applied on a limited scale to non-model species. Simple methods for preparing subclone libraries from individual cosmids have been available for some time (reviewed in Adams *et al.* 1994) and have been applied to MHC-bearing cosmids from two songbird species (Hess *et al.* 2000; Edwards *et al.* 2000). The most popular methods begin with sonication of whole cosmid clones in a cup horn sonicator, and electrophoresis of the sonicated DNA on an agarose gel. The ragged ends of the DNA fragments are repaired and blunted by T4-DNA polymerase prior to electrophoresis in an agarose gel. Fragments of approximately 1.5–2.5 kb are then excised from the gel using a razor blade and purified from the slab using any number of kits, such as Qiagen's (Valencia, California) gel extraction kit. These fragments are then cloned into a vector such as M13 or pUC18/19; several hundred to thousands of such plasmids can

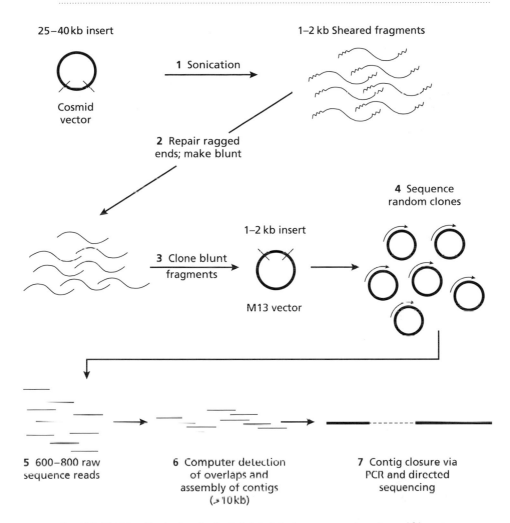

**Fig. 8.6** Diagram illustrating the logic behind shotgun sequencing of cosmid inserts.

be prepared rapidly in a 96-well plate format using kits such as Quiagen's 96 miniprep kits.

Standard primers can be used to sequence on an automated sequencer one or both ends of each subclone depending on the vector used. Modified primers can be used to minimize the amount of vector sequenced in each run as this is entirely redundant information that will be discarded. Note that some subclones will contain part of the original sCos-1 vector, which usually is not separated from the insert prior to sonication. For this reason, modified ($\approx 4\,$kb) cosmid vectors that are smaller than the typical 6–7 kb vectors sold commercially, can be constructed to minimize eventual sequencing of noninsert sequences.

Depending on the length and quality of sequence reads, 700–900 individual reads on an automated DNA sequencer are typically required to achieve the minimum number of contigs from a 30-kb cosmid clone to warrant directed sequencing and PCR approaches. The sequence coverage can be highly redundant in certain areas, but the pay-off is high confidence in the final sequence (Wilson *et al.* 1992). Powerful programs for assembly of individual sequence reads into larger contigs have recently been developed (Ewing & Green 1998; Ewing *et al.* 1998; Gordon *et al.* 1998). Additional programs such as Seqhelp (Lee *et al.* 1998) and Repeat Masker (A.F.A. Smit and P. Green, RepeatMasker at http://ftp.genome.washington.edu/RM/RepeatMasker.html) can be used to annotate and analyse the genomic features of contigs by conducting BLAST searches to databases such as Genbank, finding open-reading frames, simple sequence repeats, retroviral sequences and transposable elements, CpG islands, and the like. The rapid and increasingly automated acquisition of large segments of DNA sequence flanking MHC genes from nonmodel species provides an exciting and novel resource for further evolutionary and population genetic analysis in natural populations of vertebrates.

## 8.4  Analysis of MHC polymorphism

Once a specific MHC locus can be reliably amplified via PCR, any one of the many laboratory methods for detecting mutations and polymorphisms in DNA sequences can be employed to characterize the variation. Frequently the imprint of balancing selection on nucleotide polymorphisms can be detected, even when the locus/allele relationships are ambiguous, or when sequences from unambiguously different loci are compared (Hughes *et al.* 1994; Edwards *et al.* 1995b). When multiple loci are amplified in a PCR, different loci can sometimes be distinguished on the basis of amplified intron sequences (Ono *et al.* 1993) or upon phylogenetic analysis (Vincek *et al.* 1997), but these approaches do not work on all species (Edwards *et al.* 1998) and it is not clear how widespread their use is. In general we advocate taking the extra time and effort required to amplify single loci reliably; the number of such surveys at the sequence level in natural populations is alarmingly small, and needs to be increased before we have a general understanding of MHC variability in nature.

### 8.4.1  Laboratory techniques

In the recent past, most examinations of MHC polymorphisms in natural populations have been conducted using RFLP analyses (e.g. Yuhki & O'Brien 1990; Wittzell *et al.* 1994; von Schantz *et al.* 1996; Boyce *et al.* 1997) or sequencing of cloned PCR products (Ono *et al.* 1993; Murray *et al.* 1995). Both of these methods have drawbacks, the former in lack of resolution, the latter in potential sequencing artefacts and limited data on genotype frequencies (some alleles will be missed). Sequencing of PCR products generated from cDNA has advantages if coding sequence variability is of primary interest; amplification is generally easier than from genomic DNA because of fewer competing potential templates. However, for this method to work, RNA must be retrievable from tissues from the individuals sampled, either from rapidly snap-frozen tissues (spleen in the case of class II), blood (suitable for class I) or cell lines (e.g. Belich *et al.* 1992; Yuhki & O'Brien 1994; Edwards *et al.* 1995b; Houlden *et al.* 1996). More recently, single-stranded conformational polymorphism (SSCP) has emerged as a convenient method for resolving amplified MHC alleles (Murray *et al.* 1995; Sato *et al.* 1996; Klein *et al.* 1997; reviewed in Potts 1996). Murray *et al.* (1995) showed how individual genotypes could be reconstructed by comparing SSCP profiles of individuals with those for known, previously cloned alleles. Denaturing gradient gel electrophoresis (DGGE) is another powerful method for examining MHC diversity. A good example of this approach as applied to the MHC of natural populations is in the work of Miller (Miller & Withler 1997), who resolved over 30 alleles at a salmonid class I locus in north-west populations of Pacific Salmon (see also Miller *et al.* 1997). In general, indirect approaches to mutation detection, despite the drawbacks of each (Taylor 1997), have great promise for jump-starting large-scale descriptions of MHC variability in nature.

For well-studied systems, such as the human MHC, rapid, nonradioactive, oligonucleotide-based typing methods have been available for a number of years (Erlich *et al.* 1993). These methods utilize immobilized oligonucleotides corresponding to different allelic motifs in PBR domains, combinations of which define known alleles in the human population. Hybridization of amplified PBRs to these immobilized probes in a dot blot format allows one to determine the allele(s) of an individual by the pattern of hybridization. In principle, new alleles consisting of new combinations of previously encountered motifs can be discovered by their aberrant hybridization pattern or lack of hybridization to one or more sets of probes corresponding to a particular PBR subdomain. However, the method does not permit direct identification and characterization of novel alleles. This typing system has not been tested on species other than humans but in principle it could be easily extended to any species for which a preliminary survey of MHC diversity at the sequence level has been performed. It allows rapid typing of literally hundreds of individuals in a short timespan, and could be easily adapted for use in the field.

### 8.4.2  Statistical analysis of MHC polymorphisms

Because of the effects of balancing selection on patterns of sequence variability, MHC variability and evolution has spawned a large literature in theoretical population genetics (Takahata 1990; Takahata & Nei 1990; Takahata & Satta

1998). In one sense, the phylogenetic and population genetic analysis of MHC genes is no different than for other DNA sequences. Typically, a high ratio of nonsynonymous to synonymous substitutions per site ($d_N/d_S$) at PBR codons, calculated according to standard methods with available software packages (Nei & Gojobori 1986; Kumar *et al.* 1993), is used as a measure of the effect of balancing selection, as originally shown for MHC by Hughes & Nei (1988, 1989). More attention should be paid to maximum likelihood codon-based methods for estimating $d_N$ and $d_S$ (Goldman & Yang 1994; Nielsen & Yang 1998), and how this ratio varies over sites, because such methods and attention to such details yield more accurate estimates of evolutionary dynamics than do older methods. Use of these new methods is particularly critical if comparisons of dynamics between genes are to be made and if estimates of selection will be gleaned from such statistics (Edwards & Hedrick 1998). Phylogenetic trees of MHC sequences are typically built using standard methods. However the presence of recombination and gene conversion in the PBRs of MHC genes makes phylogenetic analysis trickier than for other genes. To date most studies have separately analysed different regions defined on a priori assumptions of previously encountered recombinants and polymorphism (Gyllensten *et al.* 1991; McAdam *et al.* 1994). New methods for detecting recombination and gene conversion potentially permit discovery of such independently evolving regions prior to or during phylogenetic analysis (Grassly & Holmes 1997); M. Kuhner, J. Yamato, J. Felsenstein, unpublished information), thereby making prior assumptions of evolutionary pattern unnecessary.

## 8.5 Evolution and complexity of MHC class II B genes: examples from songbirds

We are currently investigating the molecular evolution and population genetics of avian MHC class II B genes using cosmid cloning and large-scale sequencing approaches. Because little is known of the long-term evolutionary dynamics of MHC genes in birds, and because knowledge of these dynamics is important for characterizing MHC allelic diversity in natural populations, we are interested in both genomic and transcribed MHC sequences. The avian MHC, although less well characterized than the mammalian MHC, is known to be approximately one-fifteenth the size of the mammalian MHC (Trowsdale 1995). In chickens the MHC contains approximately one-third the number of expressed genes (Zoorob *et al.* 1993). Additionally, the avian MHC genes examined contain smaller introns and are composed of a higher GC base content (Kaufman 1995). These characteristics, and the success of characterizing the chicken MHC using cosmid methods (Kaufman *et al.* 2000), suggested to us that characterization of the MHC of other birds using similar approaches was warranted.

To date, most studies of the avian MHC have focused on the complex in chickens or in related Galliformes (Guillemot *et al.* 1988; Jarvi & Briles 1992; Wittzell *et al.* 1994). Genomic copy numbers of MHC class II genes are largely unknown outside of the order Galliformes other than for a single songbird, the Bengalese finch from the family Estrildidae (Vincek *et al.* 1995), making genomic characterization a priority for novel bird species. Kaufman (1995; Kaufman & Salomonsen 1997) has hypothesized that the MHC of chickens, and of birds in general, is 'minimal essential', functioning with fewer genes and noncoding regions in a smaller genomic space, a pattern that could result in more intense selective pressures on MHC genes and that can readily be tested by investigation of new species.

Our investigations have thus far focused on three common North American species, house finches (Fringillidae: *Carpodacus mexicanus*), red-winged blackbirds (Icteridae: *Agelaius phoeniceus*) and scrub jays (Corvidae: *Aphelocoma caerulescens*). These species represent divergent lineages within the true songbirds (suborder Passeri, order Passeriformes) and exhibit ecological, behavioural and epidemiological traits that might benefit from interrogation by MHC probes (Woolfenden & Fitzpatrick 1984; Hill 1991; Beletsky 1996; Dhondt *et al.* 1998). Before the relationship between traits of interest and MHC genotypes can begin to be examined (Edwards & Potts 1996; Apanius *et al.* 1997; Thursz *et al.* 1997), it is necessary to obtain MHC sequences and understand how gene complexity varies across distantly related taxa.

We have applied a series of protocols, including Southern blotting, cosmid cloning, library screening, PCR and sequencing to analyse MHC class II B genes of house finches. These protocols are found in the Appendix to this chapter. Together they provide powerful methods for studying the evolution of MHC genes in nonmodel organisms.

### 8.5.1  Interspecific variation in MHC class II B gene number

To study variation in copy number of MHC class II B genes among songbirds, total genomic DNA from the three songbirds was digested, Southern-blotted and hybridized to different MHC class II B probes. With probe 3, there is one to two dark, strongly hybridizing bands per individual in every species, and more notably, the number of lighter, weakly hybridizing bands is much greater in the red-winged blackbirds and Florida scrub jay than in the house finch (Fig. 8.5). With probe 2, the hybridization pattern seen is remarkably similar, except for a small (<2 kb) strongly hybridizing fragment in the red-winged blackbird only observed with probe 2 (Fig. 8.5).

The number of strongly hybridizing bands was similar between the three species, whereas the number of weakly hybridizing bands differed substantially between species, regardless of whether six-base cutter restriction

**Table 8.2** Mean number of strongly and weakly hybridizing bands observed in hybridization of MHC class II B probes to three songbirds. Overall mean is based upon all scored individuals, which was 4–6 (house finch), 1–3 (red-winged blackbird) and 1–4 (Florida scrub jay). Thus, the overall mean is not necessarily the average of each column

| Restriction enzyme | Probe used* | House finch | | Red-winged blackbird | | Florida scrub jay | |
|---|---|---|---|---|---|---|---|
| | | Strong bands | Weak bands | Strong bands | Weak bands | Strong bands | Weak bands |
| *Pst*I | 2 | 1.0 | 3.33 | 1.00 | 8.66 | 0.00 | 12.5 |
| *Pst*II | 3 | 1.0 | 3.16 | 1.33 | 11.3 | 1.00 | 7.66 |
| *Pvu*II | 2 | 1.2 | 4.80 | 0.00 | 4.00 | 0.00 | 5.33 |
| *Pvu*II | 3 | 1.2 | 2.20 | 1.00 | 3.00 | 1.00 | 4.33 |
| *Taq*I | 2 | 1.5 | 3.00 | 0.00 | 7.00 | 0.00 | 4.00 |
| *Taq*I | 3 | 1.0 | 4.33 | 1.00 | 8.00 | 0.00 | 6.00 |
| *Alu*I | 2 | 1.0 | 0.33 | 5.00 | 4.00 | 2.66 | 7.00 |
| *Alu*I | 1 | 1.0 | 2.00 | 3.00 | 14.00 | 4.00 | 9.25 |
| Overall mean | | 1.1 | 2.80 | 0.96 | 8.30 | 1.30 | 7.10 |

\* See Fig. 8.4

enzymes or four-base cutter restriction enzymes such as *Alu*I were used (Table 8.2). The house finch displays a relatively small number (mean, 2.8) of weakly hybridizing bands, while numbers of weakly hybridizing bands are much higher in the red-winged blackbird (mean, 8.3) and in the Florida scrub jay (mean, 7.1).

### 8.5.2  Cosmid cloning of house finch MHC class II B genes

To analyse the house finch MHC in further detail, a cosmid library was constructed and screened. This initial screening identified 18 putative colonies, which were plated and rescreened with the house finch MHC probe 1. Three strongly hybridizing cosmids were identified and chosen for further characterization. To estimate the fraction of house finch class II B restriction fragments cloned in the two cosmids, cloned DNA and genomic DNA, from which the house finch library was made, were digested with a number of enzymes and hybridized with $^{32}$P radiolabelled probe 1 (Fig. 8.7). Several hybridizing fragments observed in the genomic DNA were also detected in the cloned DNA; for example, the strongly hybridizing 2.8 kb *Pst*I fragment of genomic DNA, also observed in Fig. 8.5, was present in cosmid clone 10A as well, indicating that it was cloned (Fig. 8.7). Additionally, at least two weakly hybridizing *Pst*I genomic fragments were cloned (Fig. 8.7). Overall, 15 of the 18 unique genomic fragments that hybridized to the MHC probe were cloned (Table 8.3). Assuming that all MHC genes in the house finch hybridize at low stringency to the probe, then it is likely that approximately 83% of the house finch MHC class II B genes have been cloned. These data also suggest that some of the

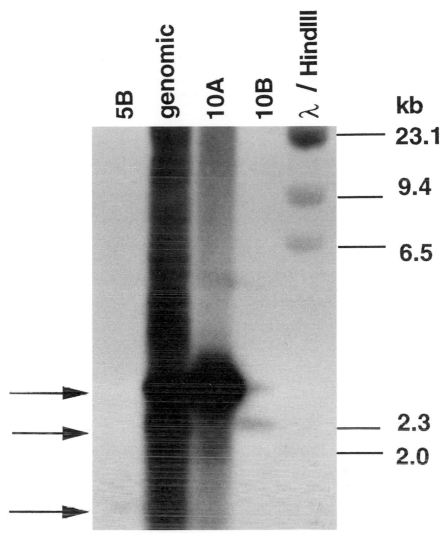

**Fig. 8.7** Comparison of the contents of house finch cosmid clones 5B, 10A and 10B with MHC-hybridizing fragments in genomic DNA of the same individual from which the library was made. All DNAs were cut with *Pst*I. All three cosmid clones possess fragments that correspond with genomic fragments hybridizing to the class II B probe (arrows).

weakly hybridizing fragments are physically linked on the same clones as the strongly hybridizing fragments.

### 8.5.3 Sequence analysis of class II B exons

To confirm that the cosmid clones 10A and 5B contained class II B genes, segments of exons 2 and 3 were amplified, cloned, and cycle-sequenced. Three

**Table 8.3** Characterization of two cosmid clones using single and double digests. The columns do not necessarily add up to the totals because hybridizing restriction fragments observed in both a single and double digest were counted only once

| Restriction enzyme(s) | Total bands in clone 5 | Total bands in clone 10 | Total bands in genomic DNA | Number of genomic bands found in clone 5 or 10 (%) |
|---|---|---|---|---|
| *Bam*I | 7 | 13 | 5 | 4 (80) |
| *Eco*RI | 6 | 10 | 4 | 3 (75) |
| *Pst*I | 7 | 12 | 3 | 3 (100) |
| *Eco*RI / *Pst*I | 7 | 12 | 5 | 4 (80) |
| *Eco*RI / *Bam*I | 7 | 14 | 7 | 6 (86) |
| Total | 25 | 44 | 18 | 15 (83) |

clones of exon 2 and seven clones from exon 3 proved to contain MHC sequences (Fig. 8.8). The three identical exon 2 sequences cloned from a single amplification product aligned without gaps to previously cloned sequences from house finches (Edwards *et al.* 1995b) (Fig. 8.8a). While the corrected sequence divergence of the cloned exon 2 sequence from the Finch 2.1 sequence cloned from genomic DNA is only 3.3%, the cloned exon 2 sequence and the Finch 1.1 sequence cloned from cDNA show 26.6% divergence. These results suggest that MHC class II B sequences that are highly divergent and probably duplicated in ancestral species exist in the house finch genome, and that we have cloned more than one locus in the three cosmid clones.

In contrast to the exon 2 sequence, two distinct exon 3 sequences were obtained from different clones and showed 10.3% sequence divergence (Fig. 8.8b). Finally, unlike the exon 2 sequences, the exon 3 sequences obtained could be aligned to previously published sequences only by invoking alignment gaps. At this time, it cannot be determined whether these gaps suggest functional or nonfunctional genes.

### 8.5.4  MHC genes in a weakly hybridizing fragment of blackbirds

To determine whether the weakly hybridizing fragments on Southern blots (Fig. 8.5) contained MHC genes, a cosmid clone (number 10) was isolated from a red-winged blackbird whose insert corresponded to one of the many weakly hybridizing bands (Fig. 8.5). Large-scale sequencing methodologies were used to determine the primary structure of a 21-kb contig on this cosmid. The MHC gene on cosmid number 10, which was designated as *Agph-DAB2*, is similar in structure and sequence to a previously cloned MHC gene from blackbirds, *Agph-DAB1* (Fig. 8.9; Edwards *et al.* 1998, 1999). There is considerable base compositional variation along this 21-kb segment and a preliminary

(a) exon 2

```
               1         11        21        31        41        51        61        71        80
10.2     GGTCAGGTTCGTGGACAGGTTCATCTACAACCGGGAGCAGTTCCTGATATTGGACAGCGACGTCGGGGTGTACGTGGGGT
Genomic  ..................................................C.........G....................
cDNA     ...............CA...........C....G..G..C................G..CAC...
Chicken  .........ATC....C...GAA........C.....A.GC.CAC..C..........G...AAA.........CTG

               81        91        101       111       121       131       141       151
10.2     TCATCGCCTATGGGGAGATGAATGCCAAGCGCTCTAACAGGCACCGGTTCACATGAGTAGTACTACCGGCGCTTTGGTG
Genomic  ...C.C.............G.
cDNA     ...C.C..........TAT............CTG.....GC.....CCATAC.....A.A..A..GC....
Chicken  AT.CAC.GCTG.T....CC.C.A..TG.ATA..GG.....CA..G.C.AGTTT.......A..CGAAT.AA.GAA...
```

(b) exon 3

```
               1         11        21        31        41        51        61
58.3     CCAGATCCAGGTGAGGTGGTTCCAGGGCCAGCAGGAGCTCTCGGAGCACGTGGTGGCCACCGACG
10A.3    .....C.......................A..C..A...........A...A
cDNA     ..CG.............................
Chicken  GG.G..G.....A.......T.AA.GG..G....GAGA......G.......T....G....

               71        81        91        101       111       121
58.3     TGGTCCCCAACGGGGACTGGACTACCAGCTCCTGGAGCTGCTGGAAACTCTCCCCCGGCGC
10A.3    .........A......................C.C.A....A.G..
cDNA     ...........G...T....--GC..GG........
Chicken  ..A.G.AG.........G....G.G..T.G.......G...-CGT..G........
```

Fig. 8.8 DNA sequences of exons 2 and 3 of class II B genes obtained from cosmid clones and compared with previously published house finch sequences and those of chicken. (a) The single sequence (10.2) obtained from cosmid 10.2) are presented in Edwards *et al.* (1995a), whereas the chicken sequence corresponds to a class II B allele from the B12 haplotype. (b) Exon 3 sequences amplified from cosmid 5B (5B.3) and 10A (10A.3), previously cloned finch sequence (1.1) (Edwards *et al.* 1995b), and a chicken sequence (B12 haplotype). Divergences among sequences were calculated according to the method of Jin & Nei (1990), with a gamma rate parameter of α = 0.5.

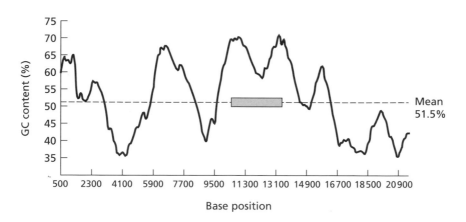

**Fig. 8.9** Sliding window analysis of GC content over an ~ 21 kb contig containing a red-winged blackbird MHC class II B gene obtained via shotgun sequencing of a cosmid clone. The window size was 1000 bp and was moved by 100 bp before each calculation of GC content. The approximate location of the class II B gene (*Agph-DAB2*) is indicated by the shade box. The high GC content near the beginning of *Agph-DAB2* possibly indicates the presence of a CpG island.

analysis of the contents of this segment using SeqHelp (Lee *et al.* 1998) uncovered several potential open-reading frames. These results confirm the presence of an MHC gene on this weakly hybridizing *Pst*I restriction fragment in the genome of red-winged blackbirds, and suggest that weakly hybridizing genomic fragments in other species may also contain MHC sequences.

### 8.5.5  Diversity of MHC class II B genes in songbirds

We have compared interspecific patterns in Southern blot band complexity among three songbird species. The observation that more bands were found in scrub jay and red-winged blackbird than in the house finch suggests that there may be more copies of the MHC class II B gene, or at least more MHC-like sequences in scrub jay and blackbird, possibly as a result of recent gene duplications or deletions. In the house finch, the strongly hybridizing band corresponds to a single MHC class II gene which we now know is a pseudogene (Hess *et al.* 2000). Although their search for class II B genes in house finches by PCR was not exhaustive, Edwards *et al.* (1995a,b) cloned up to two sequences from individual house finches, a result consistent with our Southern blot data. Using Southern blots, Vincek *et al.* (1995) estimated that there were four MHC II B genes in the Bengalese finch, which belongs to a different clade of birds from the house finch. Using phylogenetic analysis of partial PBR domains, Vincek *et al.* (1997) estimated there were four loci in Darwin's finches (Geospizidae). Wittzell *et al.* (1999) have documented complex multi-locus MHC Class I haplotypes in songbirds and a snake.

We expected that if there were differences in the number of bands detected per species, the house finch would have the highest number because this was the species from which the probe was derived. However, we found the opposite result, suggesting that the differences in band number are not solely a result of differential hybridization. Although we confirmed that at least one weakly hybridizing band in red-winged blackbird DNA contains an MHC gene, it is still possible that some of the weakly hybridizing bands in blackbirds or finches represent non-MHC DNA. While this result might be expected when very distantly related species are used for probes (Gibbs *et al.* 1991), the known divergence in conserved domains between expressed MHC sequences in different songbird species is not high (7–12% for exon 3; Edwards *et al.* 1995a,b). Paradoxically, weakly hybridizing MHC bands in birds are frequently more polymorphic than strongly hybridizing bands and, perhaps harbouring a greater proportion of functional loci (D. Hasselquist, personal communication). Lastly, our library results suggest that the weakly hybridizing bands are physically linked to strongly hybridizing bands on cosmid clones, suggesting that these bands may correspond to MHC or MHC-like sequences, as opposed to unrelated genes. If the differences in band number between species represent differences in number of functional MHC genes, then there could be functional consequences at both the immunological and behavioural levels, given the documented role of the MHC in the immune response and kin recognition (Potts *et al.* 1991; Apanius *et al.* 1997). Further MHC characterization will be necessary before we can begin to examine how MHC genotype may affect phenotype and fitness in a natural population.

## 8.6 Conclusion

MHC genes clearly have an important role to play in the molecular ecology of vertebrate species. They are fascinating from the point of view of the evolution of multigene families, and their variability can have important consequences for phenotypic variation of traditional interest to behavioural ecologists (Edwards & Hedrick 1998). Part of the challenge of bringing the power of MHC genes to molecular ecology is our ignorance of appropriate locus-specific priming sites adjacent to functional exons of interest, and the apparent close relationship between different genes in the same genomes, particularly in nonmammalian species, making characterization of individual functional MHC loci challenging. The long-term mode of evolution clearly has an important impact on the approach to cloning MHC genes. Approaches presented here for characterizing transcribed and genomic MHC sequences should aid in a better understanding of the evolution and ecology of MHC genes in natural populations.

## Acknowledgements

We would like to thank D. Garrigan, D. Hasselquist, J. Longmire, M. March and D. Slade for technical assistance and helpful discussion. M. Avery, G. Hill and E. Gray contributed to the jay, finch and blackbird samples, respectively. J. Klein and Academic Press kindly granted permission to modify a figure (our Fig. 8.1) which originally appeared in Klein *et al.* (1997). We thank D. Garrigan, C. Hess, and H. Hoekstra for constructive comments on the manuscript, and B. Murray for assistance compiling primer sequences in Table 8.1. Our research has been supported by the US National Science Foundation.

## Appendix

### Probes and primers

All probes were amplified by PCR from a partial cDNA clone (number HF# 2.1; Edwards *et al.* 1995a) containing a house finch MHC class II B gene, and phenol purified. Fifty nanograms of bacterial glycerol suspensions containing the above house finch clone were amplified using PCR in 25 µL reactions of 150 µmol dNTPs, 2 µmol of each primer, 1 ×PCR buffer, and 0.63 U of *Taq* polymerase (Boehringer Mannheim, Indianapolis, Indiana). Amplification conditions consisted of 30 s at 94 °C, 30 s at 55 °C, and 30 s at 72 °C for 30 cycles. Four primers were used to generate probes (Fig. 8.4). Probe 1, used to screen the cosmid library, consisted of 60 bp of the leader sequence (exon 1), 270 bp of exon 2, 279 bp of exon 3, and 45 bp of the transmembrane domain (exon 4). Probe 2 contains all of exon 2, with an additional 60 bp of exon 1, and 108 bp of exon 3. Probe 3 contains all of exon 3, with 51 bp of exon 2, and 45 bp of exon 4. Probes 1 and 2 overlap by 26% of their sequence length (172 bp). The locations and sizes of probes 1–3, and the primer sequences used to generate these probes, are shown in Fig. 8.4 and legend.

### Isolation of high molecular weight genomic DNA

High molecular weight DNA was isolated from 0.5 g of tissue or from 300 µL of blood. Minced tissue or blood was placed into a lysis buffer containing 100 mmol Tris pH 8, 100 mM EDTA, 0.5% SDS, and 0.5 mg mL$^{-1}$ proteinase K, and rotated overnight at 37 °C. DNA was phenol-extracted, placed into dialysis tubing with a pore size of 12–14 000 Da (The Spectrum Comparies, Laguna Hills, California), and dialysed overnight in 1 × TE (10 mM Tris, 1 mM EDTA).

## Southern blot hybridization

Approximately 10 µg of genomic DNA was digested overnight with *Pvu*II, *Pst*I, *Taq*I or *Alu*I and electrophoresed in a 1% agarose gel at 28 V for approximately 18 h (15.2 V hours cm⁻¹ of electrode separation). The DNA was transferred to a positively charged nylon membrane (Hybond N+ Amersham Pharmacia Biotech, Piscataway, New Jersey) over 2 days by capillary blotting and then fixed by baking at 80 °C for 2 h. The probe was randomly labelled with ³²P using the Stratagene (La Jolla, California) Prime-It II kit. After prehybridizing for 1 h, the probe was hybridized to the membrane overnight at 65 °C in rotating rotisserie tubes containing 0.5 M sodium phosphate buffer pH 7.2, 7% SDS, 1% bovine serum albumin, and 1 mM EDTA. Membranes were washed twice in 0.5 M sodium phosphate buffer pH 7.2, 7% SDS, 1 mM EDTA and washed last in 400 mM phosphate buffer pH 7.2, 1% SDS, 1 mM EDTA at 65 °C. These hybridizations and washes were considered to be at low stringency. Autoradiographs were prepared in a −80 °C freezer by exposing the membranes for 2–6 days to Hyperfilm-MP (Amersham Pharmacia Biotech, Piscataway, New Jersey) with one intensifying screen. Densitometric analysis of bands was not performed because differences in band intensity were generally discernible on the original autoradiographs. After accounting for the quantity of DNA in each lane, most bands could be categorized as either dark, strongly hybridizing bands or light, weakly hybridizing bands.

## Cosmid library preparation

Following Longmire *et al.* (1993), four micrograms of genomic house finch DNA were partially digested with *Sau*3A-1. Partially digested house finch DNA was dephosphorylated, and ligated overnight into sCos vector arms (Stratagene). Ligations were packaged into lambda phage using Gigapack Gold II (Stratagene), which was used to infect *Escherichia coli* bacteria strain DH 5α strain (Gibco). Ligations sufficient to produce approximately 170 000 colony-forming units were packaged and plated, achieving approximately a 4.5 × genomic representation, assuming an average insert size of 40 000 bp and a haploid genome size of 1.5 × 10⁹ bp (Tiersch & Wachtel 1991).

## Library screening and clone characterization

Replicate nylon membranes (Hybond N+) were lifted and screened using the ECL direct labelling chemiluminescence kit (Amersham Pharmacia Biotech, Piscataway, New Jersey) per manufacturer's instructions. Positive clones were chosen and isolated after the second screening. To estimate the fraction of MHC genes cloned, the cosmid clones and genomic DNA were cut with restriction enzymes in either single or double digests. DNA fragments were separated using electrophoresis, transferred to a nylon membrane via capillary blotting and hybridized to the $^{32}$P-radiolabelled probe 1.

## DNA sequencing

To verify that cosmid clones contained MHC DNA, parts of exons 2 and 3 were amplified from each clone with primers 305 and 306 (exon 2; Edwards *et al.* 1995a) and 3F and 3R (exon 3), respectively. Multiple amplified products were cloned into TA vectors and cycle-sequenced as described (Edwards *et al.* 1995a). Although this protocol can produce heteroduplexes (Borriello & Krauter 1990; L'abbe *et al.* 1992), our primary purpose in sequencing was for verification of the presence of MHC sequences rather than detailed analysis.

## References

Adams, M.D., Fields, C. & Venter, J.C. (1994) *Automated DNA Sequencing and Analysis*. Academic Press, New York.

Apanius, V., Penn, D., Slev, P.R., Ruff, L.R. & Potts, W.K. (1997) The nature of selection on the major histocompatibility complex. *Critical Reviews in Immunology* **17**, 179–224.

Arnaiz-Villena, A., Martinez-Laso, J., Alvarez, M. *et al.* (1997) Primate Mhc-E and -G alleles. *Immunogenetics* **46**, 251–256.

Ayala, F.J., Escalante, A., O'Huigin, C. & Klein, J. (1994) Molecular genetics of speciation and human origins. *Proceedings of the National Academy of Sciences USA* **91**, 6787–6794.

Baker, R.J., Longmire, J.L., Maltbie, M., Hamilton, M.J. & Van den Bussch, R.A. (1997) DNA synapomorphies for a variety of taxonomic levels from a cosmid library from the new world bat *Macrotus waterhousii*. *Systematic Biology* **46**, 579–589.

Bartl, S. & Weissman, I.L. (1994) Isolation and characterization of major histocompatibility complex class II B genes from the nurse shark. *Proceedings of the National Academy of Sciences USA* **91**, 262–266.

Bartl, S., Baish, M.A., Flajnik, M.F. & Ohta, Y. (1997) Identification of class I genes in cartilaginous fish, the most ancient group of vertebrates displaying an adaptive immune response. *Journal of Immunology* **159**, 6097–6104.

Beck, S., Abdulla, S., Alderton, R.P. *et al.* (1996) Evolutionary dynamics of non-coding sequences within the class II region of the human MHC. *Journal of Molecular Biology* **255**, 1–13.

Beletsky, L. (1996) *The Red-Winged Blackbird: the Biology of a Strongly Polygynous Songbird*. Academic Press, San Diego.

Belich, M.P., Madrigal, J.A., Hildebrand, W.H., Williams, R.C., Luz, R., Petzl-Erler, M. & Parham, P. (1992) Unusual HLA-B alleles in two tribes of Brazilian Indians. *Nature* **357**, 326–329.

Bergström, T.P., Josefsson, A., Erlich, H.A. & Gyllensten, U. (1998) Recent origin of HLA-DRB1 alleles and implications for human evolution. *Nature Genetics* **18**, 237–242.

Betz, U.A., Mayer, W.E. & Klein, J. (1994) Major histocompatibility class I genes of the coelacanth *Latemeria chalumnae*. *Proceedings of the National Academy of Sciences USA* **91**, 11065–11069.

Black, F.L. & Hedrick, P.W. (1997) Strong balancing selection at HLA loci: evidence from segregation in South American families. *Proceedings of the National Academy of Sciences USA* **94**, 12452–12456.

Borriello, F. & Krauter, K.S. (1990) Reactive site polymorphism in the murine protease inhibitor gene family is delineated using a modification of the PCR reaction (PCR + 1). *Nucleic Acids Research* **18**, 5481–5487.

Boyce, W.M., Hedrick, P.W., Muggli Cockett, N.E., Kalinowski, S., Penedo, M.C., & Ramey, R.R. II (1997) Genetic variation of major histocompatibility complex and microsatellite loci: a comparison in bighorn sheep. *Genetics* **145**, 421–433.

Dhondt, A.A., Tessglia, D.L. & Slothower, R.L. (1998) Epidemic mycoplasmal conjunctivitis in House Finches from eastern North America. *Journal of Wildlife Diseases* **34**, 265–280.

Dixon, B., van Erp, S.H.M., Rodrigues, P.N., Egberts, E. & Stet, R.J.M. (1995) Fish major histocompatibility complex genes: an expansion. *Developmental and Comparative Immunology* **19**, 109–133.

Dover, G.A. (1982) Molecular drive: a cohesive mode of species evolution. *Nature* **299**, 111–116.

Edwards, S.V. & Hedrick, P.W. (1998) Evolution and ecology of MHC molecules: from genomics to sexual selection. *Trends in Ecology and Evolution* **13**, 305–311.

Edwards, S.V. & Potts, W.K. (1996) Polymorphism of genes in the major histocompatiblility complex: implications for conservation genetics of vertebrates. In: *Molecular Genetic Approaches in Conservation* (eds T.B. Smith & R.K. Wayne), pp. 214–237. Oxford University Press, Oxford.

Edwards, S.V., Grahn, M. & Potts, W.K. (1995a) Dynamics of Mhc evolution in birds and crocodilians: amplification of class II genes with degenerate primers. *Molecular Ecology* **4**, 719–729.

Edwards, S.V., Wakeland, E.K. & Potts, W.K. (1995b) Contrasting histories of avian and mammalian Mhc genes revealed by class II B sequences from songbirds. *Proceedings of the National Academy of Sciences USA* **92**, 12200–12204.

Edwards, S., Gasper, J. & Stone, M. (1998) Genomics and polymorphism of Agph-DAB1, an Mhc class II B gene in Red-winged Blackbirds (*Agelaius phoeniceus*). *Molecular Biology and Evolution* **15**, 236–250.

Edwards, S.V., Hess, C., Gasper, J. & Garrigan, D. (1999) Toward an evolutionary genomics of the avian Mhc. *Immunological Reviews* **167**, 119–132.

Ennis, P.D., Zemmour, J., Salter, R.D. & Parham, P. (1990) Rapid cloning of HLA-A,B cDNA by using the polymerase chain reaction: frequency and nature of errors produced in amplification. *Proceedings of the National Academy of Sciences USA* **87**, 2833–2837.

Erlich, H., Bugawan, T., Begovich, A. & Scharf, S. (1993) Analysis of HLA class II polymorphism using polymerase chain reaction. *Archives of Pathology and Laboratory Medicine* **117**, 482–485.

Evans, G.A., Lewis, K. & Rothenberg, B.E. (1989) High efficiency vectors for cosmid micro cloning and genomic analysis. *GENE* **79**, 9–20.

Ewing, B. & Green, P. (1998) Base-calling of automated sequencer traces using PHRED. II. error probabilities. *Genome Research* **8**, 186–194.

Ewing, B., Hillier, L.D., Wendl, M.C. & Green, P. (1998) Base-calling of automated sequencer traces using PHRED. II. Accuracy assessment. *Genome Research* **8**, 175–185.

Fernandez-Soria, V.M., Morales, P., Castro, M.J. *et al.* (1998) Transciption and weak expression of *HLA-DRB6*: a gene with anomalies in exon 1 and other regions. *Immunogenetics* **48**, 16–21.

Figueroa, F., O'hUigin, C., Tichy, H. & Klein, J. (1994) The origin of the primate Mhc-DRB genes and allelic lineages deduced from the study of prosimians. *Journal of Immunology* **152**, 4455–4465.

Figueroa, F., Ono, H., Tichy, H., O'hUigin, C. & Klein, J. (1995) Evidence for insertion of a new intron into an Mhc gene of perch-like fish. *Proceedings of the Royal Society of London B* **259**, 325–330.

Frohman, M.A. (1988) Rapid production of full-length cDNAs from rare transcripts: amplification using a single gene-specific oligonucleotide primer. *Proceedings of the National Academy of Sciences USA* **85**, 8998–9002.

Garrigan, D. & Edwards, S.V. (1999) Polymorphism across an exon–interon boundary in an avian Mhc class II B gene. *Molecular Biology and Evolution* **16**, 1599–1606.

Geraghty, D.E., Koller, B.H., Hansen, J.A. & Orr, H.T. (1992) The HLA class I gene family includes at least six genes and twelve pseudogenes and gene fragments. *Journal of Immunology* **149**, 1934–1946.

Germain, R.N. & Margulies, D.H. (1993) The biochemistry and cell biology of antigen processing and presentation. *Annual Review of Immunology* **11**, 403–450.

Gibbs, H.L., Weatherhead, P.J., Boag, P.T., White, B.N., Tabak, L.M. & Hoysak, D.J. (1990) Realized reproductive success of polygynous red-winged blackbirds revealed by DNA markers. *Science* **250**, 1394–1397.

Gibbs, H., Boag, P., White, B., Weatherhead, P. & Tabak, L. (1991) Detection of a hypervariable DNA locus in birds by hybridization with a mouse MHC probe. *Molecular Biology and Evolution* **8**, 433–446.

Goldman, N. & Yang, Z. (1994) A codon-based model of nucleotide substitution for protein-coding DNA sequences. *Molecular Biology and Evolution* **11**, 725–736.

Gordon, D., Abajian, C. & Green, P. (1998) Consed: a graphical tool for sequence finishing. *Genome Research* **8**, 195–202.

Goulder, P., Price, D., Nowak, M., Rowland Jones, S., Phillips, R. & McMichael, A. (1997) Co-evolution of human immunodeficiency virus and cytotoxic T-lymphocyte responses. *Immunological Reviews* **159**, 17–29.

Graser, R., Vincek, V., Takami, K. & Klein, J. (1998) Analysis of zebrafish Mhc using BAC clones. *Immunogenetics* **47**, 318–125.

Grassly, N.C. & Holmes, E.C. (1997) A likelihood method for the detection of selection and recombination using nucleotide sequences. *Molecular Biology and Evolution* **14**, 239–247.

Grimsley C., Mather, K.A. & Oben, C. (1998) HLA-H: a pseudogene with increased variation due to balancing selection neighboring loci. *Molecular Biology and Evolution* **15**, 1581–1588.

Grossberger, D. & Parham, P. (1992) Reptilian class I major histocompatibility complex genes reveal conserved elements in class I structure. *Immunogenetics* **36**, 166–174.

Guillaudeux, T., Janer, M., Wong, G.K.S., Spies, T. & Geraghty, D.E. (1998) The complete genomic sequence of 424,015 bp at the centromeric end of the HLA class I region: gene content and polymorphism. *Proceedings of the National Academy of Sciences USA* **95**, 9494–9499.

Guillemot, F., Billault, A., Pourquie, O., Behar, G., Chausse, A.-M., Zoorob, R., Kreibich, G. & Auffrey, C. (1988) A molecular map of the chicken major histocompatibility complex: the class II genes are closely linked to the class I genes and the nucleolar organizer. *EMBO Journal* **7**, 2775–2785.

Gyllensten, U.B. & Erlich, H.A. (1988) Generation of single-stranded DNA by the polymerase chain reaction and its application to direct sequencing of the HLA-DQA locus. *Proceedings of the National Academy of Sciences USA* **85**, 7652–7656.

Gyllensten, U.B. & Erlich, H.A. (1989) Ancient roots for polymorphism at the HLA-DQ locus in primates. *Proceedings of the National Academy of Sciences USA* **86**, 9986–9990.

Gyllensten, U.B., Lashkari, D. & Erlich, H.A. (1990) Allelic diversification at the class II DQB locus of the mammalian major histocompatibility complex. *Proceedings of the National Academy of Sciences USA* **87**, 1835–1839.

Gyllensten, U., Sundvall, M. & Erlich, H. (1991) Allelic diversity is generated by intra-exon exchange at the DRB locus of primates. *Proceedings of the National Academy of Sciences USA* **88**, 3686–3690.

Gyllensten, U., Bergstrom, T., Josefsson, A. *et al.* (1994) The cotton-top tamarin revisited: Mhc class I polymorphism of wild tamarins, and polymorphism and allelic diversity of the class II DQA1, DQB1, and DRB loci. *Immunogenetics* **40**, 167–176.

Hashimoto, K., Nakanishi, T. & Kurosawa, Y. (1990) Isolation of carp genes encoding major histocompatibility complex antigens. *Proceedings of the National Academy of Sciences USA* **87**, 6863–6867.

Hashimoto, K., Nakanishi, T. & Kurosawa, Y. (1992) Identification of a shark sequence resembling the major histocompatibility complex class I a3 domain. *Proceedings of the National Academy of Sciences USA* **89**, 2209–2212.

Hedrick, P.W. (1994) Evolutionary genetics of the major histocompatibility complex. *The American Naturalist* **143**, 945–964.

Hedrick, P.W., Klitz, W., Robinson, W., Kuhner, M.K. & Thomson, G. (1991) Population genetics of HLA. In: *Evolution at the Molecular Level* (eds R. Selander, A. Clark & T. Whittam), pp. 248–271. Sinauer Associates, Sunderland, Massachusetts.

Hess, C.M., Gasper, J., Hoekstra, H.E. & Hill C.E. (2000) MHC Class II pseudogene and genomic signature of a 32-kb cosmid in the House Finch (*Canpodacus mexicanus*). *Genome Research*, in press.

Hill, A.V.S., Allsopp, C.E.M., Kwiatowski, D. *et al.* (1991) Common West African HLA antigens are associated with protection from severe malaria. *Nature* **352**, 595–600.

Hill, A.V., Elvin, J., Willis, A.C. *et al.* (1992) Molecular analysis of the association of HLA B53 and resistance to severe malaria. *Nature* **360**, 434–439.

Hill, G.E. (1991) Plumage color is a sexually selected indicator of male quality. *Nature* **350**, 337–339.

Houlden, B.A., Greville, W.D. & Sherwin, W.B. (1996) Evolution of MHC class I loci in marsupials: characterization of sequences from Koala (*Phascolarctos cinereus*). *Molecular Biology and Evolution* **13**, 1119–1127.

Hughes, A.L. & Nei, M. (1988) Pattern of nucleotide substitution at major histocompatibility complex class I loci reveals overdominant selection. *Nature* **335**, 167–170.

Hughes, A.L. & Nei, M. (1989) Nucleotide substitution at major histocompatibility complex class II loci: evidence for overdominant selection. *Proceedings of the National Academy of Sciences USA* **86**, 958–962.

Hughes, A.L. & Nei, M. (1993) Evolutionary relationships of the classes of major histocompatibility complex genes. *Immunogenetics* **37**, 337–346.

Hughes, A.L., Hughes, M.K., Howell, C.Y. & Nei, M. (1994) Natural selection at the class II major histocompatibility loci of mammals. *Philosophical Transactions of the Royal Society of London B* **345**, 359–367.

Jarvi, S.I. & Briles, W.E. (1992) Identification of the major histocompatibility complex in the ring-necked pheasant, *Phasianus colchicus* *Animal Genetics* **23**, 211–220.

Jarvi, S.I., Gee, G.F., Miller, M.M. & Briles, W.E. (1992) A complex alloantigen system in Florida Sandhill Cranes, *Grus canadensis pratensis*: evidence for the major histocompatibility complex (B) system. *Journal of Heredity* **86**, 348–353.

Jin, L. & Nei, M. (1990) Limitations of the evolutionary parsimony method of phylogenetic analysis. *Molecular Biology and Evolution* **7**, 82–102.

Jin, L., Macaubas, C., Hallmayer, J., Kimura, A. & Mignot, E. (1996) Tracking the evolutionary history of a microsatellite locus located in the HLA-DQA1/DQB1 class II region: a phylogenetic approach. In: *Current Topics on Molecular Evolution* (eds M. Nei & N. Takahata), pp. 79–87. Institute of Molecular Evolutionary Genetics. The Pennsylvania State University Press, University Park, Pennsylvania.

Kandil, E., Namikawa, C., Nonaka, M., Greenberg, A., Flajnik, M.F., Ishibashi, T. & Kasahara, M. (1996) Isolation of low molecular mass polypeptide complementary DNA clones from primitive vertebrates. *Journal of Immunology* **156**, 4245–4253.

Kasahara, M., Vasquez, Z., Sato, K., McKinney, E.C. & Flajnik, M.F. (1992) Evolution of the major histocompatibility complex: isolation of class II A cDNA clones from the cartilaginous fish. *Proceedings of the National Academy of Sciences USA* **89**, 6688–6692.

Kaufman, J. (1995) A 'minimal essential Mhc' and an 'unrecognized' Mhc: two extremes in selection for polymorphism. *Immunological Reviews* **143**, 63–88.

Kaufman, J. & Salomonsen, J. (1997) The 'Minimal Essential MHC' revisited: both peptide-binding and cell surface expression level of MHC molecules are polymorphisms selected by pathogens in chickens. *Hereditas* **127**, 67–73.

Kaufman, J., Skjødt, K., Salomonsen, J., Simonsen, M., Du Pasquier, L., Parisot, R. &

Riegert, P. (1990) MHC-like molecules in some nonmammalian vertebrates can be detected by some cross-reactive xenoantisera. *Journal of Immunology* **144**, 2258–2272.

Kaufman, J., Salomonsen, J., Riegert, P. & Skjødt, K. (1991) Using chicken class I sequences to understand how xenoantibodies crossreact with MHC-like molecules in nonmammalian vertebrates. *American Zoologist* **31**, 570–579.

Kaufman, J., Salomonsen, J. & Flajnik, M. (1994) Evolutionary conservation of MHC class I and class II molecules — different yet the same. *Seminars in Immunology* **6**, 411–424.

Kaufman, J., Milne, S., Gobel, T.W.F. *et al.* (1999) The chicken B locus is a minimal – essential major histocompatibility complex. *Nature* **401**, 923–925.

Klein, D., Ono, H., O'hUigin, C., Vincek, V., Goldschmidt, T. & Klein, J. (1993) Extensive Mhc variability in cichlid fishes of Lake Malawi. *Nature* **364**, 330–334.

Klein, J. & O'hUigin, C. (1993) Composite origin of major histocompatibility complex genes. *Current Opinions in Genetics and Development* **3**, 923–930.

Klein, J., Ono, H., Klein, D. & O'hUigin, C. (1994) The accordion model of Mhc evolution. In: *Progress in Immunology*, vol. VIII (eds J. Gergely & G. Petraryi), pp. 137–143. Springer-Verlag, Heidelberg.

Klein, J., Figueroa, F., Klein, D., Sato, A. & O'hUigin, C. (1997) Major histocompatibility complex genes in the study of fish phylogeny. In: *Molecular Systematics of Fishes* (eds T.D. Kocher & C.A. Stepien), pp. 271–283. Academic Press, San Diego.

Kobari, F., Sato, K., Shum, B.P. *et al.* (1995) Exon-intron organization of Xenopus MHC class II beta chain genes. *Immunogenetics* **42**, 376–385.

Kroemer, G., Bernot, A., Behar, G. *et al.* (1990) Molecular genetics of the chicken MHC: current status and evolutionary aspects. *Immunological Reviews* **113**, 119–145.

Kumar, S., Tamura, K. & Nei, M. (1993) MEGA: *Molecular Evolutionary Genetic Analysis*, v. 1.01. The Pennsylvania State University, University Park, Pennsylvania.

L'abbe, D., Belmaaza, A., Decary, F. & Chartrand, P. (1992) Elimination of heteroduplex artifacts when sequencing HLA genes amplified by polymerase chain reaction (PCR). *Immunogenetics* **35**, 395–397.

Lee, M., Lynch, E.D. & King, M.-C. (1998) SeqHelp: a program to analyze molecular sequences utilizing common computational resources. *Genome Research* **8**, 306–312.

Lenz, L.L. & Bevan, M.J. (1997) CTL responses to H2-M3-restricted Listeria epitopes. *Immunology Reviews* **158**, 115–121.

Li, W.-H. (1997) *Molecular Evolution*. Sinaur Associates, Sunderland, Massachusetts.

Lindahl, K.F., Byers, D.E., Dabhi, V.M. *et al.* (1997) H2–M3, a full-service class Ib histocompatibility antigen. *Annual Review of Immunology* **15**, 851–879.

Longmire, J.L., Brown, N.C., Meinke, L.J. *et al.* (1993) Construction and characterization of partial digest DNA libraries from flow-sorted human chromosome 16. *Genetic Analysis, Techniques and Applications* **10**, 69–76.

McAdam, S.N., Boyson, J.E., Liu, X., Garber, T.L., Hughes, A.L., Bontrop, R.E. & Watkins, D.I. (1994) A uniquely high level of recombination at the HLA-B locus. *Proceedings of the National Academy of Sciences USA* **91**, 5893–5897.

McConnell, T.J., Godwin, U.B., Norton, S.F., Nairn, R.S., Kazianis, S. & Morizot, D.C. (1998) Identification and mapping of two divergent, unlinked major histocompatibility complex class II B genes in *Xiphophorus* fishes. *Genetics* **149**, 1921–1934.

Málaga-Trillo, E., Zaleska-Rutczynska, Z., McAndrew, B., Vincek, V., Figueroa, F.S., Sültmann, H. & Klein, J. (1998) Linkage relationships and haplotype polymorphism among cichlid Mhc class II B loci. *Genetics* **149**, 1527–1537.

Manning, C.J., Wakeland, E.K. & Potts, W.K. (1992) Communal nesting patterns in mice implicate MHC genes in kin recognition. *Nature* **360**, 581–583.

Meager, S. & Potts, W.K. (1997) A microsatellite-based MHC genotyping system for house mice (*Mus domesticus*). *Hereditas* **127**, 75–82.

MHC Sequencing Consortium (1999) Complete sequence and gene map of a human major histocompatibility complex. *Nature* **401**, 921–923.

Miller, K.M. & Withler, R.E. (1997) Mhc diversity in Pacific salmon: population structure and trans-species allelism. *Hereditas* **127**, 83–95.

Miller, K.M., Withler, R.E. & Beacham, T.D. (1997) Molecular evolution at Mhc genes in two populations of chinook salmon *Oncorhynchus tshawytscha*. *Molecular Ecology* **6**, 937–954.

Milne, S., Thorpe, K., Theaker, A., Kaufman, J., Trowsdale, J. & Beck, S. (1997) Chance of necessity: the evolution of the MHC gene cluster (Abstract). *Hereditas* **127**, 164.

Mizuki, N., Ando, H., Kimura, M. *et al.* (1997) Nucleotide sequence analysis of the HLA class I region spanning the 237-kb segment around the HLA-B and -C genes. *Genomics* **42**, 55–66.

Murray, B.W. & White, B.N. (1998) Sequence variation at the *Mhc DRB* loci in beluga (*Delphinapterus leucas*) and narwhal (*Monodon monoceros*). *Immunogenetics* **48**, 242–252.

Murray, B.W., Malik, S. & White, B.N. (1995) Sequence variation at the major histocompatibility complex locus DQb in Beluga whales. *Molecular Biology and Evolution* **12**, 582–593.

Nei, M. & Gojobori, T. (1986) Simple methods for estimating the numbers of synonymous and nonsynonymous nucleotide substitutions. *Molecular Biology and Evolution* **3**, 418–426.

Nei, M. & Hughes, A.L. (1992) Balanced polymorphism and evolution by the birth-and-death process in the MHC loci. In: *Proceedings of the 11th Histocompatibility Workshop and Conference*, vol. 2 (eds K. Tsuji, M. Aizawa & T. Sasazuki), pp. 27–38. Oxford University Press, Oxford.

Nei, M., Gu, X. & Sitnikova, T. (1997) Evolution by the birth-and-death process in multigene families of the vertebrate immune system. *Proceedings of the National Academy of Sciences USA* **94**, 7799–7806.

Nielsen, R. & Yang, Z. (1998) Likelihood models for detecting positively selected amino acid sites and applications to the HIV-1 envelope gene. *Genetics* **148**, 929–936.

Okamura, K., Ototake, M., Nakanishi, T., Kurosawa, Y. & Hashimoto, K. (1997) The most primitive vertebrates with jaws possess highly polymorphic MHC class I genes comparable to those of humans. *Immunity* **7**, 777–790.

Ono, H., Klein, D., Vincek, V. *et al.* (1992) Major histocompatibility class II genes of zebrafish. *Proceedings of the National Academy of Sciences USA* **89**, 11886–11890.

Ono, H., O'hUigin, C., Tichy, H. & Klein, J. (1993) Major histocompatibility complex variation in two species of cichlid fishes from Lake Malawi. *Molecular Biology and Evolution* **10**, 1060–1072.

Parham, P. & Ohta, T. (1996) Population biology of antigen presentation by MHC class I molecules. *Science* **272**, 67–74.

Paterson, S. (1998) Evidence for balancing selection at the major histocompatibility complex in a free-living ruminant. *Journal of Heredity* **89**, 289–294.

Paterson, S., Wilson, K.E. & Pemberton, J.M. (1998) Major histocompatibility complex variation associated with juvenile survival and parasite resistance in a large unmanaged ungulate population (*ovis aries L.*), *Proceedings of the National Academy of Sciences USA* **95**, 3714–3719.

Potts, W.P. (1996) PCR-based cloning across large taxonomic distances and polymorphism detection: MHC as a case study. In: *Molecular Zoology: Advances, Strategies and Protocols* (eds J.D. Ferraris & S.R. Palumbi), pp. 181–194. Wiley-Liss, New York.

Potts, W.K., Manning, C.J. & Wakeland, E.K. (1991) Mating patterns in seminatural populations of mice influenced by MHC genotype. *Nature* **352**, 619–621.

Primmer, C.R., Raudsepp, T., Chowdhary, B.P.M., Møller, A.P. & Ellegren, H. (1997) Low frequency of microsatellites in the avian genome. *Genome Research* **7**, 471–482.

Puel, A., Groot, P.C. Lathrop, M.G., Demant, P.E. & Mouton, D. (1998) Mapping of genes controlling quantitative antibody production. *Immunogenetics* **47**, 326–331.

Radtkey, R.R., Becker, B., Miller, R.D., Riblet, R. & Case, T.J. (1996) Variation and evolution of class I Mhc in sexual and parthenogenetic geckos. *Proceedings of the Royal Society of London B* **263**, 1023–1032.

Rowen, L., Koop, B. & Hood, L. (1996) The complete 685-kilobase DNA sequence of the human b T cell receptor locus. *Science* **272**, 1755–1762.

Ruvulo, M. (1996) A new approach to studying modern human origins: hypothesis testing with coalescence time distributions. *Molecular Phylogenetics and Evolution* **5**, 202–219.

Sammut, B., Laurens, V. & Tournefier, A. (1997) Isolation of Mhc class I cDNAs from the axolotl *Ambystoma mexicanum*. *Immunogenetics* **45**, 285–294.

Sato, A., Figueroa, F., O'hUigin, C., Reznick, D.N. & Klein, J. (1996) Identification of major histocompatibility complex genes in the guppy, *Poecilia reticulata*. *Immunogenetics* **43**, 38–39.

Sato, K., Flajnik, M.F., Du Pasquier, L., Katagiri, M. & Kasahara, M. (1993) Evolution of the MHC: isolation of class II b chain cDNA clones from the amphibian *Xenopus laevis*. *Journal of Immunology* **150**, 2831–2843.

Satta, Y., Kupfermann, H., Li, Y.-J. & Takahata, N. (1999) Molecular clock and recombination in primate MHC genes. *Immunological Reviews* **167**, 357–379.

Schneider, S., Vincek, V., Tichy, H., Figueroa, F. & Klein, J. (1991) MHC class II genes of a marsupial, the red-necked wallaby (*Macropus rufogriseus*): identification of new gene families. *Molecular Biology and Evolution* **8**, 753–766.

Schonbach, C., Vincek, V., Mayer, W.E., Golubic, M., O'hUigin, C. & Klein, J. (1994) Multiplication of Mhc-DRB5 loci in the orangutan for the evolution of DRB haplotypes. *Mammalian Genome* **5**, 405–415.

Schwaiger, F.-W., Weyers, E., Epplen, C., Brün, J., Ruff, G., Crawford, A. & Epplen, J.T. (1993) The paradox of MHC-DRB exon/intron evolution: α-helix and β-sheet encoding regions diverge while hypervariable intronic simple repeats coevolve with b-sheet codons. *Journal of Molecular Evolution* **37**, 260–272.

Seddon, J.M. & Baverstock, P.R. (2000) Evolutionary lineages of *RT1.Ba* in the Australian *Rattus*. *Molecular Biology and Evolution* **17**, 768–772.

She, J.-X., Boehme, S., Wang, T.W., Bonhomme, F. & Wakeland, E.K. (1990) The generation of MHC class II gene polymorphism in the genus Mus. *Biological Journal of the Linnaean Society* **41**, 141–161.

She, J.-X., Boehm, S., Wang, T.W., Bonhomme, F. & Wakeland, E.K. (1991) Amplification of MHC class II gene polymorphism by intra-exonic recombination. *Proceedings of the National Academy of Sciences USA* **88**, 453–457.

Sheldon, B.C. & Verhulst, S. (1996) Ecological immunology: Costly parasite defences and trade-offs in evolutionary ecology. *Trends in Ecology and Evolution* **11**, 317–321.

Shum, B.P., Avila, D., Du Pasquier, L., Kasahara, M. & Flajnik, M.F. (1993) Isolation of a classical MHC class I cDNA from an amphibian: evidence for only one class I locus in the Xenopus MHC. *Journal of Immunology* **151**, 5376–5386.

Slade, R.W. (1992) Limited MHC polymorphism in the southern elephant seal: implications for MHC evolution and marine mammal population biology. *Proceedings of the Royal Society of London B* **249**, 163–171.

Sültmann, H., Mayer, W.E., Figueroa, F., O'hUigin, C. & Klein, J. (1993) Zebrafish Mhc class II alpha chain-encoding genes: polymorphism, expression, and function. *Immunogenetics* **38**, 408–420.

Takahata, N. (1990) A simple genealogical structure of strongly balanced allelic lines and trans-specific evolution of polymorphism. *Proceedings of the National Academy of Sciences USA* **87**, 2419–2423.

Takahata, N. (1993) Allelic genealogy and human evolution. *Molecular Biology and Evolution* **10**, 2–22.

Takahata, N. & Nei, M. (1990) Allelic genealogy under overdominant and frequency-dependent selection and polymorphism of major histocompatibility complex loci. *Genetics* **124**, 967–978.

Takahata, N. & Satta, Y. (1997) Evolution of the primate lineage leading to modern humans: phylogenetic and demographic inferences from DNA sequences. *Proceedings of the National Academy of Sciences USA* **94**, 4811–4815.

Takahata, N. & Satta, Y. (1998) Footprints of intergenic recombination on silent nucleotide diversity at HLA loci. *Immunogenetics* **47**, 430–441.

Taylor, G.R. (1997) *Laboratory Methods for the Detection of Mutations and Polymorphisms in DNA*. CRC Press, New York.

Thursz, M.R., Thomas, H.C., Greenwood, B.M. & Hill, A.V.S. (1997) Heterozygote advantage for HLA class-II type in hepatitis B virus infection. *Nature Genetics* **17**, 11–12.

Tiersch, T.R. & Wachtel, S.S. (1991) On the evolution of genome size of birds. *Journal of Heredity* **82**, 363–368.

Troutt, A.B., McHeyzer-Williams, M.G., Pulendran, B. & Nossal, G.J.V. (1992) Ligation-anchored PCR: a simple amplification technique with single-sided specificity. *Proceedings of the National Academy of Sciences USA* **89**, 9823–9825.

Trowsdale, J. (1995) 'Both bird and man and beast': comparative organization of MHC genes. *Immunogenetics* **41**, 1–17.

van Erp, S.H.M., Dixon, B., Figueroa, F., Egberts, E. & Stet, R.J.M. (1996) Identification and characterization of a new major histocompatibility complex class I gene in carp (*Cyprinus carpio* L.). *Immunogenetics* **44**, 49–61.

Vincek, V., Klein, D., Graser, R.T., Figueroa, F., O'hUigin, C. & Klein, J. (1995) Molecular cloning of major histocompatibility complex class II B gene cDNA from the Bengalese finch, *Lonchura striata*. *Immunogenetics* **42**, 262–267.

Vincek, V., O'hUigin, C., Satta, Y., Takahata, N., Boag, P.T., Grant, P.R., Grant, B.R. & Klein, J. (1997) How large was the founding population of Darwin's finches? *Proceedings of the Royal Society of London B* **264**, 111–118.

von Schantz, T., Wittzell, H., Goransson, G., Grahn, M. & Persson, K. (1996) MHC genotype and male ornamentation: genetic evidence for the Hamilton-Zuk model. *Proceedings of the Royal Society of London B* **263**, 265–271.

Watkins, D.I., Chen, Z.W., Hughes, A.L., Evans, M.G., Tedder, T.F. & Letvin, N.L. (1990) Evolution of the MHC class I genes of a New World primate from ancestral homologues of human non-classical genes. *Nature* **346**, 60–63.

Watkins, D.I., McAdam, S.N., Liu, X. *et al.* (1992) New recombinant HLA-B alleles in a tribe of South American Amerindians indicate rapid evolution of MHC class I loci. *Nature* **357**, 329–333.

Westerdahl, H., Wittzell, H. & von Schantz, T. (1998) Polymorphism and transcription of *Mhc* class I genes in a passerine bird, the great reed warbler. *Immunogenetics* **49**, 158–170.

Wilson, R.K., Koop, B.F., Chen, C., Halloran, N., Sciammis, R. & Hood, L. (1992) Nucleotide sequence analysis of 95 kb near the 3′ end of the murine T-cell receptor a/d chain locus: strategy and methodology. *Genomics* **13**, 1198–1208.

Wittzell, H., von Schantz, T., Zoorob, R. & Auffrey, C. (1994) Molecular characterization of three Mhc class II B haplotypes in the ring-necked pheasant. *Immunogenetics* **39**, 395–403.

Wittzell, H., Bernot, A., Auffrey, C. & Zoorob, R. (1999) Concerted evolution of two Mhc class II B loci in pheasants and domestic chickens. *Molecular Biology and Evolution* **16**, 479–490.

Wittzell, H., Madsen, T., Westendahl, H., Shine, R. & von Schantz, T. (1999) MHC variation in birds and reptiles. *Genetica* **104**, 307–309.

Woolfenden, G.E. & Fitzpatrick, J.W. (1984) *The Florida Scrub Jay: Demography of a Cooperative-breeding Bird*. Princeton University Press, Princeton, New Jersey.

Yuhki, N. & O'Brien, S.J. (1990) DNA variation of the mammalian major histocompatibility complex reflects genomic diversity and population history. *Proceedings of the National Academy of Sciences USA* **87**, 836–840.

Yuhki, N. & O'Brien, S.J. (1994) Exchanges of short polymorphic DNA segments predating speciation in feline major histocompatibility complex class I genes. *Journal of Molecular Evolution* **39**, 22–33.

Zoorob, R., Bernot, A., Renoir, D.M., Choukri, F. & Auffrey, C. (1993) Chicken major histocompatibility complex class II B genes: analysis of interallelic and interlocus sequence variance. *European Journal of Immunology* **23**, 1139–1145.

# CHAPTER 9

# DNA-Fragment Markers in Plants

## CAROL RITLAND AND KERMIT RITLAND

### 9.1 Molecular markers in plants

The use of molecular markers in plant population biology and evolution has a rich history (Cruzan 1998). Perhaps its impetus dates from Mendel's use of conspicuous morphological mutants in peas to demonstrate principles of inheritance. During the 1950s and 1960s, the U.C. Davis group of R.W. Allard, S.K. Jain, C. Rick and colleagues developed a paradigm for plant research in their use of simply inherited morphological markers in crop plants for estimating population genetic parameters. Their work focused on inbreeding short-lived crop plants such as tomato (*Lycopersicon*), bean (*Phaseolus*) and barley (*Hordeum*), from which morphological markers were easy to identify, and for which large-scale experiments were feasible (Jain & Allard 1960). This tradition of marker-based estimation of genetic parameters further flourished with the introduction of isozymes, fostering studies involving estimation of genetic diversity (Brown & Weir 1983), mating systems (Ritland 1983), local population structure (Adams 1983) and even gene mapping (Bernatsky & Tanksley 1986).

With the advent of DNA techniques, more loci of greater polymorphism were revealed and used in many applications. The polymerase chain reaction (PCR; Saiki *et al.* 1985; Mullis & Faloona 1987) revolutionized marker studies, providing entire classes of new markers. Workers from many disciplines of plant biology, ranging from ecology to breeding to evolution, have integrated the use of these markers in their research. In addition, the tradition of genetic data analysis with these markers has continued (Cruzan 1998).

This chapter focuses on the assay and analysis of a class of markers based upon the generation of DNA fragments, with an emphasis on those more popularly used in plants. Despite the commonality of DNA, and hence marker types, across organisms, plants can differ from animals in the usage of a particular marker; for example, a survey of Biological Abstracts reveals that 'random amplified polymorphic DNA' (RAPD, see Section 9.2.1) is 40 times more likely to appear in an abstract with the word 'plant' as opposed to 'animal', whereas 'microsatellite' is half as likely to appear with 'plant' as opposed to an 'animal'. This difference is probably a result of the great use of RAPDs in plant genome maps, as controlled crosses with large progenies are much easier with plants.

Genomic composition also can differ between plants and animals, giving rise to a preference for a marker, or even unique types of markers; for example, the organelle chloroplast genome is unique to plants and it exhibits uniparental inheritance and no evident recombination. Sequencing of this extranuclear genome has been completed for several species including corn (*zea maize*), rice (*Oryza sativa*), black pine (*Pinus thunbergii*), liver wort (*Marchantia*) and tobacco (*Nicotiana tabacum*). Genome sequences have been used to identify chloroplastic microsatellites (Powell *et al.* 1995a,b). This chapter pays special attention to situations like this where plants provide opportunities (or problems).

## 9.2 Randomly amplified polymorphic DNA

### 9.2.1 Basic methodology

Randomly amplified polymorphic DNA (RAPD) consist of fragments generated via the PCR using a randomly selected 10-base primer (Williams *et al.* 1990, 1993). The word 'random' refers to the fact that primers are chosen at random, without prior knowledge of any specific primer sites in the genome. Hence the step of cloning and identification of specific sequences is skipped. In addition, primers can be inexpensively obtained by ordering 'primer kits' from either Operon technologies (Almeda, California) or from the University of British Columbia Biotechnology lab (N.A.P.S. Unit, attn. John Hobbs).

When two complementary primer sites are within approximately 2 kb of each other, the RAPD primer will generate double-stranded DNA products equal to the length of this intervening region. The DNA fragments that represent the RAPD loci are separated via electrophoresis on modified (approx. 4%) agarose gels (Synergel ™(Diversified Biotechnology, Boston, Massachusetts)/ agarose). Alternatively, low percentage (3–5%) polyacrylamide gels can be used which can provide better resolution (Ronning & Schnell 1995). Following separation, gels are stained with ethidium bromide (EtBr) and visualized under ultraviolet light. The scoring of these gels is usually by eye and the presence/absence of bands recorded. For a more detailed description of the technique see Newbury & Ford-Lloyd (1993).

Early research with random primers determined that the best length of the primer sequence is normally 10 bp (basepairs), resulting in 3–5 'good' or interpretable bands (loci) per primer, although several times that number of faint bands are often observed. The probability of finding proximate, complementary primer sites within a genome might seem to be small, but the enormous size of genomes, plus the abundance of duplicated DNA within genomes, makes their occurrence highly probable. However, the number and quality of bands varies widely among primers and the normal protocol for

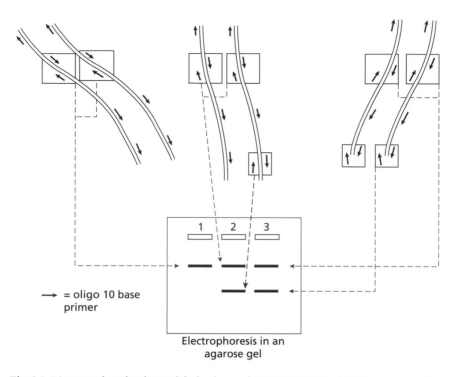

→ = oligo 10 base
primer

Electrophoresis in an
agarose gel

**Fig. 9.1** Diagram of randomly amplified polymorphic DNA (RAPD) which demonstrates the presences and absences of polymerase chain reaction (PCR) products. For extract 1 there is one PCR product. For extracts 2 and 3, there are two products; however, dominance is manifested for one of the products from extract 2 (see text).

RAPD studies is to screen several dozen primers before deciding upon the use of 5–10 primers.

Figure 9.1 illustrates the amplification of DNA and consequent representation of bands for a RAPD primer (actual gels appear somewhat like Fig. 9.3, but with fewer bands). In extract 1, both homologous chromosomal segments have a region where two complementary sites lie close together, resulting in a single band on the gel. In extracts 2 and 3, there is a second locus that amplifies a fragment of different length, discernible as a second band. However, in extract 2, just one of the two homologous alleles amplifies, yet a band of equal intensity is observed. This dominance of banded phenotype over the unbanded phenotype occurs because the PCR reaction is not limiting, e.g. the same amount of final product is obtained regardless of the number of initial copies (single versus double). In a few cases (3–5% of loci), polymorphism occurs by small insertions or deletions in the DNA lying between the primer sites, resulting in a codominant banding pattern like that found with isozymes. However, these loci are often difficult to detect because the alternative alleles usually overlap with the bands at other loci (as is typical for minisatellites).

Randomly amplified polymorphic DNAs are well known often to lack reproducibility (Levi *et al.* 1993; Perez *et al.* 1998). Any change of the PCR parameter such as annealing temperature or the amount of template could increase or decrease the number of bands. Also, because quite small amounts of DNA can be used as the template, contaminating foreign DNA can be a problem. These problems can be minimized by carefully controlling the amount and quality of DNA used for amplifications. However, even when conditions are carefully controlled, unpredictable patterns of inheritance are sometimes found, probably as a result of primer competition among segregating loci. This problem can be reduced by careful screening for 'good' primers (those that produce a few bright bands). Primers should also be tested for reliability by repeating the PCR conditions with the same DNA extract on two separate occasions, and even replicating assays among repeated extractions of the same tissue.

### 9.2.2  Applications in plants

At present, the RAPD technique is the most commonly used methodology for DNA marker studies in plants, although it is rapidly being superseded by amplified fragment-length polymorphisms (AFLPs, see Section 9.3.1). With just a minimum investment in equipment [mainly a PCR machine, an agarose gel electrophoresis system, and an ultraviolet (UV) viewer for gels], one can obtain many loci for the estimation of genetic diversity (Bai *et al.* 1997), taxonomic identification (Transue *et al.* 1994) and individual identification (Miller *et al.* 1996). Randomly amplified polymorphic DNAs serve as fingerprints for rice cultivars (Mackill 1995), clones of Sitka Spruce (*Picea sitchensis*) (Van de Ven & McNicol 1995) and the American Elm (*Ulmus americana*; Kamalay & Carey 1996). Specific chromosomes can be targeted; in corn, B chromosome markers were obtained (Gourmet & Rayburn 1996). Somatic cell hybrids can be identified, as in the potato (*Solanum tuberosum*; Baird *et al.* 1992). Randomly amplified polymorphic DNAs are also used in linkage mapping (Tulsieram *et al.* 1992), marker assisted selection, and quantitative trait locus (QTL) mapping (Deragon & Landry 1992). Markers linked to traits such as blister rust resistance (Devey *et al.* 1995) and terpenes (Plomion *et al.* 1996) have been found. Finally, many studies have characterized population structure with RAPDs; for example, marked structure was found among populations of the Tasmanian blue gum Eucalyptus (*Eucalyptus globulus*) (Nesbitt *et al.* 1995), and buffalograss (*Buchloe dactyloides*) was shown to have little variation within populations but large divergence among populations (Huff *et al.* 1993).

One positive feature of the RAPD technique is the small amount of starting DNA required (10 ng per reaction). This allows study of problems that many other techniques cannot, for example, detecting polymorphisms between cell layers of tissue of hybrid *Chrysanthemum* plants (Wolff 1996), and nonintrusive sampling of endangered plants (Stewart & Porter 1995).

A unique feature of conifers is their possession of the 'megagametophyte' in seeds. This tissue is mitotically derived from the maternal gamete contribution to the zygote, and is haploid. This makes dominance a moot issue, and its most significant use with RAPDs to date has been the quick construction of linkage maps using the seed progeny of a single tree (e.g. Tulsieram *et al.* 1992).

### 9.2.3  Homology of bands and interspecific comparisons

While RAPDs have many applications at the population and intraspecific levels, in systematics, RAPDs can only be used for very close comparisons, and cautiously so, because of the increased chance of nonhomology of comigrating fragments (Bowditch *et al.* 1993). There are three alternative ways to verify homology of fragments, involving:
1   hybridization of cloned fragments;
2   restriction fragment-length polymorphism (RFLP) digests of fragments scooped from gels; and
3   comparative genetic mapping of fragments.
Using hybridization, Thormann *et al.* (1994) found that some fragments, scored as identical, were not homologous at the interspecific level. Hurme & Savolainen (1999) found, using the restriction pattern test, that two of 33 loci in a segregating $F_2$ were not homologous to the corresponding fragment in the $F_1$ parents, suggesting that it might be necessary to verify the homology of comigrating fragments before using them even across individuals, but this is generally unlikely to be a significant problem.

Rieseberg (1996), using comparisons of linkage maps in sunflowers, found that only 174 of 220 pairs of comigrating fragments were useful, in that their map positions were consistent among crosses. He noted that, generally, mistaken homology of bands will introduce noise into RAPD data sets. However, if errors in homology of comigrating DNA bands are random, this will have little effect on the relative similarities and on ordination results. An indirect test for homology involves comparing the results of RAPD analyses with other marker analyses; for example, Adams & Rieseberg (1998) found that while RAPD markers were very similar to RFLP markers for estimating intraspecific genetic relationships, at the interspecific level, the two marker types gave different results. Clearly, RAPDs are most suitable for population-level analyses and for elucidating relationships among sister or cryptic species.

### 9.2.4 Variants of the RAPD technique

Sometimes particular bands (loci) are fixed between clones, populations or species (primer sets are often screened for such loci), and can serve as 'diagnostic' loci in future assays. These loci can then be converted into more reliable markers known as SCARs (sequence characterized amplified regions; Bodenes *et al.* 1997). To obtain a SCAR, the RAPD band is excised from the agarose gel, then cloned and sequenced for the first 50–100 bases of both ends of the fragment. Longer, specific primers are then designed within the original RAPD sequence (note the original polymorphism must lie within the primer sequence) resulting in a primer pair that amplifies only that specific locus.

A recently devised method that gives simpler, more reliable bands, but requires blotting and probing, is the random amplified microsatellite polymorphism (RAMPO; Ramser *et al.* 1997). In this technique, loci are first amplified with one 'classic' 10-base oligonucleotide RAPD primer and a second, microsatellite-like anchored oligonucleotide microsatellite-primed PCR (MP-PCR). Next, products are electrophoresed on modified agarose or polyacrylamide gels, and then blotted and probed with oligonucleotide probes such as $(CA)_8$ or $(GA)_8$. The loci detected are much like true microsatellites, having many codominant alleles but also being quite species-specific (Ramser *et al.* 1997), but the polymorphism is not often in the simple sequence repeat (SSR) region.

## 9.3 Amplified fragment-length polymorphism

### 9.3.1 Methodology

The amplified fragment length polymorphism (AFLP) technique, introduced by Vos *et al.* (1995), involves restriction digestion of genomic DNA, then PCR amplification of a subset of these fragments. It shares many of the characteristics of RAPDs, including dominance and the appearance of many loci on one gel, but the banding patterns are more reliable, and many more bands per gel are scorable. However, DNA fragments generated by this technique differ by as little as a single base, requiring use of vertical acrylamide gels or automatic fragment analysers. Vos *et al.* (1995) describe this method in detail, and below we summarize their procedure and describe some modifications introduced by Remington (Forest Biotechnology Group, NCSU) and C. Ritland and C. Goodwillie (Genetic Data Center, UBC).

The assay for AFLPs is illustrated schematically in Fig. 9.2. The first step of this method is to digest 300–1000 ng of genomic DNA with a pair of restriction enzymes, usually *Eco*RI and *Mse*I (note this is 50–100 times more DNA than needed for RAPDs), to produce fragments with 'sticky ends' (see fragments

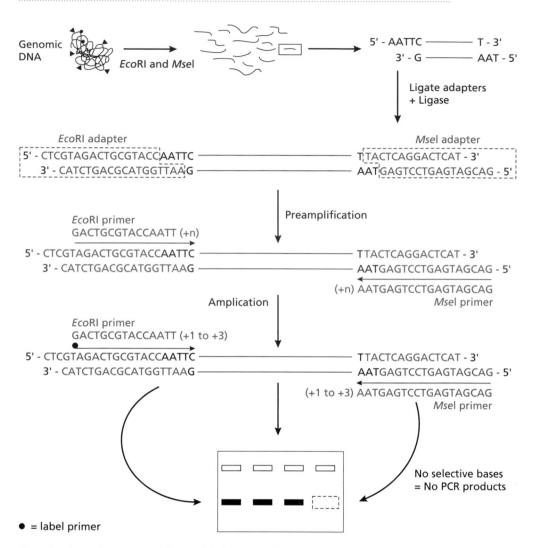

**Fig. 9.2** Schematic diagram of the amplified fragment-length polymorphism (AFLP) method (see text).

produced in upper right of Fig. 9.2). The first enzyme is a frequent cutter (usually a 4-cutter, e.g. an enzyme recognizing a sequence of 4 bp) and the other a less frequent cutter (usually a 6-cutter). Prior to the following step, the digests can be stored for a few days at 4 °C or for longer at −20 °C.

The second step is to ligate 'adaptors' to the sticky ends. These adaptors will serve as primer binding sites. A specific adaptor is attached to the 5′ end, and a second specific adaptor is attached to the 3′ end. This results in the structure following the 'ligate adaptors' stage in Fig. 9.2.

The third step is to amplify the DNA fragments using primer pairs that complement the adaptor sequences plus one to three randomly chosen nucleotides ('selective nucleotides'), added to the 3' end (beyond the restriction enzyme site). In Fig. 9.2, this is referred to as the 'amplification' step where the primer has a +1 to +3 addition of random nucleotides (we have also inserted a prior amplification step, a 'preamplification' where the addition is +$n$, to deal with large genomes, see below).

In the amplification step, as an example the primer which attaches to the *EcoRI* end may have 'AC' added, while the primer attaching to the *MseI* end may have a 'C' added. This primer pair is termed '*EcoRI*+AC/*MseI*+C', and is one example of a '+2/+1' primer combination. For every nucleotide added, the number of fragments ultimately assayed is reduced by approximately 25%, e.g. a +2/+2 combination amplifies about 1 out of 256 fragments. The number of selected nucleotides is changed until the optimal number of bands is obtained (roughly 50–100 bands or loci should appear on a 400-bp sequence gel, so that fragments of the same size, e.g. allele overlap among loci, is unlikely). Note that there are many possible selective combinations; for example, there are 256 possible +2/+2 primer combinations, giving the opportunity to detect large numbers of loci.

In Fig. 9.2, we have inserted an additional step prior to the main amplification, termed 'preamplification', which is needed for the large genomes found in many plant species, such as conifers, where even +3/+3 primer combinations give too many bands. With large genomes, one might think that one can limit the pool of selectively amplified DNA by merely increasing the number of selective nucleotides. However, Vos *et al.* (1995) found that primers with four or more added nucleotides actually suffered a loss of selectivity. As a means to select subsets of fragments beyond this limit (and to also increase template concentrations), Remington *et al.* (1999) devised an additional step prior to the main amplification, termed the 'preamplification' step (Fig. 9.2) This is merely an amplification like the above, but with somewhat shorter primer combinations, usually, +1/+1 or +2/+2 combinations.

The amplified fragments are then run out and visualized on acrylamide gels. The same methods used for visualizing microsatellite primer products are used, such as radioactive labelling, fluorescent labelling, or silver staining. In our own laboratory, we use 'tailed' infrared labels (see section 9.5.2), modified for AFLP genetic markers (details available from C. Ritland).

Figure 9.3 is a portion of an AFLP gel involving crosses between the plant species *Linanthus jepsonii* and *L. bicolor*. The first 16 lanes (in columns) are a segregating backcross of *L. jepsonii*×*L. bicolor* with *L. jepsonii*, and the next 16 lanes are F$_2$ progeny of *L. jepsonii*×*L. bicolor*. Six easily scored segregating loci in either the backcross or F$_2$ are indicated by arrows; at least three other loci are monomorphic (the actual gel is about four times longer, so approximately 30 loci can be scored with this primer combination). Generally, compared to RAPDs, the number of scorable loci per gel is several times greater with AFLPs (≈20–40 for AFLPs vs. 4–10 for RAPDs).

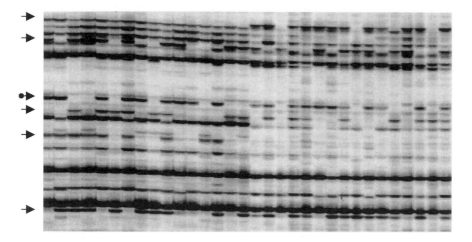

**Fig. 9.3** A sample of an amplified fragment-length polymorphism (AFLP) gel from *Linanthus* (see text). (Picture provided by Dr Carol Goodwille.)

While AFLPs are much more reproducible than RAPDs, this gel illustrates that clearly scorable loci are still somewhat of a judgement call [this picture is somewhat misleading, as scorability of any particular band can be optimized by adjusting the intensity and contrast of the 16-bit tag image format (TIF) image produced by the automated sequencer]. Interestingly, the lane marked by the dotted arrow represents a possible case of codominance, as the backcross has only two intensities, while the $F_2$ seems to have three; it may be possible to resolve codominance through the estimation of image intensities, preferably through software that takes advantage of the wide bandwidth offered by 16-bit images.

Generally, compared to RAPDs, reproducibility of AFLPs is not a major concern, as banding patterns on AFLP gels are generally quite consistent. However, the AFLP technique needs better quality DNA because restriction digestion is involved. Incomplete restriction enzyme digestion (the first step in Fig. 9.2) can be a significant problem. To check for complete digestion, DNA fragments should be electrophoresed through an 0.8% agarose gel; a 'smear' indicates digested DNA while a 'blob' near the origin indicates undigested DNA.

### 9.3.2  Applications in plants

The relatively new technique of AFLP has been enthusiastically adopted by plant workers. Table 9.1 lists the current array of studies that have used this marker technique, along with the adaptors (restriction enzymes), number of

**Table 9.1** Recent uses of amplified fragment-length polymorphisms (AFLPs) in plants, with adaptors (restriction enzymes), number of primer combinations (NPC) and numbers of polymorphic loci out of the number of total loci (NPL/NTL) detected

| Species | Adaptors | NPC | NPL/NTL | Usage | Authors |
|---|---|---|---|---|---|
| *Astragalus cremophylas* | *Eco*RI/*Mse*I | 9 | 220/352 | Measurement of genetic diversity in an endangered plant | Travis *et al.* (1996) |
| *Cocos nucifera* (coconut) | *Eco*RI/*Mse*I | 8 | 198/322 | Management of germplasm, identify parents for crossing | Perera *et al.* (1998) |
| *Cucumis melo* (melon) | *Eco*RI/*Mse*I | 109 | 221/306 | Linkage mapping including RAPD and RFLP markers | Wang *et al.* (1997) |
| *Eucalyptus urophylla* | *Eco*RI/*Mse*I | 20 | 26 used | Estimation of outcrossing rate using both RAPDs and AFLPs | Gaiotto *et al.* (1997) |
| *Glycine max* (Soybean) | *Eco*RI/*Mse*I | 12 | 87% dom. 13% codom. | High density linkage map with RFLPs, RAPDs and AFLPs | Keim *et al.* (1997) |
| *G. max* | *Eco*RI/*Mse*I | 4 | 47 polym. | Tracking parental contributions, bands and genetic changes | VanToai *et al.* (1997) |
| *Hordeum vulgare* (barley) | *Eco*RI/*Mse*I | 6 | 297 scored | Estimation of diversity and comparisons with RAPDs, RFLPs | Russell *et al.* (1997) |
| *H. vulgare* | *Eco*RI/*Mse*I | 15 | 118/1047 | Linkage mapping | Becker *et al.* (1995) |
| *H. vulgare* | *Eco*RI/*Mse*I | 24 | 21/88 | Development of AFLP markers | Qi and Lindhout (1997) |
| *H. vulgare* | *Pst*I/*Mse*I | 12 | 204/268 | Ecogeography of salt tolerance | Pakniyat *et al.* (1997) |
| *Oryza sativa* (rice) | *Eco*RI/*Mse*I | 17 | 147/529 | Compared with RAPDs for cultivar identification, linkage mapping | Mackill *et al.* (1996) |
| *O. sativa* | *Pst*I/*Mse*I | 5 | 80/400 | Linkage map anchored with RFLPs | Zhu *et al.* (1998) |
| *O. sativa* | *Pst*I/*Mse*I | 27 | 202/1048 | Linkage map, QTL map | Nandi *et al.* (1997) |
| *Phaseolus vulgaris* (bean) | *Eco*RI/*Mse*I | 16 | 110/800 | Characterizing gene pools in wild bean accessions | Tohme *et al.* (1996) |
| *Solanum tuberosum* (potato) | *Eco*RI/*Mse*I | 12 | 733/1212 | Alignment of 5 linkage maps | Rouppe van der Voort *et al.* (1997) |
| *S. tuberosum* | *Hin*dIII/*Mse*I | 108 | 6400 screen | High resolution map on chromo. V for *Phytophthora* resistance | Meksem *et al.* (1995) |
| *Triticum monococcum* (wheat) | *Eco*RI/*Mse*I | 7 | 75/288 | DNA fingerprinting of the origin of domestication for Einkorn wheat | Heun *et al.* (1997) |
| *Zea mays* (corn) | *Eco*I/*Mse*I | 6 | 209/500 | Genetic divergence between inbred lines | Marsan *et al.* (1998) |
| *Arabidopsis thaliana* | *Eco*RI/*Mse*I | 112 | 2500 screen | Saturated mapping and chromosome landing | Cnops *et al.* (1996) |
| | *Sac*I/*Mse*I | 192 | 17 used | | |

primer combinations (NPC) and numbers of polymorphic loci out of the number of total loci (NPL/NTL) detected in each study. Apart from two trees (*Eucalyptus* and *Cocos nucifera* or coconut; Perera *et al.* 1998) and an endangered species (*Astragalus cremnophylax*), all studies have involved crop plants or *Arabidopsis* (Cnops *et al.* 1996). The number of primer combinations is usually a dozen or two, but in some cases hundreds are used, indicating the vast power of the technique in terms of assayable loci. The numbers of loci detected are usually in the hundreds, and approximately 1/3–2/3 of these loci are polymorphic, although in inbreeders such as rice the proportion of polymorphic loci is much less, as has been found with other markers.

Evidently AFLP markers are apparently the next wave of DNA markers, having already been used to characterize genetic diversity (Tohme *et al.* 1996; Travis *et al.* 1996; Russell *et al.* 1997; Zhu *et al.* 1998), estimate outcrossing rates (Gaiotto *et al.* 1997; Marsan *et al.* 1998), infer the origins of domestication as in wheat (Heun *et al.* 1997), construct genetic maps (Becker *et al.* 1995; Mackill *et al.* 1996; Keim *et al.* 1997; Qi & Lindhout 1997; Rouppe van der Voort *et al.* 1997; Waugh *et al.* 1997), estimate parental contribution to germplasm (VanToai *et al.* 1997), undertake dense linkage mapping of a particular genomic area (Meksem *et al.* 1995), and screen known pieces of a chromosome for location of a quantitative trait (Cnops *et al.* 1996; Nandi *et al.* 1997; Pakniyat *et al.* 1997).

## 9.4 Interpretation and analysis of RAPD and AFLP data

### 9.4.1 Interpretation

Until the advent of RAPDs, dominance of markers was of concern in studies that utilized morphological markers and/or blood groups, or for the occasional isozyme locus containing null alleles. Randomly amplified polymorphic DNA loci usually exhibit dominance, and this enhances the sampling variance of parameter estimates, and can also induce bias of estimates (Lynch & Milligan 1994). Ignoring dominance results in an overestimate of relatedness which gets worse for closer relationship, and an underestimate of diversity which gets worse with less true diversity.

Nevertheless the predominant mode of RAPD analysis has involved band-sharing analyses, using approaches originally developed primarily in ecology. The first step of such analyses is to compute a matrix of similarities among individuals (or accessions, or populations). A large number of similarity indices exist in the literature. Jackson *et al.* (1989) present and evaluate formula for eight of these (for ecological application); for RAPD data, the coefficients most often used include Jaccard's coefficient $[n_{11}/(n-n_{00})]$, Dice's coefficient $[2n_{11}/(n+n_{11}-n_{00})]$ and Sokal and Michener's simple matching coefficient $[(n_{11}+n_{00})/n]$, where $n_{11}$ is the number of loci sharing bands, $n_{00}$ the number of loci sharing nulls, and $n$ the total number of loci compared. However, unlike allele-based similarity measures such as Nei's genetic identity and distance (Nei 1972), there have been no attempts to justify the use of any particular similarity measure in terms of a parametric evolutionary model incorporating mutation, drift, or gene identity-by-descent.

The second step is to use cluster-analysis techniques (including several alternative phylogeny-reconstruction methods) or less often, ordination techniques such as principal-components analysis, to identify patterns in the data. These methods are applicable to any type of genetic data, and as such they are beyond the topic of this chapter.

### 9.4.2 Population structure and gene diversity

Taking dominance into account, Lynch & Milligan (1994) gave estimators based upon RAPD data for gene frequencies, heterozygosity, population subdivision and pairwise relatedness. These estimators are also valid for AFLP loci which exhibit dominance. The most fundamental estimator was that for gene frequency:

$$\hat{q} = \hat{x}^{1/2}\left\{1 - [\mathrm{Var}(\hat{x})/8\hat{x}^2]^{-1}\right.$$

where $\hat{x}$ is the proportion of unbanded individuals and $\mathrm{Var}(\hat{x}) = \hat{x}(1-\hat{x})/N$ for $N$ sampled individuals. Many population parameters, such as Nei's genetic distance and diversity statistics, can then be estimated with the $\hat{q}$. Generally, at least 20 individuals are needed to obtain an adequate estimate of $q$.

Milligan & McMurry (1993) and Lynch & Milligan (1994) cautioned that two to ten times as many individuals are required to obtain the same precision compared to codominant markers. For estimating population structure, about 22 loci were determined to be needed in *Medicago truncatula* based upon a resampling method (Bonnin *et al.* 1996), although over twice this number are normally required for good estimates.

Clark & Lanigan (1993) described a method for estimating nucleotide diversity and divergence on the basis of diploid samples assayed for RAPD loci. They noted that RAPD data can be treated as analogous to restriction fragment presence/absence, with some additional assumptions, namely that presence/absence of amplified product is a result of a single base substitution, fragment sizes are accurately assessed, populations are sampled at random, loci are in linkage disequilibrium, and the same primers are used among all individuals. Their equation 8, which is too complex to give here, is used to obtain an estimate of nucleotide diversity and divergence. However, these formulae do not apply to AFLP data, and at the current time, AFLP estimators still await development.

Given these problems with dominance of RAPDs and AFLPs, careful consideration must be given to the cost and benefits of alternative marker systems, particularly isozymes. Their codominance, ease of assay and low cost makes them most efficient for many applications, including estimation of selfing rates and gene diversity. However, relatively few loci can be detected with isozymes and sometimes entire genomes are nearly monomorphic, such as with western red cedar (*Thuja plicata*) (Copes 1981; Yeh 1988).

### 9.4.3  Mating systems

While dominant markers when assayed as population samples do not provide sufficient degrees of freedom to estimate inbreeding coefficients, when assayed in progeny arrays, dominant markers can be used to estimate outcrossing rates and inbreeding coefficients of parents (Ritland 1990). A progeny array (several individuals derived from a common parent) allows inference of parent genotype so that several arrays allow inference of parent inbreeding coefficients. In turn, there are enough degrees of freedom to estimate the percent of selfed progeny once parent genotype is inferred. This enables studies of inbreeding in genetically depauperate species such as western red cedar, where isozyme variation is negligible, but where variation can be found by screening several RAPD primers.

The high number of loci potentially assayed by the AFLP technique will be of great use in paternity and mating system studies; for example, recently, using three AFLP primer pairs that generated 125 polymorphic AFLP loci, and that despite the presence of dominance, Krauss (1999) found paternity unambiguously for 242 of 252 (96.0%) naturally pollinated progeny in a population of the plant *Persoonia mollis* (Proteaceae).

### 9.4.4  Linkage and QTL mapping

Dominance poses some problems with the construction of genetic maps and QTL mapping, mainly related to reduced statistical power. In backcrosses of heterozygotes to homozygous parents, a single RAPD marker locus is informative (shows segregation) in crosses to just the recessive parent, while a codominant marker is informative in crosses to either parent. For estimating linkage, both loci must be informative; this is one-quarter as likely with RAPDs as opposed to codominant markers. Mapping via intercrossing or selfing double heterozygotes is possible, but suffers from a similar reduction in power and from increased complexity of analysis. Williams *et al.* (1993) discuss some other applications of RAPDs to breeding.

In additionn, existing computer programs for QTL mapping (compiled at http://www.stat.wisc.edu/biosci/linkage.html) have generally neglected dominant markers, although a recent revision of the program 'QTL Cartographer', by Chris Basten, Bruce Weir and Zhao-Bang Zeng of the Department of Statistics, North Carolina State University, allows for dominance. However, if both additive and dominant effects of QTLs are the object of estimation, a single type of cross is not sufficient; a joint analysis, a simultaneous backcross/$F_2$ design, involving either one or both backcrosses, is required (J.-Z. Lin and K. Ritland, unpublished information).

Saturated marker maps can be used to study the pattern of introgression in both plant breeding and natural evolution. Using 106 mapped RFLP loci, Wang et al. (1995) documented that the genome of one species of cultivated cotton, Gossypium barbudense, consisted of about 10% of the other cultivated cotton species, G. hirsutum, and that, remarkably, 57.5% of the total introgression observed was accounted for by five specific chromosomal regions that span less than 10% of the genome.

Interestingly, saturated marker maps were used to examine introgression of chromosomal blocks in Helianthus anomalus, a wild sunflower species derived via hybridization, by Ungerer et al. (1998). In providing a new and general method for estimating the tempo of hybrid speciation and dating the origin of hybrid zones, they concluded from the size of blocks that the initial hybridization event occurred fewer than 60 generations ago.

## 9.5 Sequenced-tagged-site markers

'Sequenced-tagged-sites' or STS markers are those that reveal codominant polymorphisms in specifically targeted sequences. The target sequences are initially identified from libraries of either cDNA (messenger RNA transcripts) or genomic DNA (random pieces of the genome). This class of markers includes the earliest true molecular markers used for population level studies — RFLPs as revealed by hybridization of probes to DNA cut with restriction enzymes (e.g. Quinn & White 1987) — also termed 'anonymous single-copy sequences', as the probe is of unknown sequence. If correctly used, RFLP analysis of low-copy-number anonymous nuclear loci can be a powerful tool for intraspecific systematics, and can be used for virtually any application discussed in this chapter. Restriction fragment-length polymorphisms have seen considerable use in crop plants (and their relatives) to produce linkage maps in species such as Zea maize (corn) and tomato (Bernatsky & Tanksley 1986; Helentjaris et al. 1986; Landry et al. 1987). However, the RFLP technique is relatively laborious to develop and implement compared with the newer PCR-based methods, which we now discuss.

### 9.5.1 Microsatellites

The first hypervariable DNA markers, minisatellites (VNTRs or variable number of tandem repeats) which consist of core sequences from 30 to 150 bases repeated throughout the genome (Jeffreys *et al.* 1985) created excitement among plant workers. However, after more than a decade, their application in plants has been found to be quite limited, mainly involving testing the Jeffrey's probes in some plant species (e.g. Rogstad 1993; Sharon *et al.* 1995). They have seen some use; the champion of minisatellite fingerprinting of plants, Nybom (e.g. Nybom 1996), has found this technique of greatest utility in detecting apomixis (asexual reproduction by seeds) and for inferences in plant systematics via band-sharing analysis. Fungal geneticists have also found VNTRs (not necessarily the Jeffrey's type) to be useful in identifying clones in natural populations, which can be quite large (Smith *et al.* 1992).

Since the mid 1990s, the marker of choice in plants has been microsatellites (simple tandem repeats, STRs or SSRs, see Chapter 10) (Morgante & Oliveri 1993; Powell *et al.* 1996a). A SSR locus consists of short repeats of 1–8 bases, and each repeat number represents a different allele.

### 9.5.2 Aspects of SSR methodology

Chapter 10 describes how microsatellite primers can be found for a particular species. The flanking primers are usually highly specific to the species and one of the primers is either labelled with a radioactive label or incorporation of a radioactive nucleotide into the fragment during PCR reactions. The label of choice is usually $^{33}$P or $^{35}$S to produce discrete bands. Silver staining of denatured DNA product can produce the same effect as radioactive labelling because silver staining is a very sensitive system for single-stranded DNA.

Recently, automated DNA sequences and fragment analysers have become more affordable and many marker laboratories are adopting their technology. Products can be visualized with fluorescent labelling, and either the primer or incorporated nucleotides are labelled. With infrared labelling, another cost-effective method particularly for projects that have large sample sizes (>200), involves using a third primer (usually an M13 sequence), which is labelled and attached to one of the two original primers (the 'tailed primer') at the 5′ end (Oetting *et al.* 1995). This method also means that just one labelled primer is needed for all types of primer pairs, resulting in considerable cost savings (Oetting *et al.* 1995).

Simple sequence repeats are codominant, highly variable, and somatically stable (Morgante & Oliveri 1993) and thus are the marker of choice for those who can afford the expense of their development and assay. Indeed, the cost of finding and designing the primers, which must be undertaken for every species, does limit the use of this technique. The use of primers from one species on another ('cross-amplification') is very useful in plants, particularly for the wild relatives of the well-studied crop plants. However, the PCR protocols are often different and time must be spent modifying them, and cross-amplification usually detects fewer polymorphic loci than those of the focal species (Peakall *et al.* 1998). In addition, degeneracy of the primer binding site can result in erratic detection of alleles and even the amplification of other, false loci. Direct sequencing of amplified loci in the cross-species should be performed to verify whether the appropriate locus is indeed amplified.

Scoring of the bands can also be tricky; for dinucleotide repeats, shadow bands often appear and the 'true' bands are not always obvious. Usually the uppermost band is a true band, but in rare cases, the uppermost band is actually a slight shadow band; in this case the next uppermost band is probably the true band, especially if it is darker.

### 9.5.3 Applications of SSRs in plants

Initial surveys indicated that plants have an abundance of AT nucleotide repeats widely distributed throughout their genomes, and one microsatellite occurs about every 50 kb (Morgante & Oliveri 1993). Predominantly perfect repeats are usually found, with compound repeats relatively infrequent. Trinucleotide repeats of the motifs TAT and TCT seem to be more common than other motifs, and as in humans, trinucleotide repeats are often found in coding regions (Morgante & Oliveri 1993). However, using a database search, Lagercrantz *et al.* (1993) found that microsatellites were five times less abundant in the genomes of plants than in mammals. This, combined with the often much larger genome sizes in plants, makes the development of microsatellite markers considerably more difficult in plants.

Nevertheless, SSRs have been found and well-characterized in cultivated plants such as rice (*Oryza sativa*), soybean (*Glycine max*), sugar beet (*Beta vulgaris*), wheat (*Triticum aestivum* and *T. monococcum*), wild yam (*Dioscorea villosa*) and barley (*Hordeum*) (Akkaya *et al.* 1992; Wu & Tanksley 1993; Terauchi & Konuma 1994; Becker & Heun 1995; Ma *et al.* 1996b). They are also used to assess variation among gene pools (Powell *et al.* 1996). Markers are usually spread across the genome (Roder *et al.* 1995; Schmidt & Heslop-Harrison 1996). Currently, many forest geneticists are searching and using microsatellites on conifer species. Studies of Norway spruce (*Picea abies*) and four species of *Pinus* (*P. radiata*, *P. strobus*, *P. sylvestris* and *P. taeda*) have shown that CA and GA repeats are abundant in conifers (Smith & Devey 1994; Kostia *et al.* 1995; Echt *et al.* 1996; Echt & May-Marquardt 1997; Pfeiffer *et al.* 1997).

Initial screening for microsatellites in tropical trees was undertaken to assess their abundance (Condit & Hubbell 1991; Chase *et al.* 1996). There were abundant CA and GA repeats among five tropical trees studied. Additionally, for the tropical tree *Dryobalanops lanceolata* which is a large canopy tree, bark as well as leaves were assayed to test for any genotypic difference within an individual. The same genotype was found for both the leaves and cambium tissues on collections from different distances from the ground to the top of the tree (Terauchi 1994).

As is the case in animals, SSRs in plants can be very polymorphic, with 30 or more alleles and heterozygosities of 70% or more. This level of variation opens up new avenues of inference, such as the identification of immigrants in populations, the classification of relationship (e.g. full-sib versus parent–offspring), and fine-scale inference of gene flow and heterozygosity, some of these inferences being largely confined to plants.

When more than one marker linkage map is constructed for a species, each involving a separate cross, it is often desirable to combine separate maps into a single map. However, diallelic loci such as AFLPs and RAPDs are infrequently informative (segregate) across multiple crosses, and occasional highly informative SSRs may be desirable as 'anchor loci' to join maps; for example, Waugh *et al.* (1997) found that in three different crosses of barley, involving 234, 194 and 376 AFLP marker loci, respectively, less than 4% of the segregating loci were in common between two independent crosses, and less than 2% of the segregating loci were in common between all three crosses. Fortunately they were able to join together the three maps because of the large number of markers in these maps and smaller genome size of barley. While it is very difficult to develop the number of SSR markers needed for a saturated genetic map ($\approx 100–200$), a few widely spaced markers would be desirable for their use as anchor loci.

### 9.5.4  Chloroplast microsatellites

Complete sequences of the chloroplast genome led to the rather unexpected identification of chloroplast microsatellites (Powell *et al.* 1995a,b; Vendramin *et al.* 1996). Classical studies of chloroplast DNA variation indicated rather low levels of polymorphism and complete DNA sequences indicated few noncoding regions. Nevertheless, there are a few introns, especially in the tRNA regions, and these often contain repeats. Powell *et al.* (1995a,b) first identified mononucleotide repeats of microsatellites in these chloroplast sequences, finding mostly A or T repeats. Vendramin *et al.* (1996) designed universal primers for the detection of these loci, but only using black pine chloroplast sequence. Nevertheless, amplification success is respectable, with more than 75% success observed in the conifer genera *Abies*, *Cedrus* and *Picea*.

The application of these primers to natural populations of *Abies* has revealed larger than expected numbers of detected haplotypes which could indicate high mutation rate for these chloroplast loci (Vendramin & Ziegenhagen 1997). Even in the notoriously invariable red pine, considerable chloroplast SSR variation was found (Echt *et al.* 1998). Petit and colleagues have extensively studied chloroplast SSR variation in European oaks (*Quercus*), drawing inferences from variations at six loci to study phenomena such as postglacial colonization history (Petit *et al.* 1997).

The chloroplast is maternally inherited in most seed plants. Exceptions include the pine family (Neale & Sederoff 1989) and kiwi fruit (Testolin & Cipriani 1997). Paternally inherited SSR markers provide a unique opportunity to perform paternity analysis with considerable power, an approach currently being utilized in some conifer seed orchards (Stoehr *et al.* 1998).

Craig Newton of British Columbia Research Incorporated (Vancouver, British Columbia) has discovered larger repeat motifs in the chloroplast DNA of lodgepole pine by screening chloroplast intergenic regions for polymorphism, regardless of the sequence makeup, and by designing appropriate PCR primers accordingly when polymorphism was detected. Repeat motifs of 10 bp were found. However, attempts to detect such loci in spruce did not succeed (C. Newton, personal communication).

### 9.5.5 Cross-amplification of plant SSRs

Whitton *et al.* (1997), in the first explicit examination of SSR conservation across a broad taxonomic group of plants, the Asteraceae (sunflower family), found amplification to be mostly limited to the tribe Heliantheae. Interestingly, using *rbc*L divergences for estimates of lineage divergence, they concluded that the maximum divergence time over which SSR primer sets might amplify homologous loci was 15–30 Ma ago. By contrast, the majority of plant families have arisen at least 50 Ma ago, suggesting that independent SSR development will probably be necessary for the majority of plant genera.

Indeed, this pattern is being found in the literature. Kelly & Willis (1998) found that among 19 primer pairs designed for *Mimulus nasutus*, all consistently amplified DNA from other species within its section of the genus (Simiolus), while just 20–40% of the primers gave good products in other sections of *Mimulus*. Echt *et al.* (1999) found that 16 of 21 SSRS designed for *Pinus strobus* (a soft pine) also amplified loci in two other soft pines but not in seven hard pines nor in other conifers; similarly, SSR primers designed for *Pinus radiata* (a hard pine) did not amplify SSRs in soft pines nor in other conifers. Using SSR primers designed for soybean (*Glycine max*), Peakall *et al.* (1998) found that up to 65% of these soybean primer pairs amplified SSRs within *Glycine*, but frequently, the SSR products were short and interrupted compared with those of soybeans. Cross-species amplification outside of the *Glycine* genus was much lower (3–13%). We conclude that in contrast to animals, successful cross-species amplification of SSRs in plants is largely restricted to congeners or closely related genera.

### 9.5.6 STS markers of arbitrary coding genes

The development of STS markers from cDNA libraries is exemplified by Perry & Bousquet (1998a). Using sequences from the cDNA library of black spruce (*Picea mariana*), they designed one primer from the 3′ untranscribed region (UTR) and the second primer 350–600 bp upstream in the coding region; this increases the chance of amplifying just one member of a multigene family because of the greater divergence of the 3′ UTR among members of a gene family. Of 39 pairs of primers screened, 15 loci gave codominant length polymorphisms (other loci exhibited null alleles). Perry & Bousquet suggested that most of the length polymorphisms probably occurred in noncoding (intron or 3′) regions. Also, length differences were often large enough such that PCR products need only be run on agarose gels and visualized via ethidium bromide staining and UV light. The STS primers had an average heterozygosity of 26% and allele number of 2.8 in a range-wide sample. This is well below that of a good SSR locus (heterozygosities of 60–90%), but above that commonly found for isozymes (heterozygosities of 10–20%).

These STS primers were also tested for their ability to direct specific amplification in two individuals of each of 12 additional conifer species (Perry & Bousquet 1998b). Nearly all (95–97%) of the primers functioned well in congeneric trials (and in fact, in white spruce, *Picea glauca*, more polymorphism was found), while a lower proportion (21–33%) scored positively in other Pinaceae genera; however, outside of the Pinaceae, amplification was not successful. Based on these results, it appears that, compared to SSRs (see below), STS perform better at cross-amplification across genera. This is to be expected because the primer binding regions are more evolutionarily conserved in the regions within and around cDNA, compared to the highly labile SSR regions. However, STS markers have less variation relative to SSRs.

This type of marker is often termed an 'expressed sequence tag' (EST), and generally, they are becoming the marker-of-preference for gene mapping because, being derived from sequencing of cDNA clones, their coding portions can often be matched with existing sequences of known function (usually from *Arabidopsis*) via computer searches of gene databases for homology; hence genes of known function can be immediately mapped. However, like SSRs, it takes a good investment to develop these markers, and work has been restricted to model organisms and those of economic value. A good example of the effort being put into the major crop plant species is that of Davis *et al.* (1999), who have mapped 932 EST loci in a 1736 locus map of maize.

## 9.6 Grand comparisons and future developments

A popular application is to compare estimates of population genetic parameters among alternative nuclear markers; for example, Gaiotto *et al.* (1997) used both RAPD and AFLP markers to re-estimate outcrossing rates in *Eucalyptus urophylla* and found that both markers generated rates that were similar in magnitude to those of isozymes. This can be used as a confirmation of the inheritance and utility of markers.

Comparisons among markers also provide empirical evidence of how useful alternative markers are. Powell *et al.* (1996a) compared the utility of RFLPs, RAPDs, AFLPs and SSRs for soybean germplasm analysis, and found SSR markers to have the highest expected heterozygosity (0.60), while AFLP assays revealed the greatest number of marker loci per experiment (19 per gel). They also found that estimates of genetic similarity based on RFLPs, AFLPs and SSRs were highly correlated, indicating congruence between these assays, while RAPDs were less correlated. Russell *et al.* (1997) compared the abilities of RFLPs, AFLPs, RAPDs and SSRs to determine the genetic relationships among 18 cultivated barley accessions. While all approaches uniquely fingerprinted each accession, the four marker types differed in the amount of polymorphism detected; interestingly the highest diversity index was observed for AFLPs. Genetic similarity estimates were most similar between AFLPs and RFLPs, while those for SSRs were least similar to other markers. These contrasting studies indicate that the two current contenders for 'best' marker are AFLPs and SSRs, with the exact suitability depending on the species and problem of interest.

Amplified fragment-length polymorphisms will be most useful in studies that exploit the numbers of loci revealed; for example, linkage maps constructed using AFLPs are generally more reliable than RAPD linkage maps and more saturated than SSR linkage maps. With numerous markers, linkage analysis opens new avenues of inference. As in animals, QTLs can be mapped to positions along a chromosome. Plants pose additional, unique problems that can be addressed using linked markers; for example, they can be used for detecting recombination in natural populations where no sexual reproduction is evident, as in many fungi (Anderson & Kohn 1998). Quantitative trait locus mapping is generally easier in plants because pure-breeding parental lines can often be generated and large progeny populations usually be produced from a cross.

Simple sequence repeats are most useful in applications requiring a few, highly polymorphic markers, such as in the identification of paternity and the anchoring of marker gene maps. Highly variable markers are always informative in pedigrees about linkage. They should be very useful for studying the genetic basis of inbreeding in plants (Ritland 1996). The ability directly to observe migrants, and accurately to infer paternal alleles in offspring of known mother genotypes, provides a means of accurately examining two important parameters in conservation biology, gene flow and paternity (Chase *et al.* 1996).

A unique approach to testing the utility of various markers was employed in a 'network experiment' involving several European laboratories (Jones *et al.* 1997). They examined the reproducibility of RAPDs, AFLPs and SSRs. A standardized technique was evaluated independently in several laboratories and, not surprisingly, RAPDs proved difficult to reproduce, AFLPs were much more reproducible but not perfect, while SSR alleles were amplified by all laboratories, but small differences in their sizing were obtained. Indeed, one of the challenges of collaborative work, and for database archiving, will be to integrate molecular datasets using common size markers and naming conventions.

Finally, the combination of different types of markers into a single analysis can give improved or even novel inferences. As discussed, codominant SSRs might be combined with dominant AFLPs for anchoring AFLP marker maps. A novel inference might use differences in the inheritance mechanisms among genomes to disentangle processes that would be confounded by solely nuclear data (Ennos 1994); for example, by comparing the spatial distribution of maternally inherited mitochondrial DNA with that of paternally inherited chloroplast DNA, Latta *et al.* (1998) inferred that seed dispersal in ponderosa pine was much less than pollen dispersal.

## References

Adams, R.P. & Rieseberg, L.H. (1998) The effects of non-homology in RAPD bands on similarity and multivariate statistical ordination in *Brassica* and *Helianthus*. *Theoretical and Applied Genetics* **97**, 323–326.

Adams, W.T. (1983) Application of isozymes in tree breeding. In: *Isozymes in Plant Genetics and Breeding*, Part A (eds S.D. Tanksley & T.J. Orton), pp. 381–400. Elsevier Scientific Publications, Amsterdam.

Akkaya, M.S., Bhagwat, A. & Cregan, P.B. (1992) Length polymorphism of simple sequence repeat DNA in soybean. *Genetics* **132**, 1131–1139.

Anderson, J.B. & Kohn, L.M. (1998) Genotyping, gene genealogies and genomics bring fungal population genetics above ground. *Trends in Ecology and Evolution* **13**, 444–449.

Bai, D., Brandle, J. & Reeleder, R. (1997) Genetic diversity in North American ginseng (*Panax quinquefolius* L.) grown in Ontario detected by RAPD analysis. *Genome* **40**, 111–115.

Baird, E., Cooper-Bland, S., Waugh, R., DeMaine, M. & Powell, W. (1992) Molecular characterisation of inter- and intraspecific somatic hybrids of potato using randomly amplified polymorphic DNA markers. *Molecular and General Genetics* **233**, 469–475.

Becker, J. & Heun, M. (1995) Barley microsatellites: allele variation and mapping. *Plant and Molecular Biology* **27**, 835–845.

Becker, J., Vos, P., Kuiper, M., Salamini, F. & Heun, M. (1995) Combined mapping of AFLP and RFLP markers in barley. *Molecular and General Genetics* **249**, 65–73.

Bernatsky, R. & Tanksley, S.D. (1986) Towards a saturated linkage map in tomato based on isozymes and random cDNA sequences. *Genetics* **112**, 887–898.

Bodenes, C., Joandet, S., Laigret, F. & Kremer, A. (1997) Detection of genomic regions differentiating two closely related oak species *Quercus petraea* (Matt.) Liebl. & *Quercus robur* L. *Heredity* **78**, 433–444.

Bonnin, I., Huguet, T., Gherardi, M., Prosperi, J.-M. & Olivieri, I. (1996) High level of polymorphism and spatial structure in a selfing plant species, *Medicago truncatula* (Leguminosae), shown using RAPD markers. *American Journal of Botany* **83**, 843–855.

Bowditch, B.N., Albright, D.G., Williams, J. & Braun, M.J. (1993) The use of RAPD markers in comparative genome studies. *Methods in Enzymology* **224**, 294–308.

Brown, A.H.D. & Weir, B.S. (1983) Measuring genetic variability in plant populations. In: *Isozymes in Plant Genetics and Breeding*, Part A (eds S.D. Tanksley & T.J. Orton), pp. 219–240. Elsevier Scientific Publications, Amsterdam.

Chase, M., Kesseli, R. & Kamaljit, B. (1996) Microsatellite markers for population and conservation genetics of tropical trees. *American Journal of Botany* **83**, 51–57.

Clark, A.G. & Lanigan, C.M.S. (1993) Prospects for estimating nucleotide divergence with RAPDs. *Molecular Biology and Evolution* **10**, 1096–1111.

Cnops, G., den Boer, B., Gerats, A., van Montagu, M. & van Lijsebettens, M. (1996) Chromosome landing at the *Arabidopsis* TORNADO1 locus using an AFLP based strategy. *Molecular and General Genetics* **253**, 32–41.

Condit, R. & Hubbell, S.P. (1991) Abundance and DNA sequence of two base repeat regions in tropical tree genomes. *Genome* **34**, 66–71.

Copes, D.L. (1981) Isoenzyme uniformity in western red cedar seedlings from Oregon and Washington. *Canadian Journal of Forestry Research* **11**, 451–453.

Cruzan, M.B. (1998) Genetic markers in plant evolutionary ecology. *Ecology* **79**, 400–412.

Davis, G.L., McMullen, M.D., Baysdorfer, C. *et al* (1999) A maize map standard with sequenced core markers, grass genome reference points and 932 expressed sequence tagged sites (ESTs) in a 1736-locus map. *Genetics* **52**, 1137–1172.

Deragon, J.-M. & Landry, B.S. (1992) RAPD and other PCR-based analyses of plant genomes using DNA extracted from small leaf disks. *PCR. Methods and Applications* **1**, 175–180.

Devey, M.E., Delfino-Mix, A., Kinloch, B.B. Jr & Neale, D.B. (1995) Random amplified polymorphic DNA markers tightly linked to a gene for resistance to white pine blister rust in sugar pine. *Proceedings of the National Academy of Sciences USA* **92**, 2066–2070.

Echt, C.S. & May-Marquardt, P. (1997) Survey of microsatellite DNA in pine. *Genome* **40**, 9–17.

Echt, C.S., May-Marquardt, P., Hseih, M. & Zahorchak, R. (1996) Characterization of microsatellite markers in eastern white pine. *Genome* **39**, 1102–1108.

Echt, C.S., DeVerno, L.L., Anzidei, M. & Vendramin, G.G. (1998) Chloroplast microsatellites reveal population genetic diversity in red pine, *Pinus resinosa* Ait. *Molecular Ecology* **7**, 307–316.

Echt, C.S., Vendramin, G.G., Nelson, C.D. & Marquardt, P. (1999) Microsatellite DNA as shared genetic markers among conifer species. *Canadian Journal of Forestry Research* **29**, 365–371.

Ennos, R.A. (1994) Estimating the relative rates of pollen and seed migration among plant populations. *Heredity* **72**, 250–259.

Gaiotto, R.A., Bramucci, M. & Grattapaglia, D. (1997) Estimation of outcrossing rate in a breeding population *Eucalyptus urophylla* with dominant RAPD and AFLP markers. *Theoretical and Applied Genetics* **95**, 842–849.

Gourmet, C. & Rayburn, A.L. (1996) Identification of RAPD markers associated with the presence of B chromosomes in maize. *Heredity* **77**, 240–244.

Helentjaris, T., Slocum, M., Wright, S., Schaefer, A. & Neinhuis, J. (1986) Construction of genetic linkage maps in maize and tomato using restriction fragment length polymorphisms. *Theoretical and Applied Genetics* **72**, 761–769.

Heun, M., Schafer-Pregl, R., Klawan, D. *et al.* (1997) Site of einkorn wheat domestication identified by DNA fingerprinting. *Science* **278**, 1312–1314.

Huff, D.R., Peakall, R. & Smouse, P.E. (1993) RAPD variation within and among natural populations of outcrossing buffalograss. *Theoretical and Applied Genetics* **86**, 927–934.

Hurme, P. & Savolainen, O. (1999) Comparison of homology and linkage of random amplified polymorphic DNA (RAPD) markers between individual trees of Scots pine (*Pinus sylvestris* L.). *Molecular Ecology* **8**, 15–22.

Jackson, D.A., Somers, K.M. & Harvey, H.H. (1989) Similarity coefficients: measures of co-occurrence and association or simply measures of occurrence? *American Naturalist* **133**, 436–453.

Jain, S.K. & Allard, R.W. (1960) Population studies in predominantly self-pollinated species. I. Evidence for heterozygote advantage in a closed population of barley. *Proceedings of the National Academy of Sciences USA* **46**, 1373–1377.

Jeffreys, A.J., Wilson, V. & Thein, S.L. (1985) Individual-specific 'fingerprints' of human DNA. *Nature* **316**, 76–79.

Jones, C.J., Edwards, K.J., Castaglione, S. *et al.* (1997) Reproducibility testing of RAPD, AFLP and SSR markers in plants by a network of European laboratories. *Molecular Breeding* **3**, 381–390.

Kamalay, J.C. & Carey, D.W. (1996) Using DNA markers to identify American Elm. *American Nursery* **15**, 56–57.

Keim, P., Schupp, J.M., Travis, S.E. *et al.*(1997) A high-density soybean genetic map based on AFLP markers. *Crop Science* **37**, 537–543.

Kelly, A.J. & Willis, J.H. (1998) Polymorphic microsatellite loci in *Mimulus guttatus* and related species. *Molecular Ecology* **7**, 769–774.

Kostia, S., Varvio, S.-L., Vakkair, P. & Pulkkinen, P. (1995) Microsatellite sequences in a conifer, *Pinus sylvestris*. *Genome* **38**, 1244–1248.

Krauss, S.L. (1999) Complete exclusion of nonsires in an analysis of paternity in a natural plant population using amplified fragment length polymorphism (AFLP). *Molecular Ecology* **8**, 217–226.

Lagercrantz, U., Ellegren, H. & Andersson, L. (1993) The abundance of various polymorphic microsatellite motifs differs between plants and vertebrates. *Nucleic Acids Research* **21**, 1111–1115.

Landry, B.S., Kesseli, R.V., Farrara, B. & Michelmore, R.W. (1987) A genetic map of lettuce (*Lactuca sativa* L.) with restriction fragment length polymorphism, isozyme, disease resistance and morphological markers. *Genetics* **116**, 331–337.

Latta, R.G., Linhart, Y.B., Fleck, D. & Elliot, M. (1998) Direct and indirect estimates of seed versus pollen movement within a population of ponderosa pine. *Evolution* **52**, 61–67.

Levi, A., Rowland, L.J. & Hartung, J.S. (1993) Production of reliable randomly amplified polymorphic DNA (RAPD) markers from DNA of woody plants. *Horticultural Science* **28**, 1188–1190.

Lynch, M. & Milligan, B.G. (1994) Analysis of population genetic structure with RAPD markers. *Molecular Ecology* **3**, 91–99.

Ma, A.Q., Roder, M. & Sorrells, M.E. (1996) Frequencies and sequence characteristics of di-, tri-, and tetra-nucleotide microsatellites in wheat. *Genome* **39**, 123–130.

Mackill, D.J. (1995) Classifying Japonica rice cultivars with RAPD markers. *Crop Science* **35**, 889–894.

Mackill, D.J., Zhang, Z., Redona, E.D. & Colowit, P.M. (1996) Level of polymorphism and genetic mapping of AFLP markers in rice. *Genome* **39**, 969–977.

Marsan, P.A., Castiglioni, P., Fusari, F., Kuiper, M. & Motto, M. (1998) Genetic diversity and its relationship to hybrid performance in maize as revealed by RFLP and AFLP markers. *Theoretical and Applied Genetics* **96**, 219–227.

Meksem, K., Leister, D., Peleman, J., Zabeau, M., Salamini, F. & Gebhardt, C. (1995) A high-resolution map of the vicinity of the R1 locus on chromosome V of potato based on RFLP and AFLP markers. *Molecular and General Genetics* **249**, 74–81.

Miller, A.C.E., Brookes, C.P., Loxdale, H.D. & Cussans, G.W. (1996) Using RAPD markers to identify genets of an arable grass weed, *Arrhenatherum elatius* ssp. *bulbosum*. *Annals of Applied Biology* **129**, 71–82.

Milligan, B.G. & McMurry, C.K. (1993) Dominant vs codominant genetic markers in the estimation of male success. *Molecular Ecology* **2**, 275–283.

Morgante, M. & Olivieri, A.M. (1993) PCR-amplified microsatellites as markers in plant genetics. *Plant Journal* **3**, 175–182.

Mullis, K. & Faloona, F. (1987) Specific synthesis of DNA *in vitro* via a polymerase chain reaction. *Methods in Enzymology* **55**, 335–350.

Nandi, S., Subudhi, P.K., Senadhira, D., Manigbas, N.L., Sen-Mandi, S. & Huang, N. (1997) Mapping QTLs for submergence tolerance in rice by AFLP analysis and selective genotyping. *Molecular and General Genetics* **255**, 1–8.

Neale, D.B. & Sederoff, R.R. (1989) Paternal inheritance of chloroplast DNA and maternal inheritance of mitochondrial DNA in loblolly pine. *Theoretical and Applied Genetics* **77**, 212–216.

Nei, M. (1972) Genetic distance between populations. *American Naturalist* **106**, 283–292.

Nesbitt, K.A., Potts, B.M., Vaillancourt, R.E., West, A.K. & Reid, J.B. (1995) Partitioning and distribution of RAPD variation in a forest tree species: *Eucalyptus globulus* (Myrtaceae). *Heredity* **74**, 628–637.

Newbury, H.J. & Ford-Lloyd, B.V. (1993) The use of RAPD for assessing variation in plants. *Plant Growth Regulation* **12**, 43–51.

Nybom, H. (1996) DNA fingerprinting—A useful tool in the taxonomy of apomictic plant groups. *Folia Giobotanica and Phytotaxonimica* **31**, 295–304.

Oetting, W.S., Lee, H.K., Flanders, D.J., Wiesner, G.L., Sellers, T.A. & King, R.A. (1995) Linkage analysis with multiplexed short tandem repeat polymorphisms using infrared fluorescence and M13 tailed primers. *Genomics* **30**, 450–458.

Pakniyat, H., Powell, W., Baird, E. *et al.* (1997) AFLP variation in wild barley (*Hordeum spontaneum* C. Koch) with reference to salt tolerance and associated ecogeography. *Genome* **40**, 332–341.

Peakall, R., Gilmore, S., Keys, W., Morgante, M. & Rafalski, A. (1998) Cross-species amplification of soybean (*Glycine max*) simple sequence repeats (SSRs) within the genus and other legume genera: implications for the transferability of SSRs in plants. *Molecular Biology and Evolution* **15**, 1275–1287.

Perera, L., Russell, J.R., Provan, J., McNicol, J.W. & Powell, W. (1998) Evaluating genetic relationships between indigenous coconut (*Cocus nucifera* L.) accessions from Sri Lanka by means of AFLP profiling. *Theoretical and Applied Genetics* **96**, 545–550.

Perez, T., Albornoz, J. & Dominguez, A. (1998) An evaluation of RAPD fragment reproducibility and nature. *Molecular Ecology* **7**, 1347–1357.

Perry, D.J. & Bousquet, J. (1998a) Sequence-tagged-site (STS) markers of arbitrary genes: development, characterization and analysis of linkage in black spruce. *Genetics* **149**, 1089–1098.

Perry, D.J. & Bousquet, J. (1998b) Sequence-tagged-site (STS) markers of arbitrary genes: the utility of black spruce-derived STS primers in other conifers. *Theoretical and Applied Genetics* **97**, 735–743.

Petit, R.J., Pineau, E., Demesure, B., Bacilieri, R., Ducousso, A. & Kremer, A. (1997) Chloroplast DNA footprints of postglacial recolonization by oaks. *Proceedings of the National Academy of Sciences USA* **94**, 9996–10001.

Pfeiffer, A., Olivieri, A.M. & Morgante, M. (1997) Identification and characterization of microsatellites in Norway spruce (*Picea abies* K.). *Genome* **40**, 411–419.

Plomion, C., Yani, A. & Marpeau, A. (1996) Genetic determinism of 3d-carene in maritime pine using RAPD markers. *Genome* **39**, 1123–1127.

Powell, W., Morgante, M., Andre, C. *et al.* (1995a) Hypervariable microsatellites provide a general source of polymorphic DNA markers for the chloroplast genome. *Current Biology* **5**, 1023–1029.

Powell, W., Morgante, M., McDevitt, R., Vendramin, G.G. & Rafalski, J.A. (1995b) Polymorphic simple sequence repeat regions in chloroplast genomes: applications to the population genetics of pines. *Proceedings of the National Academy of Sciences USA* **92**, 7759–7763.

Powell, W., Morgante, M., Andre, C. *et al.* (1996a) The comparison of RFLP, RAPD, AFLP and SSR (microsatellite) markers for germplasm analysis. *Molecular Breeding* **2**, 225–238.

Powell, W., Morgante, M., Doyle, J.J., McNicol, J.W., Tingey, S.V. & Rafalski, J.A. (1996b) Genepool variation in genus *Glycine* subgenus *Soja* revealed by polymorphic nuclear and chloroplast microsatellites. *Genetics* **144**, 793–803.

Qi, X. & Lindhout, P. (1997) Development of AFLP markers in barley. *Molecular and General Genetics* **254**, 330–336.

Quinn, T.W. & White, B.N. (1987) Identification of restriction fragment length polymorphisms in genomic DNA of the lesser snow goose. *Molecular Biology and Evolution* **4**, 126–143.

Ramser, J., Weising, K., Lopez-Peralta, C., Terhalle, W., Terauchi, R. & Kahl, G. (1997) Molecular marker based taxonomy and phylogeny of Guinea yam (*Dioscorea rotundata — D. cayenensis*). *Genome* **40**, 903–915.

Remington, D.L., Whetten, R.W., Liu, B.H. & O'Malley, D.M. (1999) Construction of an AFLP genetic map with nearly complete genome coverage in *Pinus taeda*. *Theoretical and Applied Genetics* **98**, 1279–1292.

Rieseberg, L.H. (1996) Homology among RAPD fragments in interspecific comparisons. *Molecular Ecology* **5**, 99–105.

Ritland, K. (1983) Estimation of mating systems. In: *Isozymes in Plant Genetics and Breeding*, Part A (eds S.D. Tanksley & T.J. Orton), pp. 289–302. Elsevier Scientific Publications, Amsterdam.

Ritland, K. (1990) A series of FORTRAN computer programs for estimating plant mating systems. *Journal of Heredity* **81**, 235–237.

Ritland, K. (1996) Inferring the genetic basis of inbreeding depression in plants. *Genome* **39**, 1–8.

Roder, M.S., Plaschke, J., Konig, S.U. *et al.* (1995) Abundance, variability and chromosomal location of microsatellites in wheat. *Molecular and General Genetics* **246**, 327–333.

Rogstad, S.H. (1993) Surveying plant genomes for variable number of tandem repeat loci. *Methods in Enzymology* **224**, 278–294.

Ronning, C.M. & Schnell, R.J. (1995) Inheritance of random amplified polymorphic DNA (RAPD) markers in *Theobroma cacao* L. *Journal of the American Society of Horticultural Science* **120**, 681–686.

Rouppe van der Voort, J.N.A.M., van Zandvoort, P., van Eck, H.J. *et al.* (1997) Use of allele specificity of comigrating AFLP markers to align genetic maps from different potato genotypes. *Molecular and General Genetics* **255**, 438–447.

Russell, J.R., Fuller, J.D., Macaulay, M. *et al.* (1997) Direct comparison of levels of genetic variation among barley accessions detected by RFLPs, AFLPs, SSRs and RAPDs. *Theoretical and Applied Genetics* **95**, 714–722.

Saiki, R.K., Scharf, S.J., Faloona, F. *et al.* (1985) Enzymatic amplification of beta-globin sequences and restriction site analysis diagnosis of sickle cell anemia. *Science* **230**, 1350–1354.

Schmidt, T. & Heslop-Harrison, J.S. (1996) The physical and genomic organization of microsatellites in sugar beet. *Proceedings of the National Academy of Sciences USA* **93**, 8761–8765.

Sharon, D., Adato, A., Mhameed, S. *et al.* (1995) DNA fingerprints in plants using simple sequence repeat and minisatellite probes. *Hortscience* **30**, 109–112.

Smith, D.N. & Devey, M.E. (1994) Occurrence and inheritance of microsatellites in *Pinus radiata*. *Genome* **37**, 977–983.

Smith, M.L., Bruhn, J.N. & Anderson, J.B. (1992) The fungus *Armillaria bulbosa* is among the largest and oldest living organisms. *Nature* **356**, 428–431.

Stewart, C.N. Jr & Porter, D.M. (1995) RAPD profiling in biological conservation: an application to estimating clonal variation in rare and endangered *Iliamna* in Virginia. *Biological Conservation* **74**, 135–142.

Stoehr, M.U., Orvar, B.L., Vo, T.M., Gawley, J.R., Webber, J.E. & Newton, C.H. (1998) Application of a chloroplast DNA marker in seed orchard management evaluations of Douglas-fir. *Canadian Journal of Forestry Research* **28**, 187–195.

Terauchi, R. (1994) A polymorphic microsatellite marker from the tropical tree *Dryobalanops lanceolata* (Dipterocarpaceae). *Japanese Journal of Genetics* **69**, 567–576.

Terauchi, R. & Konuma, A. (1994) Microsatellite polymorphism in *Dioscorea tokoro*, a wild yam species. *Genome* **37**, 794–801.

Testolin, R. & Cipriani, G. (1997) Paternal inheritance of chloroplast DNA and maternal inheritance of mitochondrial DNA in the genus *Actinidia*. *Theoretical and Applied Genetics* **94**, 897–903.

Thormann, C.E., Ferreira, M.E., Camargo, L.E.A., Tivang, J.G. & Osborn, T.C. (1994) Comparision of RFLP and RAPD markers to estimate genetic relationships within and among cruciferous species. *Theoretical and Applied Genetics* **88**, 973–980.

Tohme, J., Gonzalez, D.O., Beebe, S. & Duque, M.C. (1996) AFLP analysis of gene pools of a wild bean core collection. *Crop Science* **36**, 1375–1384.

Transue, D.K., Fairbanks, D.J., Robison, L.R. & Andersen, W.R. (1994) Species identification by RAPD analysis of grain amaranth genetic resources. *Crop Science* **34**, 1385–1389.

Travis, S.E., Maschinski, J. & Keim, P. (1996) An analysis of genetic variation in *Astragalus cremnophylax* var. *cremnophylax*, a critically endangered plant, using AFLP markers. *Molecular Ecology* **5**, 735–745.

Tulsieram, L.K., Glaubitz, J.C., Kiss, G. & Carlson, J.E. (1992) Single tree genetic-linkage mapping in conifers using haploid DNA from megagametophytes. *Bio-Technology* **10**, 686–690.

Ungerer, M.C., Baird, S.J.E., Pan, J. & Rieseberg, L.H. (1998) Rapid hybrid speciation in wild sunflowers. *Proceedings of the National Academy of Sciences USA* **95**, 11757–11762.

Van de Ven, W.T.G. & McNicol, R.J. (1995) The use of RAPD markers for the identification of Sitka spruce (*Picea sitchensis*) clones. *Heredity* **75**, 126–132.

VanToai, T.T., Peng, J. & St. Martin, S.K. (1997) Using AFLP markers to determine the genomic contribution of parents to populations. *Crop Science* **37**, 1370–1373.

Vendramin, G.G. & Ziegenhagen, B. (1997) Characterisation and inheritance of polymorphic plastid microsatellites in *Abies*. *Genome* **40**, 857–864.

Vendramin, G.G., Lelli, L., Rossi, P. & Morgante, M. (1996) A set of primers for the amplification of 20 chloroplast microsatellites in Pinaceae. *Molecular Ecology* **5**, 595–598).

Vos, P., Hogers, R., Bleeker, M. *et al.* (1995) AFLP: a new technique for DNA fingerprinting. *Nucleic Acids Research* **23**, 4407–4414.

Wang, G.L., Dong, J.M. & Paterson, A.H. (1995) The distribution of *Gossypium-hirsutum* chromatin in *Gossypium-barbadense* germplasm—Molecular analysis of introgressive plant breeding. *Theoretical and Applied Genetics* **91**, 1153–1161.

Wang, Y.H., Thomas, C.E. & Dean, R.A. (1997) A genetic map of melon (*Cucumis melo* L.) based on amplified fragment length polymorphism (AFLP) markers. *Theoretical and Applied Genetics* **95**, 791–798.

Waugh, R., Bonar, N., Baird, E. *et al.* (1997) Homology of AFLP products in three mapping populations of barley. *Molecular and General Genetics* **255**, 311–321.

Whitton, J., Rieseberg, L.H. & Ungerer, M.C. (1997) Microsatellite loci are not conserved across the Asteraceae. *Molecular and Biological Evolution* **14**, 204–209.

Williams, J.G.K., Kubelik, A.R., Livak, K.J., Rafalski, J.A. & Tingey, S.V. (1990) DNA polymorphisms amplified by arbitrary primers are useful as genetic markers. *Nucleic Acids Research* **18**, 6531–6535.

Williams, J.G.K., Hanafey, M.K., Rafalski, J.A. & Tingey, S.V. (1993) Genetic analysis using random amplified polymorphic DNA markers. *Methods in Enzymology* **218**, 704–740.

Wolff, K. (1996) RAPD analysis of sporting and chimerism in chrysanthemum. *Euphytica* **89**, 159–164.

Wu, K.-S. & Tanksley, S.D. (1993) Abundance, polymorphism and genetic mapping of microsatellites in rice. *Molecular and General Genetics* **241**, 225–235.

Yeh, F.C. (1988) Isozyme variation of *Thuja plicata* (Cupressaceae) in British Columbia. *Biochemical Systems in Ecology* **16**, 373–377.

Zhu, J., Gale, M.D., Quarrie, S., Jackson, M.T. & Bryan, G.J. (1998) AFLP markers for the study of rice biodiversity. *Theoretical and Applied Genetics* **96**, 602–611.

# CHAPTER 10

# Microsatellites: Evolutionary and Methodological Background and Empirical Applications at Individual, Population and Phylogenetic Levels

KIM T. SCRIBNER AND JOHN M. PEARCE

## 10.1 Introduction

The recent proliferation and greater accessibility of molecular genetic markers has led to a growing appreciation of the ecological and evolutionary inferences that can be drawn from molecular characterizations of individuals and populations (Burke *et al.* 1992; Avise 1994). Different techniques have the ability to target DNA sequences which have different patterns of inheritance, different modes and rates of evolution and, concomitantly, different levels of variation. In the quest for 'the right marker for the right job', microsatellites have been widely embraced as the marker of choice for many empirical genetic studies. The proliferation of microsatellite loci for various species and the voluminous literature compiled in very few years associated with their evolution and use in various research applications, exemplifies their growing importance as a research tool in the biological sciences.

The ability to define allelic states based on variation at the nucleotide level has afforded unparalleled opportunities to document the actual mutational process and rates of evolution at individual microsatellite loci. The scrutiny to which these loci have been subjected has resulted in data that raise issues pertaining to assumptions formerly stated, but largely untestable for other marker classes. Indeed this is an active arena for theoretical and empirical work. Given the extensive and ever-increasing literature on various statistical methodologies and cautionary notes regarding the uses of microsatellites, some consideration should be given to the unique characteristics of these loci when determining how and under what conditions they can be employed.

This chapter is intended to serve as a general but comprehensive review of microsatellite DNA loci. First, characteristics of microsatellites in eukaryotic genomes and aspects of their molecular evolution are described, which are germane to the understanding of levels of variability, the size and frequency distributions of alleles in natural populations, and of statistical methodologies proposed for their use in various research contexts. Second, background information is provided on how microsatellites may be acquired, either from existing genetics databases, the utility of heterologous polymerase chain reaction (PCR) primers for use in species of varying degrees of taxonomic relationship, and how microsatellites are obtained *de novo* through cloning efforts. Included is a series of internet sites where we and colleagues have compiled citations for publications describing microsatellite PCR primers. Third, a brief narrative is presented regarding PCR, locus characterizations and allele scoring. There is also a general review of the various statistical approaches that have been described for individual-level, population-level and phylogenetic analyses. Finally, a general review is given of the various novel applications of microsatellites in different areas of the biological sciences. Readers are directed to several additional excellent reviews of microsatellites that are also available in the areas of conservation biology (Bruford *et al.* 1996), genetic relatedness and paternity (Strassmann *et al.* 1996), population genetics and evolutionary biology (Bruford & Wayne 1993; Jarne & Lagoda 1996), and phylogenetic inference (Goldstein & Pollock 1997).

## 10.2 Characteristics of microsatellites

Microsatellite or simple sequence loci constitute part of a group of loci known as variable number of tandem repeat (VNTR) loci (Nakamura *et al.* 1987). Early research found that DNA regions composed of certain simple short-sequence motifs (e.g. poly (GT) or poly (GCG)) were found in much higher frequency than were random sequences of similar length, and these sequences were reported to occur ubiquitously in eukaryotic genomes (Hamada *et al.* 1982; Tautz & Renz 1984). These regions, which are often referred to as simple sequence repeat (SSR) or simple tandem repeat (STR) loci, have long been recognized as a major source of genetic variation (Tautz *et al.* 1986; Tautz 1989; Weber & May 1989).

The microsatellite repeat array typically comprises 10–50 copies of a short repeat motif (1–10 base pairs (bp), usually 2–5 bp; Tautz 1989; Weber 1990; see Fig. 10.1a, b for examples of two size variants of the same locus). The other well-known class of VNTR loci are minisatellites which differ from microsatellites in size (10–100 bp) and number of tandem repeats. These two classes of loci also differ greatly in terms of methods of characterization and detection, as well as inferred mode of evolution. Excellent background information pertaining to minisatellites can be found in Bruford *et al.* (1992), Burke *et al.* (1996), and Chapter 6.

### 10.2.1  Description, prevalence and distribution

Microsatellite loci can be classified on the basis of the repeat motif length (i.e.

G A T C     G A T C

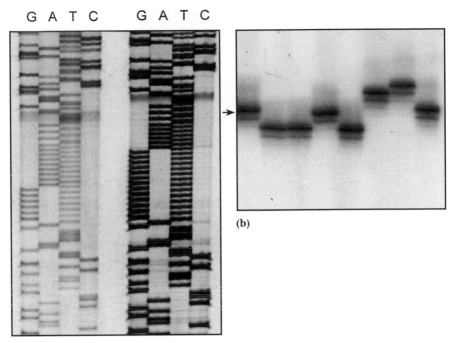

(b)

(a)

**Fig. 10.1** Microsatellite sequence and polymerase chain reaction (PCR) variation at the waterfowl sex-linked locus *Sfiu*1 (Fields & Scribner 1997). (a) Sequences of two individuals showing variation within the microsatellite repeat motif. (b) Autoradiograph of PCR products for eight individuals showing allele variation as a result of repeat length variation.

dinucleotide, trinucleotide, tetranucleotide, etc.) and motif contiguity (e.g. perfect or interrupted); for example, Weber (1990) in a study of human $(CA)_n$ loci found that the majority of microsatellite loci were perfect repeat sequences, while a lesser number were imperfect, containing one or more interruptions. Still fewer loci were found to contain compound repeat sequences with adjacent but different motifs.

Characteristics of the repeat motif (type, length and contiguity) appear to affect the rate of mutation and levels of allelic variation. Interruptions within the core sequence seem to stabilize arrays of repeats, rendering levels of polymorphism for microsatellites with interrupted repeats less variable than loci with pure repeats (Richards & Sutherland 1994; Pepin *et al.* 1995; Petes *et al.* 1997). Levels of allelic diversity are also correlated with repeat length (Weber 1990), as loci with longer repeats are generally more polymorphic than loci composed of short motifs (Beckmann & Weber 1992; but see Valdes *et al.* 1993).

Generalizations regarding the distribution and abundance of different microsatellite motifs must be cast in light of several sources of potential bias. Researchers are typically searching for highly polymorphic markers and certain motifs (frequently dinucleotide (CA) or (GA)) are preferentially screened. Furthermore, there is a bias in genetic databases as cDNAs, partial gene sequences and regions at or near suspected coding regions may not be indicative of the true representation and prevalence of different repeat motifs in the genome. Unlike minisatellites or tandemly repeated satellite DNA, predominantly found in telomeric/subtelomeric regions or in heterochromatin near centromeres, microsatellites are generally believed to be relatively uniformly or randomly distributed with high frequency in eukaryotic genomes (Dib *et al.* 1996; Dietrich *et al.* 1996).

The prevalence and distribution of various microsatellite motifs varies greatly across taxa; for example, size and density of microsatellite repeats vary in the human genome (Jurka & Pethiyagoda 1995) with abundance described as high as 1 locus per 6 kb in humans and the rat (Beckmann & Weber 1992). Stallings *et al.* (1991) estimated the occurrence to be 1 in every 30 kb for (GT) repeats across several species of mammals. Beckmann & Weber (1992) observed that $(A)_n$ sequences were the most abundant in rats and humans, followed by $(CA)_n$, $(A2-3N)_n$ (where N is T, G, or C), $(GA)_n$ and $(AT)_n$. The abundance of microsatellites in fish is probably quite high and average repeat length is thought to be larger than in mammals (Booker *et al.* 1994). The genome copy number in birds has also been described as being lower than for mammals (Primmer *et al.* 1997), possibly because of the fact that the avian genome contains relatively less noncoding DNA than most mammals and that avian SINE/LINE elements do not terminate in poly(A) tails, which are known to provide a source for the evolution of tandem simple sequences in mammals. Database searches suggest that microsatellites are five times less abundant in the genomes of plants than in animals (Lagercrantz *et al.* 1993).

### 10.2.2  Function of microsatellite loci

The function and evolutionary significance of microsatellite sequences is unknown. Early proposals speculated that $(CA)_n$ blocks served as hotspots for recombination (Pardue *et al.* 1987) or gene conversion events (Flanagan *et al.* 1984). Alternatively, $(CA)_n$ tracts have been implicated in gene regulation (Hamada *et al.* 1984; Berg *et al.* 1989). Rather than having a particular biological function, it seems more likely that such sequences arise and are maintained as a consequence of mechanisms such as DNA slippage during replication (Levinson & Gutman 1987). Indeed, the distribution and abundance of other STR elements would suggest that all such motifs arise through the same mechanisms.

### 10.2.3 Are microsatellites selectively neutral?

Several loci have been described as occurring within or adjacent to expressed gene regions (Litt & Luty 1989; Schwaiger *et al.* 1993; Burnett *et al.* 1995; Sutherland & Richards 1995; Crouau-Roy *et al.* 1996) and might therefore be under selection. For loci present within noncoding regions (i.e. introns), the extent to which these loci are under selection is unclear. Loci in noncoding regions may be more prone to selection because of the potential effects of genetic hitchhiking and background selection (Charlesworth *et al.* 1993). Trinucleotide microsatellites have been implicated in causing the onset of several inherited human diseases (see reviews in Ashley & Warren 1995; Karlin & Burge 1996). However, it is assumed that the great majority of microsatellite loci lie outside of genes and are selectively neutral.

### 10.2.4 Mode and rate of mutation

Mutation rates have been estimated directly on the basis of family pedigrees (Weber & Wong 1993) or indirectly using allele frequency distributions or linkage disequilibria by making assumptions of evolutionary effective population size (e.g. Chakraborty *et al.* 1997). Mutation rates for microsatellites are among the highest reported, with rates estimated at $10^{-2}-10^{-5}$ per haploid genome per generation (Dallas 1992; Weber & Wong 1993; Dib *et al.* 1996), a rate 2–3 times greater than that described for protein allozymes. Levinson & Gutman (1987) proposed that polymerase slippage during DNA replication was the most likely cause of microsatellite motif mutation. This mechanism was subsequently supported by *in vitro* experiments (Schlötterer & Tautz 1992; Strand *et al.* 1993).

Through direct observations of mutations at 28 loci on a single chromosome Weber & Wong (1993) found that average mutation rates for tetranucleotide loci were nearly four times those for dinucleotides, an observation corroborated by Zahn & Kwiatkowski (1995). Chakraborty *et al.* (1997) found that dinucleotide repeats have mutation rates 1.5–2.0 times higher than tetranucleotides and nondisease-related trinucleotide repeats. Nondisease-causing trinucleotide had mutation rates intermediate between dinucleotide and tetranucleotide repeats. Differences in conclusions regarding rates of mutation for various microsatellite motif types is perhaps best explained by the fact that the distribution of allele size caused by mutation is not drastically different across di-, tri- and tetranucleotide loci when large numbers of loci are investigated. Alternatively, there may be constraints on the number of repeats at a given locus and these constraints may be stricter for different repeat classes (see Nauta & Weissing 1996).

High mutation rates do not erase information about ancestral states.

Empirical studies have shown that most microsatellite mutations result in the insertion or deletion of one or two repeat units (Valdes *et al.* 1993; Weber & Wong 1993; Di Rienzo *et al.* 1994). Di Rienzo *et al.* (1994) found that the allelic distributions more closely followed expectations of a two-step step-wise mutation model, whereby most mutations were small and incremental in size with occasional saltational events of larger magnitude. Such incrementally small changes are consistent with assumptions of the step-wise mutation model (SMM; Shriver *et al.* 1993; Valdes *et al.* 1993). The SMM states that allele sizes can vary over an infinite size range. However, it is clear that the number of allele states at microsatellite loci are finite and possibly constrained (Bowcock *et al.* 1994; Garza *et al.* 1995). *In vivo* studies of pedigrees show a 2:1 bias in favour of repeat gain (Weber & Wong 1993; Banchs *et al.* 1994), and coupled with high mutation rates, suggest that microsatellites show a tendency to increase over time, perhaps to some threshold size (Garza *et al.* 1995). Garza *et al.* (1995) proposed a model of microsatellite evolution that incorporates bias in the mutational process in the form of a 'restoring force' by which small alleles tend to mutate upwards and large alleles mutate downwards. Rose & Falush (1998), in a systematic study of the complete yeast genome, found that the propensity for simple sequence tandem arrays to expand or contract appears to be a threshold process affecting repeats of approximately eight nucleotides or more. Therefore, the authors conclude that the microsatellite expansion threshold observed could represent a structural constraint on the accuracy of the mismatch repair system as proposed by Strand *et al.* (1993).

Numerous other studies have been directed at mode and rate of mutation. Amos *et al.* (1996) offered the finding that mutation rate is a function of the difference in size between the two alleles in a given individual. Evolutionary rates at the same locus may vary across different taxa (Rubinsztein *et al.* 1995), although this finding is disputed (e.g. Ellegren *et al.* 1995a). Glenn *et al.* (1996) reported that high allelic diversity was negatively correlated with the number of taxa in which the locus was amplified. The lack of locus amplification in taxa distant from the original species implies sequence evolution in the regions flanking the repeat motif. Glenn *et al.* (1996) suggested that although the mechanisms of mutation within both the unique flanking sequence and the repeat motif probably differ, the rates of these mutations or mechanisms of repair appear to be correlated.

Regardless of the specific mutational model or mechanism assumed, the high rate of mutation is of concern in application of data to population genetic analyses (see below; Slatkin 1995). Measures of population differentiation will be affected by rates of mutation, degree of allele size constraint and population size and history (Nauta & Weissing 1996; Feldman *et al.* 1997; see discussion below).

### 10.2.5 Homoplasy

High mutation rates can result in convergence of allele size (i.e. identity in state) in different populations or taxa independent of common ancestry. As a consequence of the mutational mechanism and high mutation rates, allele length variants potentially consist of a mixture of alleles identical by descent, and of alleles identical in state (i.e. of the same length) resulting from convergence, parallelism or reversion. Further, similar sized alleles may not be derived from a common ancestor, as the size and sequence of the repeat motifs and the flanking sequences surrounding the motif can both vary even within a species (B. Ely, personal communication).

Recent studies have shown that size homoplasy is common at microsatellite loci (Estoup *et al.* 1995; Shriver *et al.* 1995; Garza & Freimer 1996; Angers & Bernatchez 1997), indicating that alleles of a given size may not be identical by descent: a confounding mechanism for many statistical summaries, especially relatedness calculations (e.g. $R_{xy}$; Queller & Goodnight 1989). The size of an electromorph may not be correlated directly with repeat copy number. Although microsatellite-containing regions in different individuals may have identical nucleotide sequences, they may not share a common ancestor (e.g. Jin *et al.* 1996). Estoup *et al.* (1995), comparing sequenced microsatellite electromorphs of the same size across a spectrum of species evolutionary relations, suggested that interrupted repeats may be less susceptible to homoplasy and, as such, would be more appropriate for investigating population differentiation and evolutionary relationships between relatively distantly related populations.

### 10.3  How to find or generate microsatellite loci

### 10.3.1  Searches of molecular databases

The expanded size and greater accessibility of DNA sequence data through molecular genetics databases such as GENBANK (http://www.ncbi.nlm.nih.gov/) and EMBL (http://www.embl-heidelberg.de/) allows researchers to search for microsatellite-containing regions in taxa of interest (e.g. Moore *et al.* 1992; Van Lith & Van Zutphen 1996). For organisms which have been extensively studied or which are the focus of genome initiatives, such as *Drosophila* (Schug *et al.* 1998), humans (Dib *et al.* 1996), mouse (Dietrich *et al.* 1996), cattle (Barendse *et al.* 1994) and other agriculturally important species (Moran 1993), numerous loci already exist and can be accessed at associated web sites. Availability of microsatellite primers can be queried at MICRO-SAT (microsat@sfu.ca), the Museum of Natural History (http://www.nmnh.si.edu), the Australian Biological Research Network (http://www.abren.csu.edu.au) and Australia's Latrobe University (http://www.latrobe.edu.au), the TCG microsatellite home page (http://www.gator.biol.sc.edu/msats) and the author's web pages (http://www.fw.msu.edu/labs/scribner/index.html or http://abscweb.wr.usgs.gov/research/genetics/heterologous_primershtm).

## 10.3.2 Use of heterologous PCR primers

Repeat regions and the sequences flanking these motifs are often highly conserved across species differing in degree of taxonomic relationship (e.g. mammals: Moore *et al.* 1991; Fredholm & Wintero 1995; Kemp *et al.* 1995; Engel *et al.* 1996; waterfowl: Fields & Scribner 1997; Buchholtz *et al.* 1998; fish: Rico *et al.* 1996; cetaceans: Schlötterer *et al.* 1991; turtles: FitzSimmons *et al.* 1995). Typically, the likelihood that a taxa will amplify for a particular locus declines with increasing phylogenetic distance from the species in which the locus was cloned (e.g. Primmer *et al.* 1996; Scribner *et al.* 1996). Both Glenn *et al.* (1996) and Primmer *et al.* (1996) also observed that the more polymorphic loci were more likely to be variable within more distantly related taxa.

In compiling information from 158 references that characterized microsatellite loci in one or more species (see author's website above), we noted which primers were tested in other species, genera, families and taxonomic orders. In 89 studies that examined the utility of primers in species other than those from which loci were cloned (but within the same family), 71 (80%) found at least one locus to be variable (>2 alleles). In 74 studies where loci were tested in other genera (within the same family), 50 (68%) found at least one polymorphic locus. In an additional 31 studies that screened loci in taxa of different families, 18 (58%) found at least a single variable locus. Lastly, heterologous primers proved variable in one of four studies examining loci in taxa of a different taxonomic order.

Several studies explicitly addressed the degrees of microsatellite locus conservation as a function of phylogenetic divergence among taxa. Moore *et al.* (1991) found that 56% of 48 bovine primers amplified a similar locus in sheep (divergence 15–25 Ma), 6.2% worked in horses (80–100 Ma), while none gave a product in humans. Stallings *et al.* (1991) found six of 20 $(GT)_n$ loci were polymorphic for both humans and rodents, four were polymorphic in primates and humans and five were polymorphic for humans and rodents. Schlötterer *et al.* (1991) found that 11 microsatellite loci were conserved across Cetaceans which diverged 35–40 Ma BP. FitzSimmons *et al.* (1995) showed a high degree of locus conservation in marine turtles which have been separated evolutionarily for over 300 Ma. Whitton *et al.* (1997) found that across one plant family (Asteraceae), the utility of microsatellite loci decreased precipitously for species whose divergence time is over 14.5–29 Ma.

The cross-species performance of heterologous primers is difficult to generalize across taxa. One notable attempt to do so was carried out by Primmer *et al.* (1996). In a study of passerine birds, these authors estimated that 50% of markers may reveal polymorphisms in species up to a DNA–DNA hybridization $\Delta T_mH$ value (Sibley & Ahlquist 1990; Chapter 5) of five separating it from the species from which the loci were cloned. The approximate divergence time corresponding to a 50% success in detecting polymorphisms was estimated to be 11 Ma for passerines and 23 Ma for nonpasserines. Conversion of $\Delta T_mH$ values into species divergence times necessitates different calibrations of the molecular clock in different lineages. Primmer *et al.* (1996) offer that the mammalian data also fit the empirical relationship observed between performance and evolutionary distance for birds, suggesting that general guidelines could plausibly be extended to other taxa.

### 10.3.3  Cloning new microsatellites

If microsatellite loci are not available, novel microsatellites can be generated using standard cloning methodologies (e.g. Dracopoli *et al.* 1995). Most techniques are derivations of a common theme which was described by Rassmann *et al.* (1991). Here, we describe a general protocol and briefly highlight literature in the area of microsatellite library enrichment, which can dramatically improve the efficiency of microsatellite cloning efforts.

First, a partial genomic library is constructed for the species of interest through digestion with one or more four-base-cutting restriction enzymes, such as *Alu*I, *Hae*III, *Rsa*I, *Sau*3A1. Use of more than one enzyme is preferable as the location of restriction sites can be nonrandom (e.g. *Sau*3AI in salmonid fish; Scribner *et al.* 1996), and decreased efficiency can result with single enzyme digestion. The digested DNA is separated in agarose media and the fraction of DNA between approximately 300 and 700 bp is recovered. The size-selected fraction is cleaned, dephosphorylated and ligated into a digested, dephosphorylated vector using standard methods (e.g. Sambrook *et al.* 1989). Choice of the restriction enzyme for ligation depends on the enzyme used in the genomic DNA digestion and on the characteristics of the poly cloning site(s) of the vector (e.g. blunt-end cutting *Sma*I or *Bam*HI). Cloning vectors can be either phage-type (e.g. M13, lambda lgt10) or various plasmids or phagemids (e.g. pUC18, pBluescriptII KS±). The ligation reaction is subsequently transformed into an appropriate *Escherichia coli* host strain. It should be noted that some care should be taken in selection of an appropriate bacterial host, as the stability of DNA containing highly repetitive elements can be compromised even when using recombinant-deficient strains of bacteria. Genomic libraries should be relatively large (5000–10 000 recombinants), as the proportion of recombinants containing microsatellites is typically fairly low (<0.5–2.0%), depending on the genome copy number, which can vary greatly among taxa, and on the basis of the repeat motif (see above).

Transformation efficiencies into *E. coli* strains can be enhanced by varying the vector to insert molar ratios and through electroporation. The transformed cells are then plated at low densities onto agar containing appropriate selective media (e.g. ampicillin) and colour selection, if appropriate (e.g. IPTG/X-gal). Filter replicates of each plate are made using nylon or nitrocellulose membranes. Membranes are probed using isotope- (e.g. $(\gamma\text{-}^{32}P)ATP$) labelled or nonisotopic (e.g. alkaline phosphatase-conjugated probes) simple-sequence polymers (e.g. $(CA)_n$ or $(GA)_n$). Putative positive clones can then be identified using a variety of detection methods such as autoradiography or chemiluminescence.

The number of positive clones identified per unit effort can be increased through library enrichment procedures (see review in Hammond *et al.* 1999). Various methods include the use of PCR directly to amplify ligation products (Grist *et al.* 1993), use of PCR in combination with novel mutant *E. coli* strains (Ostrander *et al.* 1992), employing repeat analysis pooled isolation and detection (RAPID cloning; Koob *et al.* 1998), or use of streptavidin-coated beads (Kandpal *et al.* 1994; Kijas *et al.* 1994). Such enrichment techniques report efficiencies ranging from 40 to 90% of recombinants. For some species groups, associations exist between specific microsatellite motifs and other repetitive genomic DNAs (e.g. SINE elements in artiodactyls; Kaukinen & Varvio 1992) and library enrichment techniques have been described based on these associations (Band & Ron 1996). Library enrichment can result in some level of redundancy (i.e. the same locus may be cloned several times). It is thus best to check all repeat-containing sequences for duplication.

Many putative positive recombinants will not be informative as they prove to be false positives, lack sufficient flanking sequence to construct PCR primers, or will be of insufficient length (i.e. <10 repeats) or contiguity to be polymorphic. Several additional screens can be employed before clones are sequenced to eliminate further uninformative recombinants from final selection. DNAs from known microsatellite-containing clones of varying size (e.g. $(CA)_{10}$, $(CA)_{25}$) can be used as positive controls together with several negative controls. Strong positives identified through this process can then be selected for subsequent sequencing. Levels of locus polymorphism are generally related to microsatellite repeat number. Using oligo repeats of long length (>10) in the screening process at high stringency can increase the probability of obtaining informative loci.

## 10.4 Locus characterization

### 10.4.1 Primer development and optimization

Polymerase chain reaction primer sequences within flanking regions adjacent to microsatellite loci can be developed using a variety of software (e.g. OLIGO; Rychlik & Rhoads 1989). Several criteria can be employed to maximize PCR product yield and clarity. First, primers should be designed outside a minimum distance from the repeat motif, as placing primers close to (within 10–20 bp) the motif may increase the propensity for PCR stutter (see below). Second, primers should be of sufficient size and complexity (G-C content) so that annealing temperatures are moderately stringent and to ensure that primers will anneal to a single site. Primers can also be designed with a higher A-T content on the 3′ end to ensure that the primer completely anneals prior to polymerase-mediated extension. Polymerase chain reaction primers are typically designed to amplify regions from 100 to 250 bp in length. Longer amplification products require longer gel running times to resolve unambiguously and may be more difficult to score. However, this consideration will vary depending on the detection system used (e.g. autoradiograph versus automated systems).

When developing multiple loci, there are several factors to consider that can ultimately increase efficiency. Primers for different loci can be designed with similar annealing temperatures and PCR amplification conditions that facilitate the coamplification or multiplexing of loci in the same PCR reaction (see Olsen et al. 1996; O'Reilly et al. 1996). If this is a desirable strategy, then PCR primers for different loci must be designed to amplify products of different length so that alleles do not overlap during electrophoresis.

### 10.4.2 DNA sources and extraction

The PCR-based methods including analyses of microsatellites have introduced novel sampling strategies and have facilitated research in new areas as a result of the ability to extract DNA from minute and often degraded sample sources. Examples of tissues used in PCR-based DNA studies include plucked and shed hair (Morin et al. 1993; Taberlet et al. 1997a), avian feathers (Ellegren 1992; Pearce et al. 1997) and nest materials (Pearce et al. 1997), skin from wild animals (Larsen et al. 1996; Richard et al. 1996), museum specimens (Ellegren 1991; Mundy et al. 1997a), plant seeds (Dawson et al. 1997), sperm (Zhang et al. 1994; Gertsch & Fjerdingstad 1997), fish scales (Nielsen et al. 1997), faecal material (Wasser et al. 1996; Reed et al. 1997), predatory bird stomach contents (e.g. food habits analysis; Scribner & Bowman 1998), bone (Hagelberg & Sykes

1989), whale baleen (Rosenbaum *et al.* 1997), and urine (Marklund *et al.* 1996). Several protocols also exist for direct PCR from tissues without first extracting the DNA from other cellular material (McCusker *et al.* 1992).

Use of nontraditional tissue material raises three main areas of concern. First, contamination (Pearce *et al.* 1997) can result from unintentional mixing of individual-specific tissues during collection as a result of logistics or species biology; for example, avian feathers sampled from a nest may consist of a mixture of both adults and offspring. Second, PCR may not consistently amplify both microsatellite alleles when DNA quantity is low (Gagneaux *et al.* 1997). Typically between 50 and 200 ng of DNA is used per PCR amplification, although Van Oorschot & Jones (1997) were able to amplify microsatellite loci with as little as 2 ng of human DNA. However, samples that yield minute amounts of DNA (e.g. hairs; Gagneaux *et al.* 1997) can give inconsistent PCR results and thus caution is warranted. Finally, the time and cost involved in developing sufficient protocols for DNA extraction and testing sampling assumptions and PCR products for genotype error is frequently great (see Ramos *et al.* 1995; Taberlet & Waits 1998). Studies using noninvasive sampling of minute or degraded tissue sources are strongly encouraged to test sample sources against blood or tissue samples from the same animal or conduct replicate DNA extractions and PCR before proceeding with data analyses (see Pearce *et al.* 1997; Taberlet *et al.* 1997b; Taberlet & Waits 1998).

It is important that extraction techniques be efficient, inexpensive and provide adequate quantities of DNA that will remain stable for appropriate periods of time. For samples that are known to yield high quantities of DNA (e.g. blood and tissue) standard phenol-chloroform-ethanol precipitation methods (Sambrook *et al.* 1989) may be sufficient. Various other kits are available (Pure Gene, Gentra, Inc., Minneapolis, Minnesota, USA) which do not rely on organic solvents and also produce good yields of DNA. For consistency across samples, extraction procedures can combine additional washing and concentration features (e.g. QiaGen kits, QiaGen, Inc., Valeneia, California, USA). These and other protocols can be critical in situations where contaminants (e.g. heparin, pigments, lipids, etc.) inhibit Taq polymerase. DNA may be extracted using several sample sources as above, using, for example, microconcentration tubes or by boiling samples in 5% Chelex (BioRad Hercules, California, USA).

### 10.4.3  The polymerase chain reaction

The most critical elements to optimize during initial testing of loci are the PCR primer annealing temperature, dinucleotide triphosphate (dNTP) to primer concentration ratios, and $MgCl_2$ concentration. Optimal annealing temperatures generally range from 48 °C to 60 °C, with lower temperatures offering less stringent conditions and greater chance for nonspecific PCR priming and product generation. Primer and dNTP concentrations also require empirical optimization and will vary across loci and occasionally across species. Polymerase chain reactions are typically carried out in 10–30 µL volumes. Many of our microsatellite loci use dNTPs in concentrations of 200 µmol and 10 pmol each primer (e.g. Scribner *et al.* 1996). Polymerase chain reaction buffers can also alter product yield. A standard 10× PCR buffer might include 100 mM Tris HCL (pH 8.0–8.5), 500 mM KCl, and 15 mM MgCl$_2$. Additional matrix-stabilizing agents, such as gelatine and bovine serum albumin (BSA) and nonionic detergents (e.g. IGEPAL (Sigma, Inc.), Triton X-100 (Sigma, Inc., St. Louis, Missouri, USA), Tween-20 (BioRad) in 0.01% working concentrations) can be included in the buffer and may increase PCR product yield for some loci.

### 10.4.4 Analysis of microsatellite polymorphisms

Microsatellite polymorphisms are scored on the basis of allele size using a variety of gel media methods to visualize microsatellite PCR products and size alleles. Gel media can resolve PCR products in either single-stranded or double-stranded condition. Double-stranded products are typically resolved using various intercalating dyes including ethidium bromide (Dallas *et al.* 1995), Syber green (FMC, Inc., Rackland, Maine, USA) and silver staining (Klinkicht & Tautz 1992). Double-stranded products are typically run on native polyacrylamide gels (e.g. Miller & Kapuscinski 1996) or high concentration agarose gels (e.g. White & Kusukawa 1997). However, for a variety of reasons, results are not as clear as obtained with single-strand detection systems because of the greater propensity for PCR products with dinucleotide microsatellites to exhibit artefactual banding patterns. If microsatellites are visualized as double-stranded products, heteroduplex products often form as a PCR artefact (White *et al.* 1992; Wilkin *et al.* 1993) and can confound allele scoring. Furthermore, many dinucleotide repeat loci are prone to PCR stutter (see below), which can also complicate scoring. Finally, intercalating dyes often require considerable effort in the standardization of allele scoring.

Improved PCR product resolution can be obtained with denaturing sequencing systems and one of three amplification methods. First, (γ-$^{32}$P)ATP or (γ-$^{33}$P)ATP end-labelled primers can be used, whereby one primer from each locus pair is end-labelled using T4 polynucleotide kinase. Polymerase chain reactions are mixed with a denaturing mix (e.g. 95% formamide, 20 mM ethylenediaminetetraacetic acid (EDTA), 0.05% bromophenol blue, 0.05% xylene cyanol) and heated for approximately 5 min at 95 °C before loading onto the gel. Second, radioactive $^{35}$S or $^{32}$P nucleotide triphosphates (NTPs) can be incorporated directly into the PCR reaction and thus into each sirond (e.g. Strassmann *et al.* 1996). Background artefactual products may be more pronounced as nonallelic products from both strands are detected. Finally, automated sequencing technology (e.g. Applied Biosystems 373, 377 (Foster City, California) or the Hitachi, Inc. FM BIO II (Alameda, California)) and product-specific software (e.g. Genotyper (Applied Biosystems Inc. 1994) and Genescan (Applied Biosystems Inc. 1993)) can be employed. In this case, primers are labelled at the 5' ends with dyes that fluoresce at different wavelengths when the gel is scanned by a laser. Many loci can be run simultaneously with this system, even when alleles overlap in size. Results from fluorescent-based semiautomated procedures are quite comparable to those based on radioactive labelling (Schwengel *et al.* 1994), although automated techniques allow for more rapid data delivery and more efficient database management. Once a detection system is in place, loci can be screened for variation and checked for Mendelian inheritance patterns and for presence of null alleles with known pedigree samples.

### 10.4.5 PCR artefacts

The most common type of PCR artefact associated with microsatellite loci is a ladder of fragments of smaller and decreasing size compared to the primary allele. These bands are particularly evident in dinucleotide repeats and such 'stutter' or 'shadow' bands are believed to be the result of slipped-strand mispairing during Taq-mediated DNA replication (Hauge & Litt 1993). Typically, intensity of stutter bands decreases monotonically with increasing difference in size from the primary allele product. Stutter banding can be a hindrance to the scoring of alleles of similar size unless care is taken to interpret fragment intensities properly. Typically a primary allele band will be darker than a stutter band. If, for example, an individual is heterozygous for repeat-length variants of similar size, then a lower band will be more intense than the upper allele, signifying that the lower band is an additional allele and not a sub-band. Each microsatellite locus has a characteristic signature and even the most difficult banding patterns can easily be mastered with practice. Many artefacts actually aid in the scoring of genotypes across a gel. Stutter can, at times, be controlled by adjusting annealing temperatures, decreasing the amount of Taq, decreasing the number of PCR cycles, or by addition of additives such as dimethyl sulphoxide (DMSO; in working concentrations of 0.05–5.0% in the PCR cocktail), nonionic detergents (e.g. NP-40, Tween-20, etc., in working concentrations of 0.01%), or matrix-stabilizing compounds (e.g. gelatine and BSA, in working concentrations of 0.01%).

Under certain conditions, Taq polymerase can induce nucleotide additions (typically dATP), onto the 5' end of the synthesized strand, during the PCR process. This typically is seen as faint additional sub-bands 1 bp larger than either of the primary alleles. This process can be related to DNA template quality and to PCR reaction conditions, and can, to some extent, be addressed by primer modifications (Brownstein & Smith 1996) and shorter extension times.

### 10.4.6  Scoring microsatellite genotypes

Microsatellites are codominant loci and can be treated in a manner comparable to allozymes. The degree of resolution is directly a function of the gel matrix and labelling system used. Fluorescent and radioactive-labelled primers facilitate the scoring of alleles with single base pair resolution. The addition of in-lane standards, sequencing standards and individuals of known genotype on each gel also aids in consistent scoring. As such, the relative size (and thus interallelic relationships based on differences in the tandem repeat number) can be easily inferred. Such relationships are critical for many population and phylogenetic applications (see below).

If double-stranded PCR products are used to score genotypes, it is virtually impossible to accurately and precisely assign base pair sizes to alleles. Alleles are scored by relative mobilities qualitatively as ordered character states. Often, scoring ambiguities necessitate that alleles be binned (e.g. as per single-locus minisatellites; Budowle *et al.* 1991; Weir 1992).

Considerable care should be employed to verify allele homology across gels and studies. Stringent quality control is still absolutely necessary, as subtle shifts in fragment mobilities between different gels and, more critically, between different laboratories, need to be accounted for. This often necessitates running individuals in a side-by-side manner across gels to standardize scoring. Scoring can be particularly troublesome for loci with high allelic diversities, particularly when alleles differ greatly in size. Typical size standards for subsequent allele scoring include an M13 control sequence reaction and individual samples of known genotype (see Fig. 10.2).

Additional complications may arise because of the presence of nonamplifying or null alleles that are a result of mutations in one of the priming sites (Callen *et al.* 1993; Pemberton *et al.* 1995). The presence of null alleles can potentially be more of a problem when primers are used for species other than the species in which the marker was cloned. The probability of random mutations in the sequences flanking the repeat motif is often related to the degree of evolutionary differences among taxa.

### 10.5  Statistical methodologies and analyses of microsatellites

Given the proliferation of literature on microsatellite evolution and of

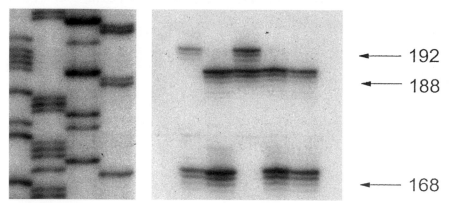

**Fig. 10.2** Mendelian inheritance of microsatellite alleles. M corresponds to the maternal genotype and offspring are numbered 1–4. N corresponds to the negative control. Individual allele assignments (in base pairs) are shown at right. Allele 188 is the paternal allele. An M13 size standard sequence is shown on the left.

cautionary notes regarding their use, some background is warranted before examples of specific applications are presented. Using microsatellites, researchers have the ability to examine polymorphisms at the molecular level and thus can address issues of evolutionary rate and mode to a degree not previously possible. Analytical methodologies have been developed explicitly to incorporate this information. Researchers will continue to refine statistical estimators which measure the degree of population or individual relatedness and test for specific biases, and for the fit of assumptions regarding statistical tests. Issues being debated are similar to those of other markers which for decades have been the standards of empirical population genetics (i.e. allozymes; Nei *et al.* 1983). However, the importance of various aspects of the molecular evolution could formerly be discussed mainly at a theoretical level. This is extremely positive, although perhaps daunting for empirical biologists. In the following sections we address several aspects of the current literature which have direct bearing on the use of microsatellites at individual, population and phylogenetic levels.

### 10.5.2 Individual-based analyses

Several statistical estimators have been used with microsatellite data to examine questions of relatedness. Estimators that are used commonly in a behavioural ecological context (e.g. probability of identity, probability of paternal exclusion and of parentage exclusion; Selvin 1980; Bruford *et al.* 1992; Table 10.1) are commonly derived using microsatellite loci. Estimates of relatedness (e.g. $r_{xy}$; Queller & Goodnight 1989; Goodnight & Queller 1996) are commonly employed with microsatellite loci. Similarly, estimates of gene correlations (Sugg *et al.* 1996) among individuals within social groups or subpopulations can be made using approaches commonly employed for other less polymorphic markers (e.g. estimates of coancestry using allozymes; Scribner *et al.* 1993).

Patterns of relatedness within and between populations can provide valuable sources of inference. In Table 10.1 we provide an example data set where genotypic data are used to construct frequency histograms of pair-wise relatedness. Heterogeneity in the distributions of $r_{xy}$ (e.g. Fig. 10.3) can be easily tested between populations (e.g. using Kruskal–Wallis tests). Alternatively, profiles within a population can be tested against simulated distributions of varying degrees of assumed individual relatedness (Goodnight & Queller 1996). Pairwise estimates of relatedness can also be useful to test specific hypotheses; for example, relatedness may be a function of individual proximity or the proximity of other breeding individuals. Analyses which use pairwise data are typically conducted using permutation tests, where observed estimates are compared to random or null distributions generated by random permutations of the data.

At the individual level, mutation rate and mode should not have an effect on the use of microsatellites in a forensics context (barring laboratory errors). Nor does the rate or mode of mutation affect their utility in various behavioural ecological contexts (i.e. detection of parentage; National Research Council

**Table 10.1** Example of the use of multilocus microsatellite genotypic data to assign parentage using data from Group 1†

| ID | (P)‡ | Age | Sex | Locus 1 | Locus 2 | Locus 3 | Locus 4 |
|----|------|-----|-----|---------|---------|---------|---------|
| 1 | | Adult | F | 105/105 | 153/157 | 139/147 | 210/216 |
| 2 | | Adult | F | 105/107 | 159/159 | 147/151 | 216/216 |
| 3 | | Adult | F | 105/109 | 143/157 | 145/147 | 212/212 |
| 4 | | Adult | M | 105/111 | 155/157 | 137/147 | 212/216 |
| 5 | | Adult | M | 105/105 | 143/155 | 145/147 | 214/216 |
| 6 | | Adult | M | 105/107 | 155/155 | 139/147 | 212/214 |
| 7 | (1/4) | Juvenile | | 105/105 | 157/157 | 137/139 | 210/212 |
| 8 | (1/5) | Juvenile | | 105/105 | 153/155 | 139/145 | 210/216 |
| 9 | (2/5) | Juvenile | | 105/105 | 143/159 | 145/147 | 214/216 |
| 10 | (2/5) | Juvenile | | 105/105 | 155/159 | 145/151 | 216/216 |

†Data for 8 loci are presented in Appendix 10.1.
‡Putative parents of each offspring inferred from observing inheritance of alleles.

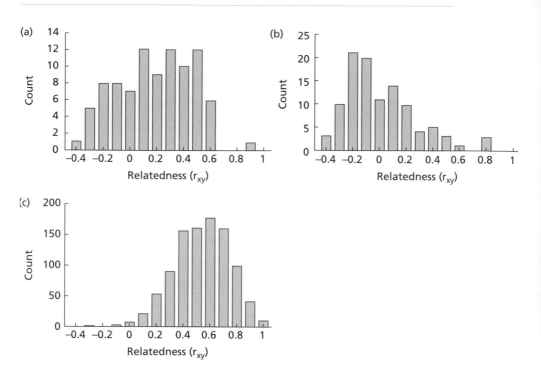

**Fig. 10.3** Histograms of pairwise relatedness ($r_{xy}$) after Queller & Goodnight (1989) with mean (±SD (standard deviation)) for each group. (a) Histogram of group 1 when all parents and offspring are considered independent (mean ± SD = 0.137 ± 0.277). (b) Group 2 (−0.043 ± 0.255). (c) Histogram of 1000 simulated pairs under the hypothesis that the average group relatedness is that of full sibs (0.489 ± 0.205).

1992). We assume that levels of homoplasy (identity in state or allele size independent of common ancestry) are minor within populations. Convergence in size resulting from reversals or the existence of alleles of similar size but with heterogeneous flanking sequence certainly occur, although presumably at low levels and not at all loci. In cases where identity is to be inferred, convergence could slightly elevate probabilities of identity. In a similar manner, if microsatellite loci are used to infer paternity or maternity, probabilities of false paternity and similar measures may be similarly affected, but in a conservative manner (i.e. the researcher would not falsely exclude individuals although the power of the test may diminish slightly). However, homoplasy would not be expected for each of the loci assayed. Many loci should be scored to estimate paternity, maternity or interindividual relatedness with accuracy and precision (Blouin *et al.* 1996). As such, convergence at a single or a few loci may slightly bias estimates of relatedness but not to a great degree. The existence of null alleles can also bias results and should be tested.

One of the more exciting recent developments in individual-based analyses involves tests which assign probabilities of an individual being derived from alternative source populations (e.g. breeding locales) based on the individual's multilocus genotypes and probabilities of occurrence given the allele frequencies (and assuming Hardy–Weinberg equilibrium genotypic frequencies) in each of the possible source populations. This 'Assignment test' (reviewed in Waser & Strobeck 1998) has wide application in forensic science, population ecology and natural resources (i.e. fisheries and wildlife) management. For example, Pearce et al. (2000) applied microsatellite and mitochondrial DNA to an evaluation of morphology used to identify subspecies of harvested Canada geese (*Branta canadensis*). The authors demonstrated that these genetic markers differentiated areas that comprise an admixed wintering group. Their analysis also raised questions about the use of morphological charactecistics to classify and monitor harvested populations at different supspecies.

### 10.5.3 Analyses at the level of populations

Researchers are frequently interested in addressing questions of population breeding structure or of departures from random mating. Estimating the deviation of genotypic frequencies from those expected under conditions of Hardy–Weinberg equilibrium provide one means of testing for departures from random mating. Given the high allelic diversities typically seen for microsatellite loci, particularly when populations are characterized using low or moderate sample sizes, large sample-size statistics such as chi-square are usually not appropriate. Researchers typically use Fisher's Exact Test, which may be calculated using one of several statistical packages (e.g. genetic data analysis (GDA): Lewis & Zaykin 1998; GENEPOP: Raymond & Rousset 1995). Estimates of multilocus gametic disequilibrium can also be of importance and can be calculated using several programs (e.g. GDA, LINKDOS: Garnier-Gere & Dillmann 1992).

Various population and phylogenetic-level statistical methodologies have been advocated or specifically developed for use with microsatellite loci (Tables 10.2 and 10.3). Estimates of the degree of population subdivision and of the partitioning of genetic variation within and among populations are typically conducted using $F$-statistics (e.g. $F_{ST}$ or $\theta$; Weir & Cockerham 1984; Weir 1996). A variety of population genetics statistical packages are available (Table 10.3). Differences in repeat scores within and between populations have been used to obtain other estimates of degree of population subdivision (e.g. $R_{ST}$: Slatkin 1995; $\rho_{ST}$: Rousset 1996; Table 10.3). Nielsen (1997) drawing on earlier work (Griffiths & Tavare 1994) used coalescence theory to derive likeli-

**Table 10.2** Individual and population-level statistics combined for three hypothetical populations (see text and Appendix 10.1 for data)

| Locus | Alleles (*n*) | $F_{st}$ | $R_{st}$ | Single locus P(I)† | Cumulative P(I)† | Single locus Pi‡ | Cumulative Pi‡ |
|---|---|---|---|---|---|---|---|
| 1 | 5 | −0.006 | 0.008 | 0.343 | 0.343 | 0.544 | 0.544 |
| 2 | 4 | 0.013 | 0.004 | 0.368 | 0.126 | 0.584 | 0.318 |
| 3 | 9 | −0.003 | 0.007 | 0.150 | 0.019 | 0.266 | 0.008 |
| 4 | 6 | 0.029 | 0.048 | 0.121 | $2.31 \times 10^{-3}$ | 0.108 | $9.18 \times 10^{-3}$ |
| 5 | 6 | 0.049§ | −0.014 | 0.171 | $2.83 \times 10^{-4}$ | 0.090 | $8.33 \times 10^{-4}$ |
| 6 | 6 | 0.025 | −0.016 | 0.060 | $4.87 \times 10^{-5}$ | 0.257 | $2.14 \times 10^{-4}$ |
| 7 | 7 | 0.022 | 0.123§ | 0.078 | $2.94 \times 10^{-6}$ | 0.114 | $2.46 \times 10^{-5}$ |
| 8 | 8 | 0.025 | 0.001 | 0.078 | $2.32 \times 10^{-7}$ | 0.040 | $1.0 \times 10^{-6}$ |

†Probability of identity after Bruford *et al.* (1992).
‡Probability of parental (male or female) nonexclusion after Selvin (1980).
§ Significance of $F_{st}$ and $R_{st}$; $P < 0.05$ (Bonferroni correction applied).

hood estimates of population genetic parameters (i.e. θ) as well as parameters of mutational models (i.e. one-step vs. two-step).

Differences in repeat scores within and between populations have also been used to obtain estimators of divergence (i.e. genetic distance; Table 10.3) between populations (average squared distance (ASD): Goldstein *et al.* 1995a; $(\delta\mu)^2$: Goldstein *et al.* 1995b). Bowcock *et al.* (1994) defined a distance measure ($D_{AS}$) between pairs of individuals based on the proportion of shared alleles. More recently Feldman *et al.* (1997) developed a distance measure that is reported to recover the linear relationship of population divergence with time even when ranges in allele size are constrained. Several additional measures of genetic distance have also recently been advocated (e.g. $D_{SW}$: Shriver *et al.* 1995; $D_{LR}$: Paetkau *et al.* 1997). Statistics which provide an assessment of population relationships (or tree topology; e.g. Fig. 10.4) and degree of divergence (or branch lengths) will vary in their accuracy and precision as a function of the length of time since divergence. An excellent review of methods of statistical inference at the population and phylogenetic levels is provided by Goldstein & Pollock (1997).

Microsatellites have not been widely used in studies of systematics but a number of studies have used either sequence information in the flanking regions (e.g. Schlötterer *et al.* 1991; Zardoya *et al.* 1996) or allelic variation in the repeat motif (e.g. Bowcock *et al.* 1994) to reconstruct phylogenies. Use of microsatellite loci in populations or higher taxonomic groupings may be problematical (see discussion in this section).

Evolutionary forces can have a pronounced effect on population allele composition and interpopulation differentiation for all genetic loci. As mutation rates affect only that probability of fixation and not the rate of fixation

**Table 10.3** Statistical methodologies for the analysis of population subdivision and genetic differentiation using microsatellite loci

| Statistic | Reference | Program | Reference |
|---|---|---|---|
| *Population subdivision/geographic variation* | | | |
| $F_{st}$ or $\theta$ | Weir & Cockerham 1984 | FSTAT | Goudet 1995 |
| | Weir 1996 | GDA | Lewis and Zaykin 1998 |
| $R_{st}$ | Slatkin 1995 | RstCalc | Goodman 1997 |
| | | GENEPOP | Raymond & Rousset 1995 |
| $\rho_{st}$ | Rousset 1996 | AMOVA | Michalakis & Excoffier 1996 |
| | | ARLEQUIN | Michalakis & Excoffier 1996 |
| *Genetic distance/ phylogenetic analysis* | | | |
| Examples of earlier distance measures | | | |
| Nei's $D_S$ | Nei 1972 | MEGA | Kumar *et al.* 1993 |
| | | PHYLIP | Felsenstein 1989 |
| | | BIOSYS-1 | Swofford & Selander 1981 |
| | | GDA | Lewis & Zaykin 1998 |
| Chord distance | Cavalli-Sforza & Edwards (1967) | BIOSYS-1 | Swofford & Selander 1981 |
| $F_{st}$-based | Reynolds *et al.* 1983 | AMOVA | Michalakis & Excoffier 1996 |
| *Examples of microsatellite-based distance measures* | | | |
| $ASD$ | Goldstein *et al.* 1995a | MICROSAT | Minch 1996 |
| $(\delta\mu)^2$ | Goldstein *et al.* 1995b | MICROSAT | Minch 1996 |
| $D_{AS}$ | Bowcock *et al.* 1994 | MICROSAT | Minch 1996 |
| $D_{SW}$ | Shriver *et al.* 1995 | — | — |
| $D_{LR}$ | Paetkau *et al.* 1997 | — | — |

of neutral alleles, there should be no greater likelihood for finding high inter-population variance in allele frequency for loci with high mutation rates (e.g. microsatellites and single-locus minisatellites) than at polymorphic protein-coding loci. Other variables such as population size, degree of reproductive isolation and the amount of time since populations have been separated are of importance. However, a series of issues argue that caution be used when microsatellite loci are used in population or phylogenetic analyses.

Research has found that the mutational model by which most genetic distance measures are based (the infinite alleles model (IAM)) is not appropriate for microsatellite data at all evolutionary scales of divergence. These loci mutate in a manner consistent with the step-wise mutation model. Further empirical evidence (Kwiatkowski *et al.* 1992; Weber & Wong 1993) suggested that mutations sometimes produce nucleotide repeat patterns which require two or more step changes. Thus various researchers (Valdes *et al.* 1993; Di Rienzo *et al.* 1994; Garza *et al.* 1995) proposed models that allow changes of >1 step with some probability. Based on analyses of genetics databases researchers further found evidence for constraints in allele size leading to

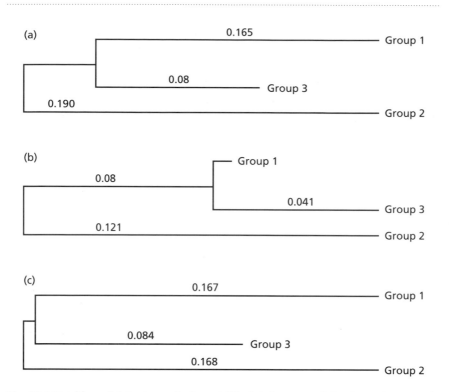

**Fig. 10.4** Neighbour-joining trees of microsatellite data for groups 1–3 using various genetic distance metrics. (a) Proportion of shared alleles (Bowcock *et al.* 1994); (b) $\delta\mu^2$ (Goldstein *et al.* 1995b), and (c) Cavalli-Sforza & Edwards' (1967) chord distance.

further theoretical work (e.g. Garza *et al.* 1995). The development of inferential methods which define population relationships (e.g. Table 10.3) have been described based on explicit models of microsatellite evolution to address these issues.

The rate of mutation for microsatellite loci is considerably higher than that observed for loci such as protein allozymes. Mutation can play a diversifying role in populations which have been isolated from one another for considerable periods of time, as the probability of fixation of mutations in populations is directly a function of the rate of mutation to novel alleles. Conversely, mutation can also play a homogenizing role, as high rates of mutation can lead to convergence (homoplasy) independent of common ancestry, which may counter the diversifying effects of random genetic drift. Furthermore, when the range of target alleles is limited, mutation will lead to the reappearance of alleles lost in the past. Microsatellites fail to reflect separation times past some threshold (i.e. among species or subspecies) because of constraints on maximal size (Garza *et al.* 1995), high rates of mutation (Goldstein *et al.* 1995a) and because of increasing likelihood of homoplasy. These findings argue that certain population-level analyses and analyses at higher taxonomic levels should proceed with caution.

Several general points are often made when describing statistical methodologies employed for microsatellite data. In many respects, the conclusions do not differ from findings of several decades ago pertaining to allozyme data (i.e. Nei *et al.* 1983). First, genetic distance is highly sensitive to the demographic history of diverging populations. As a consequence, the reliability of population parameters will be strongly affected by bottlenecks or fluctuations in population size. Second, the probability of obtaining the correct relationships among populations depends on the number of loci surveyed. Results based on only a few (i.e. 5–10) loci may be misleading. In addition, the general utility of various statistical measures may be related to the degree of evolutionary divergence among populations. Interpopulation divergence, as quantified using the various statistical measures defined above, increases monotonically, although in a nonlinear fashion as a function of the time of population separation. The lack of linearity resulting from the mutation rate, mode and constraints is particularly a problem over long time periods. Over relatively short periods of time, drift would be expected to be the predominant force affecting population divergence in allele frequencies. Therefore, genetic distance measures, estimates of co-ancestry (e.g. Chesser 1991), or population subdivision (e.g. $F_{ST}$: Weir & Cockerham 1984) that are based on differences in allele frequency would be expected to reflect microevolutionary processes more accurately than measures that incorporate mutational processes. Mutation-based measures of population differentiation may prove more appropriate at resolving deeper branches of a population or higher taxonomic phylogeny. Nauta & Weissing (1996) using computer simulations of the SMM and incorporating constraints, investigated which distance measures yield useful estimators of phylogenetic relationships and demographic parameters. These authors propose that the utility of microsatellite loci may be limited to short divergence times and to relatively small populations, underlying the dual importance of mutation and genetic drift.

The utility of the various statistical measures have been tested using simulations and using empirical data; for example, in analyses of empirical data for populations and species of bears, Paetkau *et al.* (1997) found that, at the finest scale of divergence, statistics developed specifically to account for mutational processes performed poorly relative to other distance measures (e.g. Nei 1972; Nei 1973; a likelihood ratio estimator $D_{LR}$ defined in Paetkau *et al.* 1997). Takezaki & Nei (1996) studied the efficiencies of various genetic distance measures and tree building algorithms (UPGMA versus neighbour-joining), under different demographic scenarios (bottlenecked versus constant population size), and locus conditions (high versus low levels of heterozygosity) in phylogenetic reconstructions using computer simulations. These authors found that several distance measures ($D_A$: Nei *et al.* 1983; chord distance: Cavalli-Sforza & Edwards 1967) were efficient in obtaining the correct tree topology in many of the conditions tested.

## 10.6 Empirical applications

Several features of microsatellites render them particularly useful for ecological or evolutionary applications. First, their abundance throughout the genome, high levels of variation and statistical independence (i.e. lack of physical linkage) imply that microsatellites may provide the statistical power necessary for individual identity calculations and parentage determination (National Research Council 1992), even when the pool of potential parents is unknown. Second, the codominant pattern of inheritance and the fact that each locus is a definable unit with only two possible allelic states (homozygous and heterozygous; Fig. 10.2), permits pedigree construction and determination of parentage (Morin *et al.* 1993; Morin *et al.* 1994a), kinship (Queller *et al.* 1993; Westneat & Webster 1994) and population-level patterns (Bruford & Wayne 1993; Bruford *et al.* 1996). Third, reliance on PCR techniques precludes the need for destructive sampling, minimizes or eliminates handling and allows use of small and nontraditional tissue sources (Arnheim *et al.* 1990).

### 10.6.1 Applications in behavioural ecology

The level of genetic differentiation between populations is ultimately a function of within-population demographic and social factors. Behavioural processes, such as philopatry, dispersal, juvenile recruitment and mating system type will dictate patterns of genetic differentiation within and between populations. Microsatellite loci are able to chronicle these individual, and ultimately, population-level demographic processes, which are often the focus of field ecology studies.

Two studies involving plant seed and pollen movement have addressed issues of individual dispersal and gene flow with microsatellite loci. In a study of an endangered tropical tree (*Gliricidia sepium*), Dawson *et al.* (1997) estimated pollen-mediated gene dispersal as a function of geographical distance. Dow & Ashley (1996) additionally used microsatellite markers to examine parentage of bur oak (*Quercus marocarpa*) saplings and determine the extent of seed and pollen dispersal within the population. Both studies documented extreme cases of long-distance gene transfer through pollen and seed dispersal.

Aspects of the mating process include definition of the mating strategy (e.g.

monogamy, polyandry, etc.), mate choice, parentage (maternity and paternity) and reproductive success. In the past, direct observation has been used to gather information on reproductive success and morphological and behavioural correlates with fitness. However, direct observation is either unreliable (Keane *et al.* 1997; Lanctot *et al.* 1997) or unfeasible (Morin *et al.* 1993; Jones & Avise 1997) as a method for determining the number of offspring that can be attributed to a single parent.

Microsatellite loci can uniquely identify parents and putative offspring and thus be used in pedigree construction (Prodohl *et al.* 1996; Hansen *et al.* 1997). Offspring can be examined for the presence of nonparental allele combinations as evidence of multiple maternity or paternity (see Table 10.1). Craighead *et al.* (1995) used a series of microsatellite loci to construct pedigrees for 30 family groups of grizzly bears (*Ursus arctos*). The authors were able to estimate the maximum reproductive success of each male in the study area and augment previous data to determine the proportion of breeding-age males that actually sired offspring, shedding new light on the mating strategy. Keane *et al.* (1997) combined long-term demographic and microsatellite DNA data in order to understand mating patterns and social unit breeding structure of toque macaques (*Macaca sinica*). They concluded that effective population size may be much smaller than indicated by demographic data as males may not always breed every year and thus lifetime reproductive success may be quite low. Ellegren *et al.* (1995b) examined mate choice and extra-pair paternity among breeding pairs with control and 'handicapped' (certain primary and tail feathers removed) male pied flycatchers (*Ficedula hypoleuca*). By examining the genotypes of adults and their offspring, the authors found that pairs containing a handicapped male were not more prone to extra-pair fertilizations by another male. In a study of the mating system of Gulf pipefish (*Syngnathus scovelli*), Jones & Avise (1997) found that male pipefish were rarely the recipient of eggs from more than one female, but that separate males had received eggs from the same female. Use of microsatellite loci for paternity assessment should not be restricted to the analysis of sociobiological issues but may also aid in the development of breeding programmes in captive or small wild populations.

Behavioural studies have revealed that a variety of life-history and mating strategies are cooperative in nature. The concept of inclusive fitness has become an important theory to explain the evolution and maintenance of social and cooperative behaviour (Hamilton 1964). However, while cooperative behaviours may be observed, the evolutionary importance of kin selection rests on the tenet that a nonrandom level of genetic relatedness exists between cooperating individuals (Queller & Goodnight 1989; Westneat & Webster 1994). The applicability and statistical analysis of microsatellite data in estimation of social structure and genetic relatedness within groups have only recently been developed (Queller *et al.* 1993; Westneat & Webster 1994; Ishibashi *et al.* 1997).

The broad application of microsatellite data to estimate individual relatedness is exemplified by the diversity of animal groups examined. In a study of an extreme form of male–male cooperation in the long-tailed manakin (*Chiroxiphia linearis*), McDonald & Potts (1994) used microsatellite DNA to calculate the genetic relatedness among lekking males. The authors found that males were no more related than at the level of second cousins, arguing against a kin selection hypothesis to explain their cooperative behaviour. Through field manipulations of the ant (*Myrmica tahoensis*), Evans (1994) found that with microsatellite DNA, relatedness coefficients among colony members suggested that workers skew the production of certain sexes, and thus maintain appropriate relatedness values within a colony, and to replacement queens. Garza *et al.* (1997) found that the mound-building mouse (*Mus spicilegus*) congregates with genetic relatives to build mounds to store food for immature individuals. Relatedness is also important in resolving dominance hierarchies and reproduction strategies in wild and captive populations (Morin *et al.* 1994b; Craighead *et al.* 1995; Girman *et al.* 1997; Taylor *et al.* 1994, 1997) where maintenance of genetic variation is critical to long-term population viability.

## 10.6.2  Applications in population and conservation biology

Microsatellites have been used in numerous population and conservation biology applications and population genetic information has become an integral part of many field studies (Mundy *et al.* 1997b), breeding programmes (Wolfus *et al.* 1997), and in cases of small or declining populations where the impacts of population bottlenecks are of concern (Taylor *et al.* 1994; Bancroft *et al.* 1995). The basis of microsatellite markers in PCR has also created the opportunity to compare current populations to ancestral populations from the same locale by analysing previously sampled individuals that may exist in museums (Arnheim *et al.* 1990; Ellegren 1991) or other reference collections (Nielsen *et al.* 1997). Such reference samples permit a unique opportunity to assess environmental or other impacts on contemporary levels and partitioning of genetic variability (Taylor *et al.* 1994; Nielsen *et al.* 1997).

Microsatellites are often used in biological studies to determine the magnitude of differences between populations for better understanding of population subdivision, subspecies evolution and classification, and higher order systematics and taxonomy. Information on population subdivision is important in considering various management scenarios. Numerous studies have used microsatellite loci, often in conjunction with other genetic markers, to examine conspecific populations for genetic differentiation at both macro- and microgeographical scales (Allen *et al.* 1995; Larsen *et al.* 1996; Prodohl *et al.* 1996; Favre *et al.* 1997; Estoup *et al.* 1998). Scribner *et al.* (1998) used microsatellite loci in a mixed stock analysis of anadromous Salmonid fish and found that the accuracy and precision of microsatellite loci was comparable to allozyme loci. Allelic variation at microsatellite loci has also been used in intraspecific studies of taxonomy and subspeciation; for example, Roy *et al.* (1994) examined patterns of genetic differentiation and hybridization among wolflike canids. Buchanan *et al.* (1994) used microsatellite loci to investigate evolutionary relationships among domesticated sheep breeds.

In conclusion, microsatellites are rapidly becoming the workhorse of empirical population genetics, behavioural ecology and evolutionary biology. While the enthusiasm with which these loci have been embraced has been somewhat tempered in certain areas, they will continue to be a powerful research tool for years to come because of their high levels of variability and ease in scoring.

**Appendix 10.1** Microsatellite data for eight hypothetical loci used in calculations for Tables 10.2 and 10.3 and Figs 10.3 and 10.4

| Population | Age | Sex | Locus 1 | | Locus 2 | | Locus 3 | | Locus 4 | | Locus 5 | | Locus 6 | | Locus 7 | | Locus 8 | |
|---|---|---|---|---|---|---|---|---|---|---|---|---|---|---|---|---|---|---|
| 1 | A | F1 | 105 | 105 | 146 | 146 | 149 | 149 | 184 | 190 | 153 | 157 | 180 | 182 | 139 | 147 | 210 | 216 |
| 1 | A | F2 | 105 | 107 | 146 | 148 | 149 | 159 | 184 | 190 | 159 | 159 | 182 | 184 | 147 | 151 | 216 | 216 |
| 1 | A | F3 | 105 | 111 | 146 | 152 | 157 | 157 | 188 | 190 | 143 | 157 | 184 | 186 | 145 | 147 | 212 | 212 |
| 1 | A | M1 | 105 | 111 | 146 | 146 | 151 | 165 | 184 | 188 | 155 | 157 | 184 | 190 | 137 | 147 | 212 | 216 |
| 1 | A | M2 | 105 | 105 | 146 | 146 | 149 | 149 | 184 | 190 | 143 | 155 | 182 | 184 | 145 | 147 | 214 | 216 |
| 1 | A | M3 | 105 | 107 | 146 | 146 | 149 | 149 | 184 | 190 | 155 | 155 | 182 | 190 | 139 | 147 | 212 | 214 |
| 1 | J1 | | 105 | 105 | 146 | 146 | 149 | 165 | 184 | 188 | 157 | 157 | 180 | 190 | 137 | 139 | 210 | 212 |
| 1 | J12 | | 105 | 105 | 146 | 146 | 149 | 149 | 184 | 184 | 153 | 155 | 182 | 182 | 139 | 145 | 210 | 216 |
| 1 | J2 | | 105 | 105 | 146 | 146 | 149 | 149 | 184 | 184 | 143 | 159 | 182 | 184 | 145 | 147 | 214 | 216 |
| 1 | J2 | | 105 | 105 | 146 | 148 | 149 | 149 | 184 | 190 | 155 | 159 | 182 | 182 | 145 | 151 | 216 | 216 |
| 1 | J2 | | 105 | 107 | 146 | 146 | 149 | 159 | 184 | 190 | 155 | 159 | 182 | 182 | 147 | 147 | 214 | 212 |
| 1 | J3 | | 105 | 105 | 146 | 146 | 149 | 157 | 184 | 188 | 143 | 155 | 182 | 184 | 139 | 145 | 212 | 212 |
| 1 | J3 | | 105 | 107 | 146 | 152 | 149 | 157 | 184 | 190 | 143 | 155 | 182 | 186 | 139 | 147 | 212 | 214 |
| 1 | J3 | | 105 | 111 | 146 | 146 | 149 | 157 | 190 | 190 | 155 | 157 | 182 | 186 | 147 | 147 | 212 | 212 |
| 2 | A | | 105 | 105 | 146 | 146 | 157 | 163 | 188 | 188 | 143 | 155 | 182 | 182 | 149 | 151 | 210 | 210 |
| 2 | A | | 103 | 105 | 146 | 146 | 149 | 155 | 186 | 192 | 155 | 159 | 182 | 190 | 139 | 141 | 212 | 216 |
| 2 | A | | 105 | 105 | 146 | 148 | 157 | 159 | 184 | 184 | 143 | 155 | 182 | 182 | 145 | 151 | 210 | 218 |
| 2 | A | | 105 | 105 | 150 | 152 | 149 | 149 | 188 | 194 | 155 | 157 | 184 | 184 | 145 | 145 | 216 | 220 |
| 2 | A | | 107 | 107 | 146 | 146 | 149 | 149 | 192 | 194 | 155 | 155 | 182 | 182 | 141 | 147 | 208 | 214 |
| 2 | A | | 107 | 107 | 146 | 146 | 149 | 157 | 186 | 190 | 155 | 155 | 182 | 182 | 147 | 147 | 208 | 214 |

| | | | | | | | | | | | | | | | | | |
|---|---|---|---|---|---|---|---|---|---|---|---|---|---|---|---|---|---|
| 2 | A | 105 | 107 | 146 | 146 | 149 | 149 | 190 | 194 | 155 | 155 | 182 | 182 | 147 | 147 | 202 | 208 |
| 2 | A | 105 | 105 | 146 | 148 | 149 | 161 | 184 | 188 | 153 | 159 | 182 | 182 | 137 | 147 | 210 | 210 |
| 2 | A | 105 | 105 | 146 | 148 | 153 | 159 | 188 | 190 | 151 | 151 | 182 | 188 | 149 | 151 | 210 | 212 |
| 2 | A | 105 | 105 | 146 | 148 | 149 | 159 | 184 | 184 | 151 | 159 | 182 | 182 | 151 | 151 | 210 | 216 |
| 2 | A | 105 | 105 | 146 | 152 | 149 | 157 | 188 | 190 | 151 | 159 | 182 | 188 | 149 | 151 | 202 | 216 |
| 2 | A | 105 | 105 | 146 | 152 | 149 | 159 | 184 | 194 | 151 | 155 | 182 | 188 | 137 | 147 | 218 | 220 |
| 2 | A | 105 | 105 | 146 | 145 | 149 | 149 | 184 | 186 | 155 | 157 | 182 | 182 | 151 | 151 | 212 | 212 |
| 2 | A | 105 | 105 | 146 | 145 | 157 | 165 | 188 | 194 | 151 | 155 | 180 | 184 | 139 | 139 | 210 | 212 |
| 2 | A | 105 | 107 | 146 | 148 | 157 | 159 | 184 | 188 | 151 | 155 | 184 | 184 | 141 | 147 | 210 | 212 |
| 3 | A | 105 | 109 | 146 | 148 | 149 | 159 | 184 | 188 | 155 | 155 | 184 | 186 | 141 | 145 | 212 | 216 |
| 3 | A | 105 | 107 | 148 | 152 | 149 | 151 | 184 | 186 | 143 | 155 | 182 | 186 | 145 | 145 | 210 | 216 |
| 3 | A | 105 | 105 | 146 | 148 | 149 | 159 | 184 | 184 | 143 | 157 | 182 | 184 | 139 | 147 | 210 | 216 |
| 3 | A | 105 | 105 | 146 | 148 | 149 | 149 | 184 | 190 | 155 | 155 | 182 | 182 | 137 | 139 | 210 | 214 |
| 3 | A | 105 | 107 | 146 | 146 | 159 | 161 | 184 | 184 | 155 | 159 | 182 | 190 | 137 | 141 | 210 | 210 |
| 3 | A | 105 | 109 | 146 | 146 | 149 | 159 | 184 | 188 | 155 | 155 | 184 | 184 | 147 | 147 | 212 | 212 |
| 3 | A | 105 | 107 | 146 | 146 | 149 | 149 | 188 | 190 | 143 | 155 | 184 | 184 | 139 | 147 | 212 | 214 |
| 3 | A | 103 | 105 | 146 | 146 | 149 | 165 | 190 | 190 | 155 | 155 | 182 | 190 | 139 | 147 | 212 | 214 |
| 3 | A | 105 | 105 | 146 | 152 | 149 | 165 | 190 | 190 | 155 | 155 | 182 | 190 | 141 | 147 | 212 | 214 |
| 3 | A | 105 | 109 | 146 | 146 | 149 | 159 | 184 | 188 | 151 | 159 | 182 | 182 | 139 | 149 | 210 | 212 |
| 3 | A | 103 | 107 | 146 | 152 | 157 | 149 | 184 | 192 | 155 | 155 | 184 | 184 | 141 | 147 | 210 | 218 |
| 3 | A | 105 | 105 | 145 | 146 | 149 | 159 | 184 | 188 | 155 | 155 | 184 | 184 | 137 | 145 | 210 | 212 |
| 3 | A | 105 | 105 | 146 | 146 | 149 | 161 | 184 | 184 | 143 | 155 | 182 | 182 | 139 | 145 | 212 | 212 |
| 3 | A | 105 | 105 | 146 | 146 | 149 | 149 | 190 | 190 | 153 | 155 | 130 | 182 | 141 | 145 | 210 | 216 |
| 3 | A | 105 | 105 | 146 | 152 | 153 | 157 | 184 | 188 | 143 | 155 | 184 | 184 | 141 | 145 | 210 | 216 |

# References

Applied Biosystems Inc. (1993) *GeneScan 672 software. Users Manual.* A. Foster City, California.

Applied Biosystems Inc. (1994) *Genotyper 1.1 DNA Fragment Analysis Software. Users Manual.* A. Foster City, CA.

Allen, P.J., Amos, W., Pomeroy, P.P. & Twiss, S.D. (1995) Microsatellite variation in grey seals (*Halichoerus grypus*) shows evidence of genetic differentiation between two British breeding colonies. *Molecular Ecology* **4**, 653–662.

Amos, W., Sawcer, S.J., Feakes, R.W. & Rubinsztein, D.C. (1996) Microsatellites show mutational bias and heterozygote instability. *Nature Genetics* **13**, 390–391.

Angers, B. & Bernatchez, L. (1997) Complex evolution of a salmonid microsatellite locus and its consequences in inferring allelic divergence from size information. *Molecular Biology and Evolution* **14**, 230–238.

Arnheim, N., White, T. & Rainey, W.E. (1990) Application of PCR: organismal and population biology. *Bioscience* **40**, 171–182.

Ashley, C.T. & Warren, S.T. (1995) Trinucleotide repeat expansion and human disease. *Annual Review of Genetics* **29**, 703–728.

Avise, J.C. (1994) *Molecular Markers, Natural History and Evolution.* Chapman & Hall, New York.

Banchs, I., Bosch, A., Guimera, J., Lazar, C., Puig, A. & Estivill, X. (1994) New alleles at microsatellite loci in CEPH families mainly arise from somatic mutations in the lyphoblastoid cell lines. *Human Mutations* **3**, 365–372.

Bancroft, D.R., Pemberton, J.M., Albon, S.D. *et al.* (1995) Molecular genetic variation and individual survival during population crashes of an unmanaged ungulate population. *Philisophical Transactions of the Royal Society of London* **347** (1321), 263–273.

Band, M. & Ron, M. (1996) Creation of a SINE enriched library for the isolation of polymorphic (AGC)n microsatellite markers in the bovine genome. *Animal Genetics* **27**, 243–248.

Barendse, W., Armitage, S.M., Kossarek, L.M. *et al.* (1994) A genetic linkage map of the bovine genome. *Nature Genetics* **6**, 227–235.

Beckmann, J.S. & Weber, J.L. (1992) Survey of human and rat microsatellites. *Genomics* **12**, 627–631.

Berg, D.T., Walls, J.D., Reifel-Miller, A.E. & Grinnel, B.W. (1989) E1A-induced enhancer activity of the poly (dG-dT) poly (dA-dC) element (GT element) and interactions with a GT-specific nuclear factor. *Molecular and Cellular Biology* **9**, 5238–5243.

Blouin, M.S., Parsons, M., Lacaille, V. & Lotz, S. (1996) Use of microsatellite loci to classify individuals by relatedness. *Molecular Ecology* **5**, 393–401.

Booker, A.L., Cook, D., Bentzen, P., Wright, J.M. & Doyle, R.W. (1994) Organization of microsatellites differs between mammals and cold water teleost fishes. *Canadian Journal of Fisheries and Aquatic Sciences* **51**, 11959–11966.

Bowcock, A.M., Ruiz-Linares, A., Tomfohre, J., Minch, E. & Kidd, J.R. (1994) High resolution of human evolutionary trees with polymorphic microsatellites. *Nature* **368**, 455–457.

Brownstein, M.J. & Smith, J.R. (1996) Modulation of non-templated nucleotide addition by Taq DNA polymerase: primer modifications that facilitate genotyping. *Biotechniques* **20**, 1004–1010.

Bruford, M.W. & Wayne, R.K. (1993) Microsatellites and their application to population genetic studies. *Current Opinions in Genetics and Development* **3**, 937–943.

Bruford, M.W., Hanotte, O., Brookfield, J.F.Y. & Burke, T. (1992) Single locus and multilocus DNA fingerprinting. In: *Molecular Genetic Analysis of Populations: a Practical Approach* (ed. A.R. Hoelzel), pp. 225–269. Oxford University Press, Oxford.

Bruford, M.W., Cheesman, D.J., Coote, T. *et al.* (1996) Microsatellites and their application to conservation genetics. In: *Molecular Genetic Approaches in Conservation* (eds T.B. Smith & R.K. Wayne), pp. 278–297. Oxford University Press, Oxford.

Buchanan, F.C., Adams, L.J., Littlejohn, R.P., Maddox, J.F. & Crawford, A.M. (1994) Determination of evolutionary relationships among sheep breeds using microsatellites. *Genomics* **22**, 397–403.

Buchholtz, W.G., Pearce, J.M., Pierson, B.J. & Scribner, K.T. (1998) Dinucleotide repeat polymorphisms in waterfowl (family Anatidae): characterization of a sex-linked (Z-specific) and 14 bi-parentally inherited loci. *Animal Genetics* **29**, 323–325.

Budowle, B., Giusti, A.M., Waye, J.S. *et al.* (1991) Fixed-bin analysis for statistical evaluation of continuous distributions of allelic data from VNTR loci, for use in forensic comparisons. *American Journal of Human Genetics* **48**, 841–855.

Burke, T., Rainey, W.E. & White, T.J. (1992) Molecular variation and ecological problems. In: *Genes in Ecology* (eds R.J. Berry, T.J. Crawford & G.H. Hewitt), pp. 229–254. Blackwell Scientific Publications, Oxford.

Burke, T., Hanotte, O. & Van Pijlen, I.A. (1996) Minisatellte analysis in conservation genetics. In: *Molecular Genetic Approaches in Conservation* (eds T.B. Smith & R.K. Wayne), pp. 251–277. Oxford University Press, Oxford.

Burnett, R.C., Francisco, L.V., Derose, S.A., Storb, R. & Ostrander, E.A. (1995) Identification and characterization of a highly polymorphic microsatellite marker within the canine MHC Class I region. *Mammalian Genome* **6**, 684–685.

Callen, D.F., Thompson, A.D., Shen, Y., Phillips, H.A., Richards, R.I., Mulley, J.C. & Sutherland, G.R. (1993) Incidence and origin of null alleles in the $(AC)_n$ microsatellite markers. *American Journal of Human Genetics* **52**, 922–927.

Cavalli-Sforza, L.L. & Edwards, A.W.F. (1967) Phylogenetic analysis: models and estimation procedures. *American Journal of Human Genetics* **19**, 233–257.

Chakraborty, R., Kimmel, M., Stivers, D.N., Davison, L.J. & Deka, R. (1997) Relative mutation rates at di-, tri-, and tetranucleotide microsatellite loci. *Proceedings of the National Academy of Sciences USA* **94**, 1041–1046.

Charlesworth, B., Morgan, M.T. & Charlesworth, D. (1993) The effect of deleterious mutations on neutral molecular variation. *Genetics* **134**, 1289–1303.

Chesser, R. (1991) Influence of gene flow and breeding tactics on gene diversity within populations. *Genetics* **129**, 573–583.

Craighead, L., Paetkau, D., Reynolds, H.V., Vyse, E.R. & Strobeck, C. (1995) Microsatellite analysis of paternity and reproduction in Arctic grizzly bears. *Journal of Heredity* **86**, 255–261.

Crouau-Roy, B., Bouzekri, N., Carassi, C., Clayton, J., Contu, L. & Cambon-Thomsen, A. (1996) Strong association between microsatellites and an HLA-B, DR haplotype (B18–DR3): implication of microsatellite evolution. *Immunogenetics* **43**, 255–260.

Dallas, J.F. (1992) Estimation of microsatellite mutation rates in recombinant inbred strains of mouse. *Mammalian Genome* **3**, 32–38.

Dallas, J.F., Dod, B., Boursot, P., Prager, E.M. & Bonhomme, F. (1995) Population subdivision and gene flow in mice. *Molecular Ecology* **4**, 311–320.

Dawson, I.K., Waugh, R., Simons, A.J. & Powell, W. (1997) Simple sequence repeats provide a direct estimate of pollen-mediated gene dispersal in the tropical tree *Gliricidia sepium*. *Molecular Ecology* **6**, 179–183.

Di Rienzo, A.A., Peterson, A.C., Garza, J.C., Valdez, A.M., Slatkin, M. & Freimer, N.B. (1994) Mutational processes on simple-sequence repeat loci in human populations. *Proceedings of the National Academy of Sciences USA* **91**, 3166–3170.

Dib, C. Faure, S., Fizames, C. *et al.* (1996) A comprehensive genetic map of the human genome based on 5264 microsatellites. *Nature* **380**, 152–154.

Dietrich, W. F., Miller, J., Steen, R. *et al.* (1996) A comprehensive genetic map of the mouse genome. *Nature* **380**, 149–152.

Dow, B.C. & Ashley, M.V. (1996) Microsatellite analysis of seed dispersal and parentage of samplings in bur oak *Quercus macrocarpa*. *Molecular Ecology* **5**, 615–627.

Dracopoli, N.C., Haines J.L., Korf, B.R. *et al.* (1995) Genotyping, colony hybridization to screen for microsatellites, construction of small insert libraries enriched for short tandem repeat sequences by marker selection, and characterization of $(CA)_n$ microsatellite repeats from large insert clones. In: *Current Protocols in Human Genetics* (eds N.C. Dracopoli, J.L. Haines, B.R. Korf, D.T. Moir, C.C. Morton, C.E. Seidman, J.C. Seidman & D.R. Smith), pp. 2: 2.2.2–2.4.5. John Wiley and Sons, New York.

Ellegren. H. (1991) DNA typing of museum birds. *Nature* **354**, 113.

Ellegren, H. (1992) Cloning of highly polymorphic microsatellites in the horse. *Animal Genetics* **23**, 133–142.

Ellegren, H., Primmer, C.R. & Sheldon, B.C. (1995a) Microsatellite 'evolution': directionality or bias? *Nature Genetics* **11**, 360–362.

Ellegren, H., Lifjeld, J.T., Slagsvold, T. & Primmer, C.R. (1995b) Handicapped males and extra pair paternity in pied flycatchers: a study using microsatellite markers. *Molecular Ecology* **4**, 739–744.

Engel, S.R., Linn, R.A., Taylor, J.F. & Davis, S.K. (1996) Conservation of microsatellite loci across species of artiodactyls: Implications for population studies. *Journal of Mammalogy* **77** (2), 504–518.

Estoup, A., Tailliez, C., Cornuet, J.M. & Solignac, M. (1995) Size homoplasy and mutational processes of interrupted microsatellites in two bee species, *Apis mellifera* and *Bombus terrestris* (Apidae). *Molecular Biology and Evolution* **12**, 1074–1084.

Estoup, A., Rousset, F., Michalakis, Y., Cornuet, A.M., Adriamanga, M. & Guyomard, R. (1998) Comparative analysis of microsatellite and allozyme markers: a case study investigating microgeographic differentiation in brown trout (*Salmo trutta*). *Molecular Ecology* **7**, 339–353.

Evans, J.D. (1994) Relatedness threshold for the production of female sexuals in colonies of a polygynous ant, *Myrmica tahoensis*, as revealed by microsatellite DNA analysis. *Proceedings of the National Academy of Sciences USA* **92**, 6514–6517.

Favre, L., Balloux, F., Goudet, J. & Perrin, N. (1997) Female-biased dispersal in the monogamous mammal *Crocidura russula*: Evidence from field data and microsatellite patterns. *Proceedings of the Royal Society of London Series B Biological Sciences* **264** (1378), 127–132.

Feldman, M.W., Bergman, A., Pollock, D.D. & Goldstein, D.B. (1997) Microsatellite genetic distances with range constraints: analytic description and problems of estimation. *Genetics* **145**, 207–216.

Felsenstein, J. (1989) PHYLIP. Phylogeny Inference Package (Version 3.2). *Cladistics* **5**, 164–166.

Fields, R.L. & Scribner, K.T. (1997) Isolation and characterization of novel waterfowl microsatellites: cross-species comparisons and research applications. *Molecular Ecology* **6**, 199–202.

FitzSimmons, N.N., Moritz, C. & Moore, S.S. (1995) Conservation and dynamics of microsatellite loci over 300 million years of marine turtle evolution. *Molecular Biology and Evolution* **12**, 432–440.

Flanagan, J.C., Le Franc, M.P. & Rabbitts, T.H. (1984) Mechanisms of divergence and convergence of the immunoglobulin a1 and a2 constant region gene sequences. *Cell* **36**, 681–688.

Fredholm, M. & Wintero, A.K. (1995) Variation of short tandem repeats within and between species belonging to the Canidae family. *Mammalian Genome* **6**, 11–18.

Gagneaux, P., Boesch, C. & Woodruff, D.S. (1997) Microsatellite scoring errors associated with non-invasive genotyping based on nuclear DNA amplified from shed hair. *Molecular Ecology* **6**, 861–868.

Garnier-Gere, P. & Dillmann, C. (1992) A computer program for testing pair-wise linkage disequilibrium in subdivided populations. *Journal of Heredity* **8**, 239.

Garza, J.C. & Freimer, N.B. (1996) Homoplasy for size at microsatellite loci in humans and chimpanzees. *Genome Research* **6**, 211–217.

Garza, J.C., Slatkin, M. & Freimer, N.B. (1995) Microsatellite allele frequencies in humans and chimpanzees, with implications for constraints on allele size. *Molecular Biology and Evolution* **12**, 594–603.

Garza, J.C., Dallas, J., Duryaldi, D., Gerasimov, S., Croset, H. & Boutsot, P. (1997) Social structure of the mound-building mouse *Mus spicilegus* revealed by genetic analysis with microsatellites. *Molecular Ecology* **6**, 1009–1017.

Gertsch, P.J. & Fjerdingstad, E.J. (1997) Biased amplification and the utility of spermatheca-PCR for mating frequency studies in Hymenoptera. *Hereditas* **126**, 183–186.

Girman, D.J., Mills, M.G.L., Geffen, E. & Wayne, R.K. (1997) A molecular genetic analysis of social structure, dispersal, and interpack relationships of the African wild dog (*Lycaon pictus*). *Behavioral Ecology and Sociobiology* **40**, 187–198.

Glenn, T.C., Stephan, W., Dessauer, H.C. & Braun, M.J. (1996) Allelic diversity in alligator microsatellite loci is negatively correlated with GC content of flanking sequences and evolutionary conservation of PCR amplifiability. *Molecular Biology and Evolution* **13**, 1151–1154.

Goldstein, D.B. & Pollock, D.D. (1997) Launching microsatellites: a review of mutation processes and methods of phylogenetic inference. *Journal of Heredity* **88**, 335–342.

Goldstein, D.B., Ruiz Linares, A., Cavalli-Sforza, L.L. & Feldman, M.W. (1995a) An evaluation of genetic distances for use with microsatellite loci. *Genetics* **139**, 463–471.

Goldstein, D.B., Ruiz Linares, A., Cavalli-Sforza, L.L. & Feldman, M.W. (1995b) Genetic absolute dating based on microsatellites and the origin of modern humans. *Proceedings of the National Academy of Sciences USA* **92**, 6723–6727.

Goodman, S.J. (1997) RST CALC: a collection of computer programs for calculating unbiased estimates of genetic differentiation and gene flow from microsatellite data and determining their significance. *Molecular Ecology* **6**, 881–886.

Goodnight, K.F. & Queller, D. (1996) *Relatedness, v.5.0*. Rice University, Houston, Texas.

Goudet, J. (1995) FSTAT (Version 1.2): a computer program to calculate F-statistics. *Journal of Heredity* **86**, 485–486.

Griffiths, R.C. & Tavare, S. (1994) Sampling theory for neutral alleles in a varying environment. *Philosophical Transactions of the Royal Society of London B* **344**, 403–410.

Grist, S.A., Firgaira, F.A. & Morley, A.A. (1993) Dinucleotide repeat polymorphisms isolated by the polymerase chain reaction. *Biotechniques* **15**, 304–309.

Hagelberg, E. & Sykes, B. (1989) Ancient bone DNA amplified. *Nature* **342**, 485.

Hamada, H., Petrino, M. & Kakunaga, T. (1982) A novel repeated element with Z-DNA forming potential is widely found in evolutionarily diverse Eukaryotic genomes. *Proceedings of the National Academy of Sciences USA* **79**, 6465–6469.

Hamada, H., Petrino, M.G., Kakunaga, T., Seidman, M. & Stollar, B.D. (1984) Characterization of genomic Poly (dT-dG) Poly (dC-dA) sequences: structure, organization and conformation. *Molecular and Cellular Biology* **4**, 2610–2612.

Hamilton, W.D. (1964) The genetical evolution of social behavior. *Journal of Theoretical Biology* **7**, 1–52.

Hammond, R.L., Saccheria, I.J. & Ciofi, C. (2000) Isolation of microsatellite markers in animals. In: *Molecular Tools for Screening Biodiversity: Plants and Animals* (eds A. Karp, P.G. Isaac & D.S. Ingram). Chapman & Hall, London (In press.)

Hansen, M.M., Nielsen, E.E. & Mensberg, K.L.D. (1997) The problem of sampling families rather than populations: Relatedness among individuals in samples of juvenile brown trout *Salmo trutta* L. *Molecular Ecology* **6**, 469–474.

Hauge, X.Y. & Litt, M. (1993) A study of the origin of 'shadow bands' seen when typing dinucleotide repeat polymorphisms by the PCR. *Human Molecular Genetics* **2**, 411–415.

Ishibashi, Y., Saitoh, T., Abe, S. & Yoshida, M.C. (1997) Sex-related spatial kin structure in a spring population of grey-sided voles *Clethrionomys rufocanus* as revealed by mitochondrial and microsatellite DNA analyses. *Molecular Ecology* **6**, 63–71.

Jarne, P. & Lagoda, P.J.L. (1996) Microsatellites: from molecules to populations and back. *Trends in Ecology and Evolution* **11**, 424–429.

Jin, L., Macaubas, C., Hallmayer, J., Kimura, A. & Mignot, E. (1996) Mutation rate varies among alleles at a microsatellite locus: Phylogenetic evidence. *Proceedings of the National Academy of Sciences USA* **93**, 15285–15288.

Jones, A.G. & Avise, J.C. (1997) Microsatellite analysis of maternity and the mating system in the Gulf pipefish *Syngnathus scovelli*, a species with male pregnancy and sex-role reversal. *Molecular Ecology* **6**, 203–213.

Jurke, J. & Pethiyagoda, C. (1995) Simple repetitive DNA sequences from primates: compilation and analysis. *Journal of Molecular Evolution* **40**, 120–126.

Kandpal, R.P., Kandpal, G. & Weissman, M. (1994) Construction of libraries enriched for sequence repeats and jumping clones, and hybridization selection of region-specific markers. *Proceedings of the National Academy of Sciences USA* **91**, 88–92.

Kaukinen, J. & Varvio, S.L. (1992) Artiodactyl retroposons: association with microsatellites and use in SINEmorph detection by PCR. *Nucleic Acids Research* **20**, 2955–2958.

Karlin, S. & Burge, C. (1996) Trinucleotide repeats and long homopeptides in gene and proteins associated with nervous system disease and development. *Proceedings of the National Academy of Sciences USA* **93**, 1560–1565.

Keane, B., Dittus, W.P.J. & Melnick, D.J. (1997) Paternity assessment in wild groups of toque macaques *Macaca sinica* at Polonnaruwa Sri Lanka using molecular markers. *Molecular Ecology* **6**, 267–282.

Kemp, S.J., Hishida, O., Wambuga, J. *et al.* (1995) A panel of polymorphic bovine, ovine, and caprine microsatellite markers. *Animal Genetics* **26**, 299–306.

Kijas, J.M.H., Fowler, J.C.S., Garbett, C.A. & Thomas, M.R. (1994) Enrichment of microsatellites from the citrus genome using biotynilated oligonucleotide sequences bound to streptavidiin-coated magnetic particles. *Biotechniques* **16**, 657–662.

Klinkicht, N. & Tautz, D. (1992) Detection of simple sequence length polymorphisms by silver staining. *Molecular Ecology* **1**, 133–134.

Koob, M.D., Benzow, K.A., Bird, T.D., Day, J.W., Moseley, M.L. & Ranum, L.P.W. (1998) Rapid cloning of expanded trinucleotide repeat sequences from genomic DNA. *Nature Genetics* **18**, 72–75.

Kumar, S., Tumara, K. & Nei, M. (1993) *MEGA: Molecular evolutionary genetic analysis*, v. 1.01. The Pennsylvania State University, University Park, Pennsylvania.

Kwiatkowski, D.J., Henske, E.P., Weimer, K., Ozelius, L. & Gusella, J.F. (1992) Construction of a GT polymorphism map in human 9p. *Genomics* **12**, 229–240.

Lagercrantz, U., Ellegren, H. & Andersson, L. (1993) The abundance of various polymorphic microsatellite motifs differs between plants and vertebrates. *Nucleic Acids Research* **21**, 1111–1115.

Lanctot, R.B., Scribner, K.T., Kempenaers, B. & Weatherhead, P.J. (1997) Lekking without a paradox in the buff-breasted sandpiper. *American Naturalist* **149**, 1051–1070.

Larsen, A.H., Sigurjonsson, J., Oien, N., Vikingsson, G. & Palsboll, P. (1996) Populations genetic analysis of nuclear and mitochondrial loci in skin biopsies collected from central northeastern North Atlantic humpback whales (*Megaptera novaeangliae*): Population identity and migratory destinations. *Proceedings of the Royal Society of London Series B Biological Sciences* **263** (1376), 1611–1618.

Levinson, G. & Gutman, G.A. (1987) Slipped-strand mispairing: a major mechanism for DNA sequence evolution. *Molecular Biology and Evolution* **4**, 203–221.

Lewis, P.O. & Zaykin, D. (1998) *Genetic Data Analysis*, v. 1.0, 32–bit Computer Program. University of New Mexico, Albuquerque.

Litt, M. & Luty, J.A. (1989) A hypervariable microsatellite revealed by in vitro amplification of a dinucleotide repeat within the cardiac muscle actin gene. *American Journal of Human Genetics* **44**, 397–401.

McCusker, J., Dawson, M.T. & Noone, D. (1992) Improved method for direct PCR amplification from whole blood. *Nucleic Acids Research* **20**, 6747.

McDonald, D.B. & Potts, W.K. (1994) Cooperative display and relatedness among males in a lek-mating bird. *Science* **266**, 1030–1032.

Marklund, S., Sandberg, K. & Andersson, L. (1996) Forensic tracing of horse identities using urine samples and DNA markers. *Animal Biotechnology* **7**, 145–153.

Michalakis, Y. & Excoffier, L. (1996) A generic estimation of population subdivision using distance between alleles with special reference for microsatellite loci. *Genetics* **142**, 1061–1064.

Miller, L.M. & Kapuscinski, A.R. (1996) Microsatellite DNA markers reveal new levels of genetic variation in northern pike. *Transactions of the American Fisheries Society* **125**, 971–977.

Minch, E. (1996) *MICROSAT*, v. 1.5. Standord University Medical Centre, Standord, CA.

Moore, S.S., Sargeant, L.L., King, T.J., Mattick, J.S., Georges, M. & Hetzel, D.J.S. (1991) The conservation of dinucleotide microsatellites among mammalian genomes allows the use of heterologous PCR primer pairs in closely related species. *Genomics* **10**, 654–660.

Moore, S.S., Barendse, W., Berger, K.T., Armitage, S.M. & Hetzel, D.J.S. (1992) Bovine and ovine DNA microsatellite from the EMBL and GENBANK databases. *Animal Genetics* **23**, 463 467.

Moran, C. (1993) Microsatellite repeats in pig (*Sus domestica*) and chicken (*Gallus domesticus*) genomes. *Journal of Heredity* **84**, 274–280.

Morin, P.A., Wallis, J., Moore, J.J., Chakraborty, R. & Woodruff, D.S. (1993) Non-invasive sampling and DNA amplification of paternity, community structure, and phylogeography in wild chimpanzees. *Primates* **34**, 347–356.

Morin, P.A., Wallis, J., Moore, J.J. & Woodruff, D.S. (1994a) Paternity exclusion in a community of wild chimpanzees using hypervariable simple sequence repeats. *Molecular Ecology* **3**, 469–478.

Morin, P.A., Moore, J.J., Chakraborty, R., Jin, L., Goodall, J. & Woodruff, D.S. (1994b) Kin selection, social structure, gene flow, and the evolution of chimpanzees. *Science* **265**, 1193–1201.

Mundy, N.I., Unitt, P. & Woodruff, D.S. (1997a) Skin from feet of museum specimens as a non-destructive source of DNA for avian genotyping. *Auk* **114**, 126–129.

Mundy, N.I., Winchell, C.S., Burr, T. & Woodruff, D.S. (1997b) Microsatellite variation and microevolution in the critically endangered San Clemente Island loggerhead shrike (*Lanius ludovicianus mearnsi*). *Proceedings of the Royal Society of London Series B Biological Sciences* **264** (1383), 869–875.

Nakamura, Y., Leppert, M., O'Connell, P. *et al.* (1987) Variable number of tandem repeat (VNTR) markers for human gene mapping. *Science* **235**, 1616 1622.

National Research Council (1992) *DNA Technology in Forensic Science*. National Academy Press, Washington, District of Columbia.

Nauta, M.J. & Weissing, F.J. (1996) Constraints on allele size at microsatellite loci: Implications for genetic differentiation. *Genetics* **143**, 1021–1032.

Nei, M. (1972) Genetic distance between populations. *American Naturalist* **106**, 283–291.

Nei, M. (1973) The theory and estimation of genetic distance. In: *Genetic Structure of Populations* (ed. N.E. Morton), pp. 45–54. University Press of Hawaii, Honolulu.

Nei, M., Tajima, F. & Tateno, Y. (1983) Accuracy of estimated phylogenetic trees from molecular data. *Journal of Molecular Evolution* **19**, 153–170

Nielsen, E.E., Hansen, M.M. & Loeschcke, V. (1997) Analysis of microsatellite DNA from old scale samples of Atlantic salmon *Salmo salar*: a comparison of genetic composition over 60 years. *Molecular Ecology* **6**, 487–492.

Nielsen, R. (1997) A likelihood approach to population samples of microsatellite alleles. *Genetics* **146**, 711–716.

Olsen, J.B., Wenburg, J.K. & Bentzen, P. (1996) Semiautomated multi-locus genotyping of Pacific salmon (*Oncorhynchus* spp.) using microsatellites. *Molecular Marine Biology and Biotechnology* **5**, 259–272.

O'Reilly, P.T., Hamilton, L.C., McConnel, S.K. & Wright, J.M. (1996) Rapid analysis of genetic variation in Atlantic salmon (*Salmo salar*) by PCR multiplexing of dinucleotide and tetranucleotide microsatellites. *Canadian Journal of Fisheries and Aquatic Sciences* **53**, 2292–2298.

Ostrander, E.A., Jong, P.M., Rine, J. & Duyk, G. (1992) Construction of small-insert genomic DNA libraries highly enriched for microsatellite repeat sequences. *Proceedings of the National Academy of Sciences USA* **89**, 3419–3423.

Paetkau, D., Waits, L.P., Clarkson, P.L., Craighead, L. & Strobeck, C. (1997) An empirical evaluation of genetic distance statistics using microsatellite data from bear (Urdisae) populations. *Genetics* **147**, 1943–1957.

Pardue, M.L., Lowenhaupt, K., Rich, K. & Nordhei, A. (1987) $(dC-dA)_n \cdot (dG-dT)_n$ sequences have evolutionary conserved chromosomal locations in *Drosophila* with implications for roles in chromosome structure and function. *EMBO Journal* **6**, 1781–1789.

Pearce, J., Fields, R.L. & Scribner, K.T. (1997) Nest materials as a source of genetic data for avian ecological studies. *Journal of Field Ornithology* **68**, 471–481.

Pearce, J.M., Pierson, B.J., Talbot, S.L., Derksen, D.V., Kraege, O. & Scribner, K.T. (2000) A genetic evaluation of morphology used to identify harvested Canada geese. *Journal of Wildife Management*. In press.

Pemberton, J.M., Slate, J., Bancroft, D.R. & Barrett, J.A. (1995) Nonamplifying alleles at microsatellite loci: a caution for parentage and population studies. *Molecular Ecology* **4**, 249–252.

Pepin, L., Amigues, Y., Lepingle, A., Berthier, J.L., Bensaid, A. & Vaiman, D. (1995) Sequence conservation of microsatellite between *Bos taurus* (cattle), *Capra hiricus* (goat), and related species. Examples of use in parentage testing and phylogeny analysis. *Heredity* **74**, 53–61.

Petes, T.D., Greewell, P.W. & Dominska, M. (1997) Stabilization of microsatellite sequences by variant repeats in the yeast *Saccharomyces cerevisiae*. *Genetics* **146**, 491–498.

Primmer, C.R., Moller, A.P. & Ellegren, H. (1996) A wide-ranging survey of cross-species microsatellite amplification in birds. *Molecular Ecology* **5**, 365–378.

Primmer, C.R., Raudsepp, T., Chowdhary, B.P., Moller, A.P. & Ellegren, H. (1997) Low frequency of microsatellites in the avian genome. *Genome Research* **7**, 471–482.

Prodohl, P.A., Loughry, W.J., McDonough, C.M., Nelson, W.S. & Avise, J.C. (1996) Molecular documentation of polyembryony and the micro-spatial dispersion of clonal sibships in the nine-banded armadillo, *Dasypus novemcinctus*. *Proceedings of the Royal Society of London Series B Biological Sciences* **263** (1377), 1643–1649.

Queller, D.C. & Goodnight, K.F. (1989) Estimating relatedness using genetic markers. *Evolution* **43**, 258–275.

Queller, D.C., Strassmann, J.E. & Hughs, C.R. (1993) Microsatellites and kinship. *Trends in Ecology and Evolution* **8**, 285–288.

Ramos, M.D., Lalueza, C., Girbau, E., Perez-Perez, A., Quevedo, S., Turbon, D. & Estivill, X. (1995) Amplifying dinucleotide microsatellite loci from bone and tooth samples of up to 5000 years of age: more inconsistency than usefulness. *Human Genetics* **96**, 205–212.

Rassmann, K., Schlötterer, C. & Tautz, D. (1991) Isolation of simple-sequence loci for use in polymerase chain reaction-based DNA fingerprinting. *Electrophoresis* **12**, 113–118.

Raymond, M. & Rousset, M. (1995) GENEPOP (Version 1.2), a population genetics software for exact tests and ecumenicism. *Journal of Heredity* **86**, 248–249.

Reed, J.Z., Tollit, D.J., Thompson, P.M. & Amos, W. (1997) Molecuar scatology: the use of molecular genetic analysis to assign species, sex and individual identity to seal faeces. *Molecular Ecology* **6**, 225–234.

Reynolds, J., Weir, B.S. & Cockerham, C.C. (1983) Estimation of the coancestry coefficients: basis for a short-term genetic distance. *Genetics* **105**, 767–779.

Richard, K.R., Whitehead, H. & Wright, J.M. (1996) Polymorphic microsatellites from sperm whales and their use in the genetic identification of individuals from naturally sloughed pieces of skin. *Molecular Ecology* **5**, 313–315.

Richards, R.I. & Sutherland, G.R. (1994) Simple repeat DNA is not replicated simply. *Nature Genetics* **6**, 114–116.

Rico, C., Rico, I. & Hewitt, G. (1996) 470 million years of conservation of microsatellite loci among fish species. *Proceedings of the Royal Society of London B* **263**, 549–557.

Rose, O. & Falush, D. (1998) A threshold size for microsatellite expansion. *Molecular Biology and Evolution* **15**, 613–615.

Rosenbaum, H.C., Egan, M.G., Clapham, P.J., Brownell, R.L. Jr & Desalle, R. (1997) An effective method for isolating DNA from historical specimens of baleen. *Molecular Ecology* **7**, 677–681.

Rousset, F. (1996) Equilibrium values of measures of population subdivision for stepwise mutation processes. *Genetics* **142**, 1357–1362.

Roy, M.S., Geffen, E., Smith, D., Ostrander, E.A. & Wayne, R.A. (1994) Patterns of differentiation and hybridization in North American wolflike canids revealed by analysis of microsatellite loci. *Molecular Biology and Evolution* **11**, 553–570.

Rubinsztein, D.C., Amos, W., Leggo, J. *et al.* (1995) Microsatellite evolution-evidence for directionality and variation in rate between species. *Nature Genetics* **10**, 337–343.

Rychlik, W. & Rhoads, R.E. (1989) A computer program for choosing optimal oligonucleotides for filter hybridization, sequencing, and *in vitro* amplification of DNA. *Nucleic Acids Research* **17**, 8543–8551.

Sambrook, J., Fritsch, E.F. & Maniatis, T. (1989) *Molecular Cloning: a Laboratory Manual*, 2nd edn. Cold Spring Harbor Laboratory Press, Cold Spring Harbor, New York.

Schlötterer, C. & Tautz, D. (1992) Slippage synthesis of simple sequence DNA. *Nucleic Acids Research* **20**, 211–215.

Schlötterer, C., Amos, B. & Tautz, D. (1991) Conservation of polymorphic simple sequence loci in catacean species. *Nature* **354**, 63–65.

Schug, M.D., Wetterstrand, K.A., Gaudette, M.S., Lim, R.H., Hutter, C.M. & Aquadro, C.F. (1998) The distribution and frequency of microsatellite loci in *Drosophila melanogaster Molecular Ecology* **7**, 57–70.

Schwaiger, F.-W., Weyers, E., Epplen, C., *et al.* (1993) The paradox of MHC-DRB exon/intron evolution: α-helix and β-sheet encoding regions diverge while hypervariable intronic simple repeats coevolve with β-sheet codons. *Journal of Molecular Evolution* **37**, 260–272.

Schwengel, D.A., Jedlicka, A.E., Nanthakumar, E.J., Weber, J.L. & Levitt, R.C. (1994) Comparison of fluorescent-based semi-automated genotyping of multiple microsatellites with autoradiographic techniques. *Genomics* **22**, 46–54.

Scribner, K.T. & Bowman, T.D. (1998) Microsatellites identify depredated waterfowl remains from glaucous gull stomachs. *Molecular Ecology* **7**, 1401–1405.

Scribner, K.T., Congdon, J.D., Chesser, R.K. & Smith, M.H. (1993) Annual differences in female reproductive success affect spatial and cohort-specific genotypic heterogeneity in painted turtles. *Evolution*, **47** 1360–1373.

Scribner, K.T., Gust, J.R. & Fields, R.L. (1996) Isolation and characterization of novel microsatellite loci: cross-species amplification and population genetic applications. *Canadian Journal of Fisheries and Aquatic Sciences* **53**, 685–693.

Scribner, K.T., Crane, P.A., Spearman, W.J. & Seeb, L.W. (1998) DNA and allozyme markers provide concordant estimates of population differentiation: analyses of U.S. & Canadian populations of Yukon River fall-run chum salmon. *Canadian Journal of Fisheries and Aquatic Sciences* **55**, 1748–1758.

Selvin, S. (1980) Probability of nonpaternity determined by multiple allele codominant systems. *American Journal of Human Genetics* **32**, 276–278.

Shriver, M.D., Jin, R., Chakraborty, R. & Boerwinkle, E. (1993) VNTR allele frequency distributions under the stepwise mutation model: a computer simulation approach. *Genetics* **134**, 983–993.

Shriver, M.D., Jin, L., Boerwinkle, E., Deka, R., Ferrell, R.E. & Chakraborty, R. (1995) A novel measure of genetic distance for highly polymorphic tandem repeat loci. *Molecular Biology and Evolution* **12**, 457–462.

Sibley, C.G. & Ahlquist, J.E. (1990) *Phylogeny and Classification of Birds: a Study in Molecular Evolution*. Yale University Press, New Haven, Connecticut.

Slatkin, M. (1995) A measure of population subdivision based on microsatellite allele frequencies. *Genetics* **139**, 457–462.

Stallings, R.L., Ford, A.F., Nelson, D., Torney, D.C., Hilderbrand, C.E. & Moyzis, R.K. (1991) Evolution and distribution of $(GT)_n$ repetitive sequences in mammalian genomes. *Genomics* **10**, 807–815.

Strand, M., Prolla, T.A., Liskay, R.M. & Petes, T.D. (1993) Destabilization of tracts of simple repetitive DNA in yeast by mutations affecting DNA mismatch repair. *Nature* **365**, 274–276.

Strassmann, J.E., Solis, C.R., Peters, J.M. & Queller, D.C. (1996) Strategies for finding and using highly polymorphic DNA microsatellite loci for studies of genetic relatedness and pedigrees. In: *Molecular Zoology: Advances, Strategies, and Protocols* (eds J.D. Ferraris & S.R. Palumbi). Wiley-Liss. New York.

Sugg, D.W., Chesser, R.K., Dobson, F.S. & Hoogland, J.L. (1996) Population genetics meets behavioral ecology. *Trends in Ecology and Evolution* **11**, 338–342.

Sutherland, G.R. & Richards, R.I. (1995) Simple tandem DNA repeats and human genetics disease. *Proceedings of the National Academy of Sciences USA* **92**, 3636–3641.

Swofford, D.L. & Selander, R.B. (1981) BIOSYS-I a Fortran program for the comprehensive analysis of electrophoretic data in population genetics and systematics. *Journal of Heredity* **72**, 281–283.

Taberlet, P. & Waits, L.P. (1998) Non-invasive genetic sampling. *Trends in Ecology and Evolution* **13**, 26–27.

Taberlet, P., Camarra, J.J., Griffin, S. *et al.* (1997a) Non-invasive genetic tracking of the endangered Pyrenean brown bear population. *Molecular Ecology* **6**, 869–876.

Taberlet, P., Griffin, S., Goossens, B. *et al.* (1997b) Reliable genotyping of samples with very low DNA quantities using PCR. *Nucleic Acids Research* **24**, 3189–3194.

Takezaki, N. & Nei, M. (1996) Genetic distances and reconstruction of phylogenetic trees from microsatellite DNA. *Genetics* **144**, 389–399.

Tautz, D. (1989) hypervariablity of simple sequences as a general source for polymorphic DNA markers. *Nucleic Acids Research* **17**, 6463–6471.

Tautz, D. & Renz, M. (1984) Simple sequences are ubiquitous repetitive components of eukaryotic genomes. *Nucleic Acids Research* **12**, 4127–4138.

Tautz, D., Trick, M. & Dover, G. (1986) Cryptic simplicity in DNA is a major source of genetic variation. *Nature* **322**, 652–656.

Taylor, A.C., Sherwin, W.B. & Wayne, R.K. (1994) Genetic variation of microsatellite loci in a bottlenecked species: The northern hairy-nosed wombat *Lasiorhinus drefftii*. *Molecular Ecology* **3**, 277–290.

Taylor, A.C., Horsup, A., Johnson, C.N., Sunnucks, P. & Sherwin, B. (1997) Relatedness structure detected by microsatellite analysis and attempted pedigree reconstruction in an endangered marsupial, the northern hairy-nosed wombat *Lasiorhinus krefftii*. *Molecular Ecology* **6**, 9–19.

Valdes, A.M., Slatkin, M. & Freimer, N.B. (1993) Allele frequencies at microsatellite loci: the stepwise mutation model revisited. *Genetics* **133**, 737–749.

Van Lith, H.A. & Van Zutphen, L.M. (1996) Characterization of rabbit DNA microsatellites extracted from the EMBL nucleotide sequence database. *Animal Genetics* **27**, 387–395.

Van Oorschot, R.A.H. & Jones, M.K. (1997) DNA fingerprints from fingerprints. *Nature* **387**, 767.

Wasser, P.M. & Strobeck, C. (1998) Genetic signatures of interpopulation dispersal. *Trends in Ecology and Evolution* **13**, 43–44.

Wasser, S.K., Houston, C.S., Cadd, G.G. *et al.* (1997) Techniques for application of faecal DNA methods to field studies of ursids. *Molecular Ecology* **6**, 1091–1097.

Weber, J.L. (1990) Informativeness of $(dC-dA)_n \cdot (dG-dT)_n$ polymorphisms. *Genomics* **7**, 524–530.

Weber, J.L. & May, P.E. (1989) Abundant class of human DNA polymorphisms which can be typed using the polymerase chain reaction. *American Journal of Human Genetics* **44**, 388–396.

Weber, J.L. & Wong, C. (1993) Mutation of human short tandem repeats. *Human Molecular Genetics* **2**, 1123–1128.

Weir, B.S. (1992) Independence of VNTR alleles defined as floating bins. *American Journal of Human Genetics* **51**, 992–997.

Weir, B.S. (1996) *Genetic Data Analysis II*. Sinauer Associates, Inc, Sunderland, Massachusetts.

Weir, B.S. & Cockerham, C.C. (1984) Estimating F-statistics for the analysis of population structure. *Evolution* **38**, 1358–1370.

Westneat, D.F. & Webster, M.S. (1994) Molecular analysis of kinship in birds: interesting questions and useful techniques. In: *Molecular Ecology and Evolution: Approaches and Applications* (eds B. Schierwater, B. Steit, G.P. Wagner & R. DeSalle), pp. 91–126. Birkhäuser Verlag, Basel, Switzerland.

White, H.W. & Kusukawa, N. (1997) Agarose-based system for separation of short tandem repeat loci. *Biotechniques* **22**, 976–980.

White, M.B., Carvalho, M., Derse, D., O'Brian, S.J. & Dean, M. (1992) Detecting single-base substitutions as heteroduplex polymorphisms. *Genomics* **12**, 301–306.

Whitton, J., Rieseberg, L.H. & Ungerer, M.C. (1997) Microsatellite loci are not conserved across the Asteraceae. *Molecular Biology and Evolution* **14**, 204–209.

Wilkin, D.J., Koprivnikar, K.E. & Cohn, D.H. (1993) Heteroduplex analysis can increase the informativeness of PCR-amplified VNTR markers: application using a marker tightly linked to the COL2A1 gene. *Genomics* **15**, 372–375.

Wolfus, G.M., Garcia, D.K. & Alcivar-Warren, A. (1997) Application of the microsatellite technique for analyzing genetic diversity in shrimp breeding programs. *Aquaculture* **152** (1–4), 35–47.

Zahn, L.M. & Kwiatkowski, D.J. (1995) A 37-marker PCR-based genetic linkage map of human chromosome 9: observations on mutations and positive inference. *Genomics* **28**, 140–146.

Zardoya, R., Vollmer, D.M., Craddock, C., Streelman, J.T., Karl, S. & Meyer, A. (1996) Evolutionary conservation of microsatellite flanking regions and their use in resolving the phylogeny of cichlid fishes (Pisces: Perciformes). *Proceedings of the Royal Society of London Series B Biological Sciences* **263** (1376), 1589–1598.

Zhang, L., Leefang, E.P., Yu, J. & Arnheim, N. (1994) Studying human mutations by sperm typing: instability of CAG trinucleotide repeats in the human androgen receptor gene. *Nature Genetics* **7**, 531–535.

# CHAPTER 11

# Introns

## VICKI FRIESEN

## 11.1 Introduction

A wide variety of molecular tools have become available for the study of ecology and evolution in recent years (Avise 1994; Hillis *et al.* 1996a; and see Chapter 1). All have their strengths, but most also have limitations; for example, levels of variability detected through protein electrophoresis often are too low to be informative, especially for species breeding in previously glaciated areas or on islands (e.g. Evans 1987; Chapter 4). Mitochondrial DNA (mtDNA) is generally more variable than allozymes, but represents essentially a single supergene whose mode of inheritance is not typical of the majority of the genome (Wilson *et al.* 1985; Chapter 7). Randomly amplified polymorphic DNA (RAPDs) and microsatellite loci (Chapter 10) generally are highly variable and can be screened efficiently, but they either involve unknown genes (RAPDs) or require laborious and expensive groundwork (microsatellite loci). This chapter describes the potential of nuclear introns to provide an additional tool for studies of ecology and evolution. The function and evolution of introns, methods for assaying sequence variation in introns and results of early studies of ecology and evolution based on introns are discussed.

### 11.1.1 Molecular genetics and evolution of introns

Introns are noncoding segments of DNA that interrupt the coding sequences (exons) of nuclear genes of all eukaryotes (Fig. 11.1; Gilbert 1978; Lewin 1994; Brown *et al.* 1996). (Plant chloroplasts and mitochondrial DNA also contain introns, but these are not discussed in the present chapter.) Introns are transcribed into premessenger RNA (premRNA) along with the exons, but are spliced out by specific enzymes that complex with the premRNA as part of a ribonucleoprotein particle or spliceosome during RNA processing. The resultant messenger RNA (mRNA) is then exported to the cytoplasm for translation. As a general rule, introns begin with GU (or rarely, GC), end with AG, and contain both a 'branch site' with the sequence YUNAY (where 'Y' indicates a pyrimidine, 'N' indicates any base) located 18–40 bases from the 3' end and a polypyrimide tract upstream from the 3' splice site. Splicing begins with a transesterification reaction between the 3' end of an exon and the branch

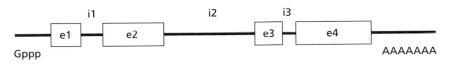

**Fig. 11.1** Structure of a typical eukaryotic gene, including introns and exons. 'Gppp' indicates the 5' cap; 'e' designates an exon; 'i' designates an intron; 'AAAAAAA' represents the poly A tail.

site, resulting in formation of a free -OH at the 3' end of the exon and a 'lariat' (loop) structure within the intron. The free -OH then undergoes transesterification with the 3' end of the intron, resulting in excision of the intron and joining of sequential exons. Recognition and processing of introns from pre-mRNAs appears to be aided by polypyrimide tracts in mammals, and high U contents in plants (Ko *et al.* 1998).

Introns range in size from ~50 bp (base pairs) to tens of thousands of bp (e.g. Brown *et al.* 1996), and sometimes contain microsatellites or exons for other loci. Their function is uncertain, but they may either provide sites for recombination between the functional domains of proteins (Gilbert 1978; de Souza *et al.* 1996; Tomita *et al.* 1996), or represent entirely nonfunctional 'junk' DNA. Two theories have been proposed for their evolution:

**1** according to the 'introns-early' hypothesis, they arose in early forms of life where they provided sites for recombination between exons, and were secondarily lost from the genomes of prokaryotes (Gilbert 1978); or

**2** according to the 'introns-late' hypothesis, introns were inserted randomly into continuous protein-coding regions in the primordial genes of eukaryotes (Cavalier-Smith 1985).

Studies aimed at differentiating between these alternatives tend to favour the introns-late hypothesis, but the evidence is not yet conclusive (e.g. Gilbert 1978; Cavalier-Smith 1985; de Souza *et al.* 1996; Cho & Doolittle 1997; Gamulin *et al.* 1997; Rzhetsky *et al.* 1997).

The potential of nuclear introns to provide molecular markers for studies of ecology and evolution results primarily from the fact that they are noncoding and essentially neutral to selection; the substitution rate for introns therefore is greater than for other single-copy nuclear DNA (Li & Graur 1991). Furthermore, because they are flanked by exons, which generally are under stabilizing selection, conserved sites are usually available for general polymerase chain reaction (PCR) primers that will amplify the desired target in a variety of species. This greatly reduces the amount of groundwork required for laboratory analyses compared to the analysis of microsatellites, and increases the versatility of introns relative to allozymes, which require high quality tissue samples. Also, because introns occur in virtually all structural genes and are scattered throughout the genome, they can provide potentially thousands of independent markers. Finally, sequence variation in introns can be determined directly, in contrast to methods such as protein electrophoresis and the analysis of restriction fragments, which do not detect a high proportion of variation (e.g. Birt *et al.* 1995); this enables application of both sequence- and frequency-based statistical analyses.

## 11.2 Methods

The most efficient method of analysing variation in introns involves DNA amplification and either direct or indirect sequence analysis. Polymerase chain reaction-based analysis of variation in introns begins with the selection of primers. To design primers, DNA sequences for a variety of species for a specific gene are extracted from a sequence database and are aligned. Introns of a size appropriate for a given problem are then located: larger introns (>1000 bp) generally are more favourable for phylogenetic analyses, because they can yield a large amount of sequence information from a single amplification (e.g. Slade *et al.* 1994; Prychitko & Moore 1997); smaller ones are more appropriate for population-level analyses, when indirect mutation-detection methods are required to screen many loci in large numbers of individuals (Lessa & Applebaum 1993; Friesen *et al.* 1997). Proteins belonging to large multigene families (e.g. immunoglobulins) and loci undergoing concerted evolution (e.g. the internal transcribed spacers of the ribosomal RNA genes) are less useful as molecular markers for ecological or evolutionary studies because more than one locus may be amplified by a single pair of primers, complicating the interpretation of radiograms. Polymerase chain reaction primers generally are designed to anneal to blocks of sequence that are conserved across species in the exons flanking the target intron (Kocher *et al.* 1989; Lessa 1992; Palumbi & Baker 1994). Intron primers are most useful if they also amplify a portion of exon, so that the identity of the amplification product can be confirmed by comparison of exon sequences with those of a reference taxon. Primers also have greatest utility if they have high melting temperatures (either by homology to GC-rich regions or through increased length) to enable amplification and sequencing at high temperatures if required, either to improve specificity or to overcome strong secondary structure in the template DNA. A variety of general PCR primers, many of which were developed originally for genome mapping, are already available through the efforts of a number of research groups (Table 11.1).

Following the selection of appropriate primers, PCR conditions generally need to be optimized to ensure both that only a single locus is amplified and that all alleles for the locus will amplify and thus to prevent null alleles (see Appendix). For phylogenetic analyses, in which sequence is required from only a few representatives of each taxon, sequence variation in introns can be

**Table 11.1** Examples of general polymerase chain reaction (PCR) primers used to amplify introns for studies of ecology and evolution†‡

| Locus | Primers (forward/reverse)§ | Taxa | Reference¶ |
|---|---|---|---|
| Aldolase A & C | TGTGCCCAGTATAAGAAGGATGG/CCCATCAGGGAGAATTTCAGGCTCCACAA | Vertebrates | 1,2 |
| Arginine kinase | GACCACCTCCGAGTCATCTCZATG/GTGCCAAGGTTGGTDGGGCA | Invertebrates | 3 |
| Creatine kinase | GACCACCTCCGAGTCATCTCZATG/CAGGTGCTCGTTCCACATGAA | Vertebrates | 3 |
| Cytochrome c | CGCTGCGCCACTGCCACAC/CATCTTGGTGCCGGGGATGTACTTCTT | Insects | 3 |
| Cytochrome c | AAGTGCGCCAGTGCCACAC/CATCTTGGTTCCAGGGATGTA | Plants | 3 |
| Elongation factor 1α | TCCGGATGGCAYGGCAAYATG/ATGTGAGCAGTGTGGCAATCCAA | Invertebrates | 4 |
| α-Enolase | TGGACTTCAAATCCCCGATGAATCCCAGC/CCAGGCACCCAGTCTACCTGGTCAAA | Vertebrates | 5 |
| Ependymin II | CCGTCGCTGCCCTCTCC/GCCACTGATCAGACAG | Salmonids | 6 |
| Glyceraldehyde-3-phosphate dehydrogenase | ACCTTTAATGCGGGTGCTGGCATTGC/CATCAAGTCCACAACACGGTTGCTGTA | Vertebrates | 5 |
| Lamin A | CCAAGAAGCAGGTGCAGGATGAGATGC/CTGCCGCCCGTTGTGCGATCTCCACCA | Vertebrates | 5 |
| Major histocompatibility complex DQA | CCGGATCCCAGTACACCATGAATTCATGG/ | Mammals | 2 |
| | CCGGATCCCAGTGCTCCACCTTGCCAGTC | Animals | 3 |
| Proto-oncogene int | AACCTTCACAACAAYGAGGC/TTGCACTCTTGICGCATYTC | | |
| ras-1 proto-oncogene | GAGCCGCGCTCACCATCCAGCTC/CATGTCTCCCATCAATCAC | Salmonids | 6 |
| S7 Ribosomal protein | TGGCCTCTTCCTGGCCGTC/AACTCGTCTGGCTTTTCGCC | Fish | 7 |
| β-Tubulin | CAGGCTGGTCAATGTGGYAAYGA/CCRTGYTCATCACTKATYACCTCCCA | Animals | 4 |

† Please check original references for notes on potential problems before using any of these primers.
‡ Additional primers, whose utilities are less general or uncertain, are given in Lessa (1992; pocket gophers *Thomomys bottae*), Lessa & Applebaum (1993; Mammalia), Slade *et al.* (1993; Vertebrata), Leicht *et al.* (1993; *Drosophila*), Palumbi (1996; various), Venta *et al.* (1996; mammals), Friesen *et al.* (1997; Vertebrata), Lyons *et al.* (1997; Mammalia), Moran *et al.* (1997; Salmonidae), Prychitko & Moore (1997; Picidae), Harris & Disotell (1998; Papionini), Slattery and O'Brien (1998; Carnivora: Felidae), Friesen *et al.* (2000; general).
§ D=G or T; I=inosine; K=G, A or T; R=A or G; Y=C or T; Z=C or G.
¶ 1, Lessa & Applebaum 1993; 2, Slade *et al.* 1993; 3, Palumbi & Baker 1994; 4, Palumbi 1996; 5, Friesen *et al.* 1997; 6, Moran *et al.* 1997; 7, Chow & Hazama 1998.

determined directly through various sequencing protocols (Chapter 3; Hillis *et al.* 1996b). For population-level investigations, in which large numbers of individuals need to be assayed, a combination of indirect methods of mutation detection and direct sequencing becomes more efficient (Chapter 3; Lessa & Applebaum 1993; Hillis *et al.* 1996b). Various researchers have screened samples for sequence variation using analyses of restriction fragment-length polymorphisms (RFLPs, e.g. Baker *et al.* 1998), temperature gradient gel electrophoresis (TGGE, e.g. Wartell *et al.* 1990), denaturing gradient gel ele trophoresis (DGGE, e.g. Lessa 1992), or analysis of single-stranded conformational polymorphisms (SSCPs, e.g. Friesen *et al.* 1997; Liu *et al.* 1999). These indirect methods greatly reduce the amount of time and money required for large-scale population surveys without necessarily compromising the sensitivity of the approach. Various researchers have found that combining the analysis of SSCPs with direct sequencing of variant genotypes provides a highly efficient method of assaying population-level sequence variation (see Appendix, e.g. Friesen *et al.* 1996, 1997; Liu *et al.* 1999; Holder *et al.* 1999). The analysis of SSCPs is based on the observations that a single-stranded fragment of DNA forms a specific secondary structure that depends on the primary sequence of bases and temperature, and that the rate at which a segment of DNA migrates through a gel in response to an electrical field depends in part on its secondary structure (Figs 11.2, 11.3). DNA fragments that differ by even a single base have slightly different secondary structures and therefore migrate at different rates (Orita *et al.* 1989).

For phylogenetic analyses, intron sequences can be subjected directly to various methods of phylogenetic reconstruction (Swofford *et al.* 1996). For population-level studies, data from introns can be analysed in any of at least three ways: (1) using classical frequency-methods (Chapter 4; Weir 1996); (2)

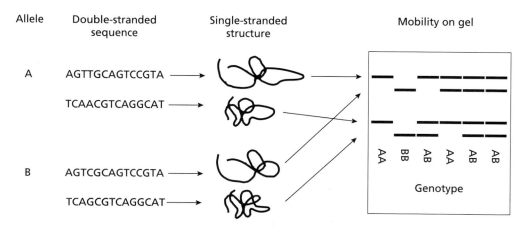

**Fig. 11.2** Diagram of a single-stranded conformational polymorphism (SSCP) gel.

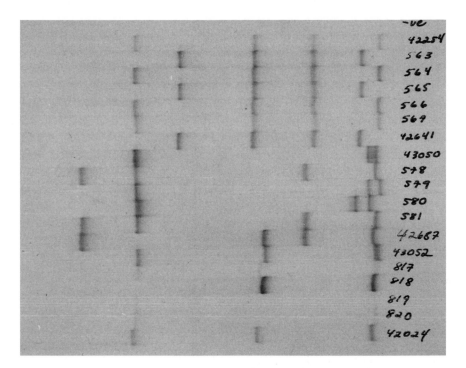

**Fig. 11.3** Example of a single-stranded conformational polymorphism (SSCP) autoradiogram, representing variation in a lactate dehydrogenase intron in three families of tree swallows (*Tachycineta bicolor*). Parents are designated by five-digit codes; chicks are designated by three-digit codes. Each band represents one strand of one allele: homozygotes are represented by two bands; heterozygotes are represented by three or four bands (depending on whether a mutation results in a shift in migration rate of one or both strands). (E. Croteau, R. Robertson, A. Moller and V.L. Friesen unpublished data.)

using sequence-based methods (Swofford *et al.* 1996); or (3) using methods that combine allele frequencies and sequences (e.g. analysis of molecular variance, Excoffier *et al.* 1992; Michalakis & Excoffier 1996; coalescence analysis, Slatkin & Hudson 1991; Rogers & Harpending 1992; Rogers *et al.* 1996; Beerli & Felsenstein 1999). The latter are the most powerful approaches, and cannot be applied easily to data from either allozymes or microsatellites.

## 11.3  Molecular evolution

Except in special cases, the use of molecular markers to study ecology and evolution assumes that substitutions are neutral to selection. It usually also requires that the identity of the target, and therefore mechanisms of molecular evolution, are known, and it sometimes assumes that substitutions fit a molecular clock with a predictable rate. To date, little has been published regarding molecular evolution in introns. Friesen *et al.* (1997) and V.L. Friesen

**Table 11.2** (a) Nucleotide sequences and inferred amino acid sequences of a lamin intron and parts of the flanking exons for marbled murrelets and a reference taxon (the chicken, *Gallus gallus*). Dots represent the intron of the reference taxon, which could not be aligned easily with that of murrelets. Sites that are underlined are sites of variation within murrelets. Site designations are positions relative to the 5′ primer. (b) Sequence variation within a lamin intron for *Brachyramphus* murrelets. 'N' is the number of individuals possessing a given allele. 'Bm' indicates an allele that occurs in marbled murrelets; 'Bb' indicates an allele that occurs in Kittlitz's murrelets; 'Bp' indicates an allele that occurs in long-billed murrelets (*Brachyramphinus perdix*). (From Friesen *et al.* 1997)

(a)

| Amino acid | R   R   V   D   A   E   N   R   L   Q   T   L   K   E   E   L | 17 |
|---|---|---|
| Chicken | TGCGCCGCGTGGACGCCGAGAACCGCCTGCAGACCCTGAAGGAGGAGTT | 49 |
| Murrelet | TGCGCCGCGTGGACGCCGAGAACCGGCTGCAGACACTGAAGGAAGAGCT | 49 |
| Amino acid | E   F   Q   K   N   I   Y   S   E | 26 |
| Chicken | GGAGTTCCAGAAGAACATTTACAGCGAG............................ | 76 |
| Murrelet | CGAATTCCAGAAGAACATCTACAGCGAGGTGAGGTGCCATGCGCGGTGCC | 99 |
| Amino acid | | |
| Chicken | ............................................................ | |
| Murrelet | GGGCTTGCGTGGGGTAACACCCCCCGTGTGTCGCATCCCCGTCCCCAAGC | 149 |
| Amino acid | E   L | 28 |
| Chicken | ...............................................GAGCT | 81 |
| Murrelet | CGATCCCACCCCGAGGTGACACTGTGTCCCCTGGCCTGTCCCCAGGAGCT | 199 |
| Amino acid | R   E   T   K   R   R   H   E   T   R | 38 |
| Chicken | GCGGGAGACCAAACGGCGGCACGAGACGCGC | 112 |
| Murrelet | GCGGGAGACCAAGCGCCGCCACGAGACCCGG | 230 |

(b)

| Genotype | Site 111111 19123356 74941412 | N |
|---|---|---|
| Bm3 | CGCCCAGG | 17 |
| Bm4 | CGCCTAGG | 5 |
| Bm5 | CGTCCAGG | 3 |
| Bm6 | CGCCCGGG | 7 |
| Bb3 | CGCTCGAA | 1 |
| Bp4 | CGCCCGGG | 1 |
| Bp3 | TACCCGGG | 2 |

and B.C. Cogdon (unpublished data) studied sequence variation in nine introns in brachyramphine murrelets. They concluded that their amplification products represented the target introns for most loci (Table 11.2); e.g. sequences of putative introns began with 'GT', ended with 'AG' and contained a potential branch site, and sequences of putative exons matched those of the reference bird almost perfectly. Substitution patterns for murrelet introns also did not differ from predictions of the neutral theory (Tables 11.2–11.4): most

**Table 11.3** Intron sizes, numbers of alleles and numbers of transitions (TI), transversions (TV) and insertions/deletions (indels) among intron alleles and cytochrome *b* genotypes for marbled murrelets (V.L. Friesen & B.C. Cogdon, unpublished data)

| Locus† | Size (bp) | Alleles (*n*) | TI (*n*) | TV (*n*) | Indels (*n*) (sizes, bp) |
|---|---|---|---|---|---|
| Aldolase B | 280+ | 4 | 0–2 | 0–2 | 0 |
| α-Enolase | 247 | 5 | 1–6 | 0–2 | 0 |
| GAPD | 351 | 14 | 0–5 | 0–2 | 0–2 (4,13) |
| Lamin | 117 | 5 | 1–2 | 0 | 0 |
| LDH | 314 | 7 | 0–5 | 0–1 | 0 |
| MPP | 220 | 4 | 1–2 | 0 | 0 |
| OD | 478 | 8 | 0–5 | 0–2 | 1 (20) |
| P40 | 250 | 8 | 1–5 | 0 | 0–2 (1,9) |
| Tropo | 1181 | 7 | 1–3 | 0 | 0 |
| Cytochrome *b* | 1045 | 13 | 1–4 | 0 | 0 |

†GAPD, glyceraldehyde-3-phosphate dehydrogenase; LDH, lactate dehydrogenase; MPP, myelin proteolipid protein; OD, ornithine decarboylase; P40, ribosomal protein 40; Tropo, tropomyosin.

**Table 11.4** Numbers of transitions (TI), transversions (TV) and insertions/deletions (indels), and percent sequence divergence (not including indels) between intron alleles and cytochrome *b* genotypes of marbled murrelets and Kittlitz's murrelets (V.L. Friesen & B.C. Cogdon, unpublished data)

| Locus† | TI (*n*) | TV (*n*) | Indels (*n*) (sizes, bp) | % Sequence divergence |
|---|---|---|---|---|
| Aldolase | 1–4 | 0–1 | 1 (1) | 0.4–1.4 |
| Enolase | 1–6 | 0–1 | 0 | 0.1–2.8 |
| GAPD | 0–3 | 1–2 | 1 (15) | 0.0–1.1 |
| Lamin | 0–5 | 0 | 0 | 0.0–4.3 |
| LDH | 1–4 | 1–2 | 0 | 0.6–1.6 |
| MPP | 2–3 | 0 | 0 | 0.9–1.4 |
| OD | 2–5 | 2–3 | 0 | 0.8–1.7 |
| P40 | 4–9 | 1–2 | 1–2 (1,9) | 2.4–4.4 |
| Tropo | 2–5 | 0 | 0 | 0.2–0.4 |
| Cytochrome *b* | 59 | 4 | 0 | 5.6 |

†GAPD, glyceraldehyde-3-phosphate dehydrogenase; LDH, lactate dehydrogenase; MPP, myelin proteolipid protein; OD, ornithine decarbodylase; P40, ribosomal protein 40; Tropo, tropomyosin.

substitutions within exons were silent; substitutions were distributed randomly within introns; transitions outnumbered transversions; substitutions described a 'star' pattern with one or two hubs (e.g. Fig. 11.4); and Tajima's D statistic did not differ significantly from 0 (D=−0.012, P>0.1), suggesting an absence of strong selective forces. Clark *et al.* (1996), Prychitko & Moore

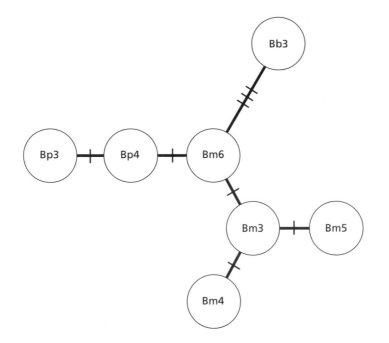

**Fig. 11.4** Substitutional relationships among lamin alleles identified in Table 11.2(b). Cross-bars indicate substitutions. (From Friesen *et al.* 1997.)

(1997) and Slade *et al.* (1998) reached similar conclusions for sequence varia-tion in various introns in *Drosophila*, woodpeckers (family Picidae) and South-ern elephant seals (*Mirounga leonina*), respectively. Divergence rates for introns in murrelets averaged $0.5\pm0.06\%$ Ma$^{-1}$ (Congdon *et al.* 2000), which is approximately one-quarter the mean for mtDNA. Rates differed signifi-cantly among loci ($F_{1,8}$, d.f.$=1,51$, $P<0.001$), being highest for the P40 intron ($0.72\pm0.04\%$ Ma$^{-1}$) and lowest for the tropomyosin intron ($0.16\pm0.01\%$ Ma$^{-1}$). Similarly, Prychitko & Moore (1997) concluded that substitutions within an intron of the β-fibrinogen gene in woodpeckers occur at a rate approximately one-quarter that of the mitochondrial cytochrome *b* gene, and Slade *et al.* (1998) estimated that substitutions in a number of introns occur at a mean rate approximately one-sixtieth that of the mitochondrial control region. Although mutation rates in introns are lower than in mtDNA, variabil-ities in the two types of loci are generally equivalent (see below), probably because the effective population size of nuclear genes is four times that of mtDNA (Wilson *et al.* 1985).

## 11.4 Applications

Because sequence analysis of nuclear introns is a relatively new tool, few examples of its application to the understanding of ecology and evolution exist. Nonetheless, results of studies published to date underline the potential of introns for investigating a wide variety of questions.

## 11.4.1 Phylogenetics

Although numerous studies have used data either on exon sequences or on intron locations in phylogenetic reconstruction, few have used intron sequences. However, intron sequences are being used increasingly to supplement sequence from mtDNA in phylogenetic analyses, and may be especially useful for determining whether phylogenies derived from mtDNA represent gene trees or taxon trees (Pamilo & Nei 1988; Moore 1995). In an exemplary study, Prychitko & Moore (1997) used sequence information from an intron in the β-fibrinogen gene, as well as the mitochondrial cytochrome *b* gene, to investigate phylogenetic relationships among woodpeckers. They found that, although overall variation in the intron was lower than in cytochrome *b*, the intron contained more phylogenetically informative sites and fewer homoplasies, and therefore was as useful as cytochrome *b* for phylogenetic reconstruction. Intron sequences have also been used to reconstruct phylogenetic relationships among felids (Slattery & O'Brien 1998) and various primates (Koop *et al.* 1989; Harada *et al.* 1995; Harris & Disotell 1998). Slade *et al.* (1994) found insufficient information in individual introns to reconstruct a phylogeny for the pinnipeds, but were able to derive a well-supported tree when they combined sequences from a number of loci.

## 11.4.2 Hybridization

To the knowledge of this author, no researchers have used nuclear introns to study hybridization, however, results from phylogenetic analyses indicate that introns can provide a rich source of species-specific markers for identification of hybrids and back-cross individuals. For example, behaviour and plumage variation suggest that marbled and Kittlitz's murrelets may hybridize in the Gulf of Alaska (K. Kuletz, unpublished data); N. Pacheco *et al.* (unpublished data) compared variation in six introns among 35 marbled murrelets and 15 Kittlitz's murrelets, and concluded that (a) none of the birds they sampled were either F1 hybrids or first generation back-crosses, and (b) the frequency of F1 hybrids and first generation back-crosses in the Gulf of Alaska must be less than 2%.

## 11.4.3 Population genetics

A few studies in which intron variation was used to measure genetic differentiation and gene flow among populations of animals have been completed recently. Congdon *et al.*'s (2000) investigation of marbled murrelets (*Brachy-ramphus marmoratus*) is noteworthy in that it enables direct comparison of results for nuclear introns, mtDNA, allozymes and microsatellites. Marbled murrelets are pursuit-diving seabirds that feed inshore and nest predominantly in large trees in old-growth forest along the Pacific coast of North America (Ralph *et al.* 1995). They are a conservation concern because their breeding habits place them in direct conflict with logging interests, and their feeding ecology renders them vulnerable to oil pollution (Ralph *et al.* 1995). They are also of interest in the study of evolutionary genetics because birds in western Alaska nest on the ground, but the genetic distinctiveness and evolutionary origin of ground-nesting murrelets are unknown. Pitocchelli *et al.* (1995) compared variation in morphometrics and mitochondrial RFLPs among murrelets nesting in Alaska; they found that tree-nesting and ground-nesting murrelets differed slightly in morphometrics but not in mtDNA. Friesen *et al.* (1996) compared variation in allozymes and mitochondrial cytochrome *b* sequences among murrelets breeding between Oregon and western Alaska and found extensive variation in cytochrome *b* genotypes, but no evidence of population differentiation. In contrast, Friesen *et al.* (1996) found significant population differentiation in allozymes (Table 11.5), but were unable to determine details of the genetic structure.

In a later analysis of a larger number of murrelets breeding between British Columbia and the western Aleutian Islands, Congdon *et al.* (2000) found between two and 18 alleles at each of three microsatellite loci, and reported weak but significant differentiation between tree-nesting and ground-nesting murrelets (Table 11.5). They also surveyed sequence variation in nine introns and found high levels of variability (Table 11.5): the percentage loci polymorphic was five times higher than for 39 allozymes; the number of alleles per locus was five times higher than for allozymes; and the number of alleles for the most variable intron was similar to the numbers of alleles found both in cytochrome *b* and in the most polymorphic microsatellite locus. Numbers of

**Table 11.5** Indices of variability, and estimates of $\Phi_{ST}$ obtained using different molecular markers for marbled murrelets (V.L. Friesen & B.C. Cogdon, unpublished information)

|  | Introns | Allozymes | Cytochrome *b* | Microsatellites |
|---|---|---|---|---|
| Number of loci surveyed | 9.00 | 39.00 | 1.00 | 3.00 |
| Percent of loci polymorphic | 100.00 | 21.00 | — | 100.00 |
| Mean number of alleles/loci | 7.10 | 1.30 | — | 7.30 |
| Maximum number of alleles/loci | 14.00 | 3.00 | 13.00 | 16.00 |
| $\Phi_{ST}$† | 0.09 | 0.09 | 0.01 | 0.03 |

†$F_{ST}$ for microsatellites.

alleles did not correlate with intron size. Both analysis of molecular variance and phylogenetic analysis based on genotype frequencies and sequence divergence among alleles indicated that murrelets from the Aleutian Islands are significantly different from those from mainland North America. The mismatch distributions (the frequencies of pairs of individuals whose genotypes differ by a given number of substitutions) for introns showed two peaks, suggesting that their sample of murrelets represents two presently or historically differentiated populations (Fig. 11.5). Estimates of gene flow based on coalescent

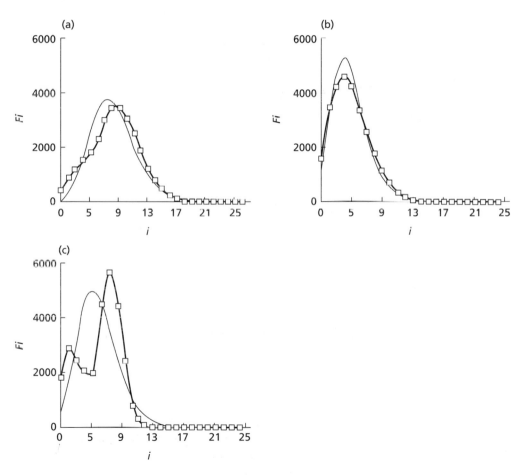

**Fig. 11.5** Distributions of pairwise sequence differences among individual marbled murrelets for: (a) all intron loci sampled; (b) introns with estimated mutation rates between 0.35 and 0.65% Ma$^{-1}$; and (c) introns with estimated mutation rates > 0.6% Ma$^{-1}$. $Fi$, frequency of a difference of magnitude $i$; $i$, number of pairwise differences between genotypes. (From Congdon *et al.* 2000.)

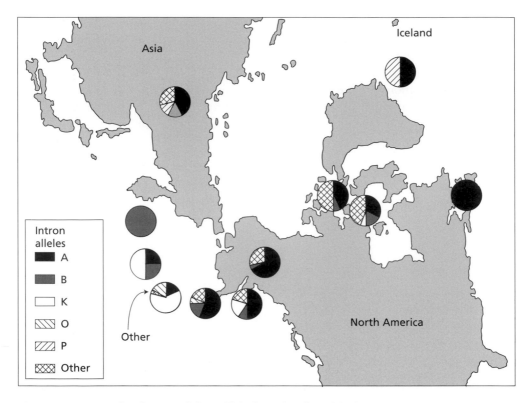

**Fig. 11.6** Frequency distributions of glyceraldehyde-3-phosphate dehydrogenase intron alleles in rock ptarmigan populations. Pie charts indicate relative frequencies of the five most common alleles; alleles with frequencies ≤2% are grouped together and designated as 'other'. From Holder *et al.* (2000).

theory suggested that net gene flow is from west to east, is restricted between murrelets in the Aleutian Islands and those in the mainland, and is not restricted between tree- and ground-nesting murrelets. Congdon *et al.* (2000) concluded that murrelets recolonized Alaska and British Columbia from at least two Pleistocene refugia, and that murrelets breeding in the Aleutian Islands are genetically isolated from those breeding in mainland North America. Analysis of allozymes, microsatellites and cytochrome *b* sequences failed to detect any of these effects (Friesen *et al.* 1996), possibly because of the lower mutation rates of allozymes, the greater susceptibility of mtDNA to population bottlenecks, and/or a greater incidence of homoplasy in the microsatellite loci.

Introns also have been used in population genetic studies of other species of vertebrates and invertebrates; for example, Holder *et al.* (2000) compared variation in a glyceraldehyde-3-phosphate dehydrogenase intron as well as in the mitochondrial control region among rock ptarmigan (*Lagopus mutus*) from throughout the Nearctic; their results suggest that ptarmigan were isolated and diverged from each other both genetically and morphologically in multiple refugia during the Pleistocene glaciations (Fig. 11.6). In this study, data from introns were less informative than those for the control region, but served to confirm the results for mtDNA. Similar results were found for a number of introns in Southern elephant seals (Slade *et al.* 1998). Palumbi & Baker (1994) and Baker *et al.* (1998) found significant population differentiation in mtDNA and an actin intron among humpback whales (*Megaptera novaengliae*), both between the Atlantic and Pacific Oceans and within the Pacific Ocean. In another study, Moran *et al.* (1997) used RFLPs in eight introns to distinguish populations of chinook salmon (*Oncorhynchus tshawytscha*) and steelhead (*O. mykiss*) from north-western North America. Intron sequences are also being used to help unravel the evolutionary history of humans (Huang *et al.* 1998). Leebens *et al.* (1998) used sequence variation in an intron of a lysozyme gene to measure gene flow within and between yucca moths (*Tegeticula yuccasella*) that pollinate different species of yuccas (*Yucca* spp.), and inferred that moths pollinating different species of hosts are genetically isolated, despite high migration among populations of moth that pollinate the same host species. Finally, introns in medflies (*Ceratitis capitata*) were found to retain more variation than either allozymes or mtDNA during population bottlenecks, and so may be more useful in tracing colonization histories (Villablanca *et al.* 1998).

### 11.4.4 Identification of gender and kinship

Introns have not yet been applied to the identification of either gender or kinship (including parentage), although results of studies to date suggest that introns have potential for addressing both of these problems. One locus (a follistatin intron) that was screened by V.L. Friesen & B.C. Congdon (unpublished information) in murrelets was found to be sex-linked, and could be used to identify gender reliably. (The universality of this intron as a sex-linked marker has not yet been assessed.) The fact that variabilities of some introns in murrelets are as high as some microsatellite loci (Table 11.5) suggests that they could be used in analyses of parentage; variation in intron II of an MHC-II-$\beta$ gene in auklets is particularly high (H.E. Walsh, unpublished information). Use of introns to assign parentage would be technically more efficient than classical DNA fingerprints, and would enable parental assignments rather than just exclusions. Introns would also provide an advantage over microsatellites in that the extensive groundwork involved in designing PCR primers for microsatellites is not necessary.

## 11.4.5 Conservation

Molecular markers have many potential applications in conservation (Avise 1994; Avise & Hamrick 1996). Most importantly, they can be used to identify cryptic species—populations of organisms that are morphologically and behaviourally similar but that are reproductively isolated. Molecular markers may also be used to estimate levels of gene flow among local populations, and thus the capacity of a species to recolonize areas where it has declined or been eliminated. They have the potential to indicate the extent of genetic differentiation of local populations, and thus may be used to identify appropriate sites for population monitoring and restoration, to prioritize conservation and restoration efforts, and to identify possible source and sink populations (Friesen 1997). Furthermore, they may be useful both for assessing the impact of natural and anthropogenic mortality and for estimating effective population sizes. Molecular markers have been used extensively in plant conservation and fisheries management (e.g. Ryman & Utter 1987; Loeschcke *et al.* 1994; Avise & Hamrick 1996), but the potential of nuclear introns is only starting to be realized; for example, Moran *et al.* (1997) used geographical variation in eight introns in two species of salmon to estimate genetic differentiation of populations for management purposes. Currently, my students and I are using introns to aid in the restoration of populations of common murres (A. Patirana, T. Birt *et al.*, unpublished information), pigeon guillemots (*Cepphus columba*; Moy *et al.*, unpublished information) and marbled and Kittlitz's murrelets (*Brachyramphus brevirostris*; Congdon *et al.*, (2000) to the Gulf of Alaska following the *Exxon Valdez* oil spill. Insofar as introns provide a sensitive and efficient tool for studies of phylogenetics, hybridization and population genetics, they should provide a powerful addition to the arsenal of molecular tools for conservation.

## 11.5 Utility of introns for studies of plants and other taxa

Plastid introns have been used extensively in phylogenetic studies of plants, but uses of nuclear introns in studies of ecology and evolution in plants are only just beginning, with marked success; for example, Liu *et al.* (1999) used SSCPs to compare sequence variation in a phosphoglucose isomerase intron in *Leavenworthia*, and found that variation in selfing and inbreeding populations was greater than in self-incompatible populations; however, the causes of these differences could not be inferred. Doyle *et al.* (1996) compared sequence variation in chloroplast DNA and two histone (H3) introns among the soybean and its congeners (*Glycine* spp.); they found that one intron did not provide phylogenetically useful information, but that the other was useful for inferring relationships among species, and clarified some relationships that could not be resolved in previous studies. Similarly, intron sequences aided in the resolution of phylogenetic relationships among species of peonies (Sang *et al.* 1997). Introns may also be useful in agriculture and industry; for example, Fulton & Brown (1997) found that the presence of an intron in *Monilinia fructicola*, a fungus causing brown rot in pome and stone fruits, provided a rapid and sensitive method for detecting the fungus in quarantined fruit. Similarly, Perlin *et al.* (1997) found that they could distinguish strains of the fungal phytopathogen *Microbotryum violaceum* using sequence variation in introns of the $\gamma$-tubulin gene in combination with electrophoretic karyoptyes and RAPDs. And de Barros *et al.* (1996) were able to differentiate strains of yeast during fermentation of wine using PCR primers based on intron splicesites. However, variation within an alcohol dehydrogenase intron in barley (*Hordeum vulgare*) was too low for use in ecological studies (Petersen & Seberg 1998).

## 11.6 Limitations of introns

Results to date suggest that introns have only a few limitations as a tool for studying ecology and evolution. On a theoretical level, introns are nuclear genes and, therefore, have a larger effective population size than mitochondrial genes. They are therefore less useful for studying recent population bottlenecks and restrictions in gene flow than is mtDNA. However, given the inherent limitations of analyses of mtDNA (e.g. it is maternally inherited so it tracks female-mediated gene flow only), combination of analyses of the two types of markers should provide a powerful approach. On a more practical level, laboratory assays of intron sequences for population-level problems involve two main difficulties. The first involves determination of sequences of individual alleles. Provided that homozygous individuals occur within a sample, unambiguous sequences can be obtained for each allele. Problems arise with loci such as the major histocompatibility complex, which are especially variable, because alleles occur mostly or entirely in the heterozygous state (see Chapter 3). In such cases, additional steps such as cloning or amplification of alleles from gel slices may be required to obtain unambiguous sequences of individual alleles. The second main problem involves interpretation of SSCP gels (Fig. 11.3). Theoretically, SSCPs should be represented by only two bands in homozygous individuals, and four bands in heterozygous individuals. In practice, most individuals are represented by larger numbers of bands. Extraneous bands appear to represent multiple stable secondary structures, reannealed products and possibly products of PCR jumping. Nonetheless, autoradiograms usually can be interpreted with practice and with the aid of additional steps such as direct sequence analysis of a few individuals or cloning.

## 11.7 Summary and future directions

Sequence variation in introns promises to provide a sensitive and efficient tool for studies of phylogenetics, hybridization and population genetics. To date, the major limitation appears to be in resolving recent evolutionary events because of the larger effective population size and lower mutation rate relative to mitochondrial and plastid DNA. Also, the efficiency with which sequence variation can be screened in large numbers of samples is lower than for microsatellites and allozymes at present. The potential of these markers for studies of gender and kinship, as well as for the conservation of biodiversity, is high but remains to be tested empirically.

## Appendix

### DNA amplification

In my laboratory, we usually conduct initial amplifications on a small number of samples in 25 µL of a cocktail containing 100 mmol Tris pH 8.0, 1.5 mmol $MgCl_2$, 50 mmol KCl, 0.2 mmol deoxyribonucleoside triphosphates (dNTPs), 0.4 µmol each primer and 0.5 units of Taq DNA polymerase (Friesen *et al.* 1997). Initial PCR temperature profiles involve denaturation at 94 °C for 90 s followed by 30 cycles of denaturation at 94 °C for 30 s, annealing at 50 °C for 30 s and extension at 72 °C for 45 s (longer for targets > 1 kb), and ending with extension at 72 °C for 3 min. Amplifications are conducted both with and without 5% dimethyl sulphoxide (DMSO), which aids in denaturation of GC-rich regions but which appears to be inhibitory for some loci. Bovine serum albumin is sometimes required if the quality of the DNA template is poor, but other adjuncts do not appear to improve the reactions significantly. Products are subjected to electrophoresis through 1.2–2.0% agarose gels in 49 mmol Tris-acetate (pH 7.5) in the presence of 10 µg mL⁻¹ ethidium bromide. Amplifications are repeated with a suite of buffers ranging from 0.5 to 3.5 mmol $MgCl_2$ and from 35 to 65 mmol KCl to obtain the sharpest and brightest bands possible. To minimize the potential for 'null' alleles in population surveys, amplifications are repeated at increasing annealing temperatures until reactions fail; an annealing temperature 6–8 °C below the temperature at which amplifications cease is used as the working temperature for further analyses.

### Analysis of SSCPs

For analysis of SSCPs, we label samples with α-³³P-dATP following optimization of PCR protocols using either of two protocols:

**1**   the radioisotope is incorporated directly into the PCR product by including 1 µCi $^{33}$P-α-dATP in each amplification reaction; or

**2**   one or both primers are end-labelled with $^{33}$P-γ-dATP prior to amplification using T4 polynucleotide kinase.

Alternatively, SSCP gels may be stained with ethidium bromide or silver, but we have found these approaches to be less satisfactory than radiolabelling. An aliquot of each PCR product is sometimes subjected to electrophoresis through agarose to verify amplification, then 2 µL are combined with 4 µL of a loading buffer containing 95% deionized formamide, 10 mM NaOH, 0.05% bromophenol blue and 0.05% xylene cyanol. Samples are heated to 95 °C for 2 min to denature the DNA, then immediately submerged in ice-water to enable the single strands to form their secondary structures. (This step is critical: if samples are not plunged into ice-water, the complementary strands will reanneal with each other rather than folding onto themselves.) Samples are subjected to electrophoresis through 25×20 cm 0.5% MDE$^R$ (J.T. Baker, Geulph, Ontario) nondenaturing acrylamide gels in 0.6×TBE (53 mmol Tris-borate, 1.2 mmol ethylenediaminetetraacetic acid (EDTA), pH=8.3) at 4 W for 4–24 h. We usually precool gels (either with a temperature regulator or in a refrigerated room) and run them at 4 °C, but resolution of SSCPs for some loci (e.g. MHC-II-β) is better at room temperature. Resolution is sometimes also improved by altering the concentration of acrylamide in the gel. We usually run a number of test gels to optimize acrylamide concentrations, temperatures and electrophoresis times (so that samples produce sharp bands and run through one-half to three-quarters of the gel), then assay all available samples under the optimal conditions. Individuals representing different alleles are then sequenced directly to identify sequence differences between alleles.

## References

Avise, J.C. (1994) Molecular markers. *Natural History and Evolution*. Chapman & Hall, New York.

Avise, J.C. & Hamrick, J.L. (eds) (1996) *Conservation Genetics*. Chapman & Hall, New York

Baker, C.S., Medrano-Gonzalez, L., Calambokidis, J. *et al.* (1998) Population structure of nuclear and mitochondrial DNA variation among humpback whales in the North Pacific. *Molecular Ecology* **7**, 695–707.

de Barros, L.M., Soden, A., Henschke, P.A. & Langridge, P. (1996) PCR differentiation of commercial yeast strains using intron splice site primers. *Applied and Environmental Microbiology* **62**, 4514–4520.

Beerli, P. & Felsenstein, J. (1999) Maximum-likelihood estimation of migration rates and effective population numbers in two populations using a coalescent approach. *Genetics* **152**, 763–773.

Birt, T.P., Friesen, V.L., Birt, R.D., Green, J.M. & Davidson, W.S. (1995) Mitochondrial DNA variation in Atlantic capelin *Mallotus villosus*: a comparison of restriction and sequence analysis. *Molecular Ecology* **4**, 771–776.

Brown, J.W.S., Smith, P. & Simpson, C.G. (1996) *Arabidopsis* consensus intron sequences. *Plant Molecular Biology* **32**, 531–535.

Cavalier-Smith, T. (1985) Selfish DNA and the origin of introns. *Nature* **315**, 283–284.

Cho, G. & Doolittle, R.F. (1997) Intron distribution in ancient paralogs supports random insertion and not random loss. *Journal of Molecular Evolution* **44**, 573–584.

Chow, S. & Hazama, K. (1998) Universal PCR primers for S7 ribosomal protein gene introns in fish. *Molecular Ecology* **7**, 1247–1263.

Clark, A.G., Leicht, B.G. & Muse, S.V. (1996) Length variation and secondary structure of introns in the *Mlc1* gene in six species of *Drosophila*. *Molecular Biology and Evolution* **13**, 471–482.

Congdon, B.C., Piatt, J.F., Martin, K. & Friesen, V.L. (2000) Mechanisms of population differentiation in marbled murrelets: historical versus contemporary processes. *Evolution* (in press)

Doyle, J.J., Kanazin, V. & Shoemaker, R.C. (1996) Phylogenetic utility of histone H3 intron sequences in the perennial relatives of soybean (*Glycine*: Leguminosae). *Molecular Phylogenetics and Evolution* **6**, 438–447.

Evans, P.G.H. (1987) Electrophoretic variability of gene products. In: *Avian Genetics* (eds F. Cooke & P.A. Buckley). pp. 105–162 Academic Press, London.

Excoffier, L., Smouse, P.E. & Quattro, J.M. (1992) Analysis of molecular variance inferred from metric distance among DNA haplotypes: application to human mitochondrial DNA restriction data. *Genetics* **131**, 479–491.

Friesen, V.L. (1997) Population genetics and the spatial scale of conservation of colonial waterbirds. *Colonial Waterbirds* **20**, 353–368.

Friesen, V.L., Baker, A.J. & Piatt, J.F. (1996) Molecular evidence for a 'new' species of alcid: the long-billed murrelet (*Brachyramphus perdix*). *Condor* **98**, 681–690.

Friesen, V.L., Congdon, B.C., Walsh, H.E. & Birt, T.P. (1997) Intron variation in marbled murrelets detected using analysis of single-stranded conformational polymorphisms. *Molecular Ecology* **6**, 1047–1058.

Friesen, V.L., Congdon, B.C., Kidd, M.G. & Birt, T.P. (1999) General PCR primers for the analysis of nuclear introns in vertebrates. *Molecular Ecology* **8**, 2141–2152.

Fulton, C.E. & Brown, A.E. (1997) Use of SSU rDNA group-I intron to distinguish *Monilinia fructicola* from *M. laxa* and *M. fructigena*. *FEMS Microbiology Letters* **157**, 307–312.

Gamulin, V., Skorokhod, A., Kavsan, V., Müller, I.M. & Müller, W.E.G. (1997) Experimental indication in favor of the introns-late theory: the receptor tyrosine kinase gene from the sponge *Geodi cydonium*. *Journal of Molecular Evolution* **44**, 242–252.

Gilbert, W. (1978) Why genes in pieces? *Nature* **271**, 501.

Harada, M.L., Cruz Schneider, M.P., Sampaio, I., Czelusniak, J. & Goodman, M. (1995) DNA evidence on the phylogenetic systematics of New World monkeys: support for the sister-grouping of *Cebus* and *Saimiri* from two unlinked nuclear genes. *Molecular Phylogenetics and Evolution* **4**, 331–349.

Harris, E.E. & Disotell, T.R. (1998) Nuclear gene trees and the phylogenetic relationships of the mangabeys (Primates: Papionini). *Molecular Biology and Evolution* **15**, 892–900.

Hillis, D.M., Moritz, B.K. & Mables, C. (eds) (1996a) *Molecular systematics*. Sinauer Associates, Sunderland, Massachussets.

Hillis, D.M., Mable, B.K., Larson, A., Davis, S.K. & Zimmer, E.A. (1996b) Nucleic acids IV: sequencing and cloning. In: *Molecular Systematics* (eds D.M. Hillis, C. Moritz & B.K. Mable), pp. 321–381. Sinauer Associates, Sunderland, Massachusetts.

Holder, K., Montgomerie, R. & Friesen, V.L. (1999) A test of the glacial refugium hypothesis using patterns of mitochondrial and nuclear DNA sequence variation in rock ptarmigan (*Lagopus mutus*). *Evolution* **53**, 1936–1950.

Huang, W., Yun-Xin, F., Chang, B.H.-J., Gu, X., Jorde, L.B. & Li, W.-H. (1998) Sequence variation in ZFX introns in human populations. *Molecular Biology and Evolution* **15**, 138–142.

Ko, C.H., Brendel, V., Taylor, R.D. & Walbot, V. (1998) U-richness is a defining feature of plant introns and may function as an intron recognition signal in maize. *Plant Molecular Biology* **36**, 573–583.

Kocher, T.D., Thomas, W.K., Meyer, A., Edwards, S.V., Pääbo, S., Villablanca, F.X. & Wilson, A.C. (1989) Dynamics of mitochondrial DNA evolution in animals: amplification and sequencing with conserved primers. *Proceedings of the National Academy of Sciences USA* **86**, 6196–6200.

Koop, B.F., Tagle, D.A., Goodman, M. & Slightom, J.L. (1989) A molecular view of primate phylogeny and important systematic and evolutionary questions. *Molecular Biology and Evolution* **6**, 580–612.

Leebens, M.J., Pellmyr, O. & Brock, M. (1998) Host specificity and the genetic structure of two yucca moth species in a yucca hybrid zone. *Evolution* **52**, 1376–1382.

Leicht, B.G., Lyckegaard, E.M.S., Benedict, C.M. & Clark, A.G. (1993) Conservation of alternative splicing and genomic organization of the myosin light-chain (*Mlc1*) gene among *Drosophila* species. *Molecular Biology and Evolution* **10**, 769–790.

Lessa, E.P. (1992) Rapid surveying of DNA sequence variation in natural populations. *Molecular Biology and Evolution* **9**, 323–330.

Lessa, E.P. & Applebaum, G. (1993) Screening techniques for detecting allelic variation in DNA sequences. *Molecular Ecology* **2**, 119–129.

Lewin, B. (1994) *Genes VI*. Oxford University Press, Oxford.

Li, W.-H. & Graur, D. (1991) *Fundamentals of Molecular Evolution*. Sinauer Associates, Sunderland, Massachusetts.

Liu, F., Charlesworth, D. & Kreitman, M. (1999) The effect of mating system differences on nucleotide diversity at the phosphoglucose isomerase locus in the plant genus *Leavenworthia*. *Genetics* **151**, 343–357.

Loeschcke, V., Tomiuk, J. & Jain, S.K. (1994) *Conservation Genetics*. Birkhäuser Verlag, Basel.

Lyons, L.A., Laughlin, T.F., Copeland, N.G., Jenkins, N.A., Womack, J.E. & O'Brien, S.J. (1997) Comparative anchor tagged sequences (CATS) for integrative mapping of mammalian genomes. *Nature Genetics* **15**, 47–56.

Michalakis, Y. & Excoffier, L. (1996) A generic estimation of population subdivision using distance between alleles with special interest to microsatellite loci. *Genetics* **142**, 1061–1064.

Moore, W.S. (1995) Inferring phylogenies from mtDNA variation: mitochondrial-gene trees versus nuclear-gene trees. *Evolution* **49**, 718–726.

Moran, P., Dightman, D.A., Waples, R.S. & Park, L.K. (1997) PCR-RFLP analysis reveals substantial population-level variation in the introns of Pacific salmon (*Oncorhynchus* spp.). *Molecular Marine Biology and Biotechnology* **6**, 315–327.

Orita, M., Iwahana, H., Kanazawa, H., Hayashi, K. & Sekiya, T. (1989) Detection of polymorphisms of human DNA by gel electrophoresis and single-strand conformation polymorphisms. *Proceedings of the National Academy of Sciences USA* **86**, 2766–2770.

Palumbi, S.R. (1996) Nucleic acids II: the polymerase chain reaction. In: *Molecular Systematics* (eds D.M. Hillis, C. Moritz & B.K. Mable), pp. 205–247. Sinauer Associates, Sunderland, Massachusetts.

Palumbi, S.R. & Baker, C.S. (1994) Contrasting population structure from nuclear intron sequences and mtDNA of humpback whales. *Molecular Biology and Evolution* **11**, 426–435.

Pamilo, P. & Nei, M. (1988) Relationships between gene trees and species trees. *Molecular Biology and Evolution* **5**, 568–583.

Perlin, M.H., Hughes, C., Welch, J. *et al.* (1997) Molecular approaches to differentiate subpopulations or *formae speciales* of the fungal phytopathogen *Microbotryum violaceum*. *International Journal of Plant Science* **158**, 568–574.

Petersen, G. & Seberg, O. (1998) Molecular characterization and sequence polymorphism of the alcohol dehydrogenase I gene in *Hordeum vulgare* L. *Euphytica* **102**, 57–63.

Pitocchelli, J., Piatt, J.F. & Cronin, M. (1995) Morphological and genetic divergence among Alaskan populations of *Brachyramphus* murrelets. *Wilson Bulletin* **107**, 235–250.

Prychitko, T.M. & Moore, W.S. (1997) The utility of DNA sequences of an intron from the β-fibrinogen gene in phylogenetic analysis of woodpeckers (Aves: Picidae). *Molecular Phylogenetics and Evolution* **8**, 193–204.

Ralph, C.J., Hunt, G., Raphael, M. & Piatt, J.F. (eds) (1995) *Ecology and conservation of the marbled murrelet.* General Technical Report PSW-GTR-152. Pacific Southwest Research Station, Forest Service, US Department of Agriculture) Albany, California.

Rogers, A.R. & Harpending, H. (1992) Population growth makes waves in the distribution of pairwise genetic differences. *Molecular Biology and Evolution* **9**, 552–569.

Rogers, A.R., Fraley, A.E., Bamshad, M.J., Watkins, W.S. & Jorde, L.B. (1996) Mitochondrial mismatch analysis is insensitive to the mutational process. *Molecular Biology and Evolution* **13**, 895–902.

Ryman, N. & Utter, F. (eds) (1987) *Population Genetics and Fishery Management.* University of Washington Press, Seattle.

Rzhetsky, A., Ayala, F.J., Hsu, L.C., Chang, C. & Yoshida, A. (1997) Exon/intron structure of aldehyde dehydrogenase genes supports the *'introns-late'* theory. *Proceedings of the National Academy of Sciences USA* **94**, 6820–6825.

Sang, T., Donoghue, M.J. & Zhang, D. (1997) Evolution of alcohol dehydrogenase genes in peonies (*Paeonia*): Phylogenetic relationships of putative nonhybrid species. *Molecular Biology and Evolution* **14**, 994–1007.

Slade, R.W., Moritz, C., Heideman, A. & Hale, P.T. (1993) Rapid assesment of single-copy nuclear DNA variation in diverse species. *Molecular Ecology* **2**, 359–373.

Slade, R.W., Moritz, C. & Heideman, A. (1994) Multiple nuclear-gene phylogenies: application to pinnipeds and comparison with a mitochondrial DNA gene phylogeny. *Molecular Biology and Evolution* **11**, 341–356.

Slade, R.W., Moritz, C., Hoelzel, A.R. & Burton, H.R. (1998) Molecular population genetics of the southern elephant seal *Mirounga leonina. Genetics* **149**, 1945–1957.

Slatkin, M. & Hudson, R.R. (1991) Pairwise comparisons of mitochondrial DNA sequences in stable and exponentially growing populations. *Genetics* **129**, 555–562.

Slattery, J.P. & O'Brien, S.J. (1998) Patterns of *M* and *X* chromosome DNA sequence divergence during the Felidae radiation. *Genetics* **148**, 1245–1255.

de Souza, S.J., Long, M., Schoenbach, L., Roy, S.W. & Gilbert, W. (1996) Intron positions correlate with module boundaries in ancient proteins. *Proceedings of the National Academy of Sciences USA* **93**, 14632–14636.

Swofford, D.L., Olsen, G.J., Waddell, P.J. & Hillis, D.M. (1996) Phylogenetic inference. In: *Molecular Systematics* (eds D.M. Hillis, C. Moritz & B.K. Mable), pp. 407–514. Sinauer Associates, Sunderland, Massachusetts.

Tomita, M., Shimizu, N. & Brutlag, D.L. (1996) Introns and reading frames: correlation between splicing sites and their codon positions. *Molecular Biology and Evolution* **13**, 1219–1223.

Venta, P.J., Brouillette, J.A., Yuzbasiyan-Gurkan, V. & Brewer, G.J. (1996) Gene-specific universal mammalian sequence-tagged sites: application to the canine genome. *Biochemical Genetics* **34**, 321–341.

Villablanca, F.X., Roderick, G.K. & Palumbi, S.R. (1998) Invasion genetics of the Mediterranean fruit fly: variation in multiple nuclear introns. *Molecular Ecology* **7**, 547–560.

Wartell, R.M., Hosseini, S.H. & Moran, C.P. (1990) Detecting base pair substitutions in DNA fragments by temperature gradient gel electrophoresis. *Nucleic Acids Research* **18**, 2699–2705.

Weir, B.S. (1996) Intraspecific differentiation. In: *Molecular Systematics* (eds D.M. Hillis, C. Moritz & B.K. Mable), pp. 385–405. Sinauer Associates, Sunderland, Massachusetts.

Wilson, A.C., Cann, R.L., Carr, S.M. *et al.* (1985) Mitochondrial DNA and two perspectives on evolutionary genetics. *Biological Journal of the Linnean Society* **26**, 375–400.

# CHAPTER 12

# Sex Identification using DNA Markers

RICHARD GRIFFITHS

## 12.1 Introduction

The visual identification of sex in a vertebrate can be difficult. It depends on the morphological difference between the sexes or identification from the evidence that an animal leaves behind; for example, male and female house sparrows can readily be told apart, but a nest full of chicks, a pile of droppings or a leg that is left under a falcon's nest all present problems. However, each of these samples contains DNA and this can be used to identify sex.

The use of DNA is, again, dependent on differences between the sexes. It relies on the genetic sex determination (SD) system, which is the switch that controls gender. The switch can be as small as a couple of genes but when this system is evolutionarily established, whole chromosomes can become specialized sex chromosomes (Charlesworth 1991). The common types of genetic SD are male heterogamety (m(ale)XY, f(emale)XX) and female heterogamety (fZW, mZZ), where the name indicates which gender contains a unique chromosome. As the two forms of sex chromosomes are equivalent (W=Y, Z−X), this chapter refers only to the XY system unless specificity is required.

Other genetic SD systems do occur. Some like mXXYY, fXXXX are convolutions of XY and, for our purposes, will be treated as such. Others like the X chromosome:autosome ratio, haplodiploidy or polyfunctional sex determination are fully described by Bull (1983). None of these possess DNA that is unique to one sex which makes gender difficult to identify using molecular markers.

In the common XY system, it is possible to trace the evolution of the sex chromosomes. They descended from a pair of autosomal chromosomes (Ohno 1967). One of the pair gained genes that made it male-determining. Unfortunately the autosomal pair usually exchange DNA during meiosis and this would break up the sex-determining region. Consequently recombination became restricted to allow chromosomal differences to be maintained between the primitive sex chromosomes (Charlesworth 1991). By a process that remains unclear, the X and Y chromosome diverged yet further until 'chromosomal sex determination' was established (Charlesworth 1996). At this stage the X chromosomes still pair and recombine in the female (XX).

They also contain a full complement of genes and function like an autosome. In some species the X and the Y become totally separate but in most a small section of the X and Y remain identical as a recombining pseudoautosomal region (Graves 1995a).

Outside the Y pseudoautosomal region, the remainder of the Y chromosome has evolved independently. This suggests that the DNA sequences should become unique to both the Y chromosome and the male, making them reliable sex-linked markers. Unfortunately the separation of the Y from the X does not require entirely new DNA sequences. Those that match the X can be juggled to a different position, Y chromosome sections removed or sequences imported from other chromosomes. Tandem repeats are particularly frequent on the Y chromosome, as a single sequence is duplicated again and again. Examples are the highly repetitive satellite DNA and the moderately repetitive mini- and microsatellites (Jones & Singh 1985; Griffiths & Holland 1990; Longmire *et al.* 1991, 1993). These exist, not only on the Y, but have close relatives throughout the genome. Other shared repeats are dispersed elements such as transposons and LINES (Charlesworth 1996; Smith *et al.* 1987). Moreover, most of the genes that the Y shared with the X chromosome degenerate, whilst those that remain can diverge from their X-linked homologues. Even so, mutation and the generation of novel sex-linked sequences does occur (Longmire *et al.* 1993; Graves 1995b) and it is these that are useful for sex identification.

In conclusion, we have to answer three questions before we can identify sex using DNA:

1   Does a species have sex-specific DNA?
2   Can sex-specific DNA be identified?
3   Can a reliable test be designed?


## 12.2  Does a species have sex-specific DNA?

To discriminate males from females, sex-specific markers must be available. Unfortunately, in all crocodiles, most turtles and some lizards sex determination is controlled by the environment so good DNA markers do not exist (Johnston *et al.* 1995; Bull 1983). The same is true of many fish that are cosexual, so an individual can function as either male or female throughout its life (Francis 1992). Despite this, some fish species do have genetically separate sexes which can be identified using DNA. Unfortunately when the sex chromosomes of a genetic SD fish, and indeed some plants, are examined microscopically, there is likely to be little or no morphological difference between the X and the Y (Francis 1992). In this situation, which is similar to that of some plants, the sexual system has 'recently' evolved so genetic, physiological and environmental factors may all have an input to the sex of an individual (Charlesworth 1991; Grant *et al.* 1994). Under such circumstances a single genetic marker is obviously not consistently sex-linked. Having said that, plants and fish are in large and variable taxa, so I expect that data on the species under investigation could reveal that DNA sexing is possible.

Further up the scale are the amphibians and the reptiles. Again these taxa show a variety of sex determination systems but these are often better established (Hillis & Green 1990). Janzen & Paukstis (1991), for example, selected a sample of 1000 of the 6000 reptiles. In a quarter of the species the sex chromosomes could be identified down the microscope by their morphological divergence. As morphology reflects sequence difference, it suggests that there are more reliable sex-specific markers available.

At the top of the scale are the birds and the mammals where each class is based on a single sex-determination system. This suggests that the system was adopted early in the development of each taxon and is now firmly in place. A karyotype analysis will confirm that sex chromosomes are present, although they may be difficult to spot in birds as a result of the large chromosome number (50–150; Christidis 1990). Additionally, a number of conserved sex-linked genes have been found that can be used to provide accurate sex identification, not just in one species but across each class.

In summary, a sexing test is reliable and easy to design when sex determination has been well established. As Table 12.1 shows, such genetic stability varies between vertebrate taxa so whatever species you choose, it is important to carry out a review of background information.

## 12.3 Can sex-specific DNA be identified?

If the study organism has a genetic sex determination system, preferably one that is well established, it is possible to search for a sex-linked genetic

**Table 12.1** Sex determination and identification in vertebrates. GSD, genetical sex determination, ESD, environmental SD, None, cosexuality. Sex-linked marker genes are more conserved in both their sequence and location than noncoding sequences. The latter are given an approximate score between 1 and 5 to indicate the number of species within that class that contain a recognized noncoding marker

| Class | Sex determination system | Established sex chromosomes | Y/W gene markers | Noncoding markers |
|---|---|---|---|---|
| Mammals† | X/Y | 100% of 4300 spp. | Yes | +++ |
| Birds‡ | W/Z | 99% of 10 000 spp. | Yes | +++ |
| Reptiles§ | GSD/ESD | ≈28% of 6000 spp. | No | + |
| Amphibians¶ | GSD | ≈2% of 3900 spp. | Yes†† | + |
| Bony fish‡‡ | GSD/ESD/none | 20 000 spp. | No | + |

†See section 12.8; ‡ see section 12.9; § Janzen & Paukstis (1991) a sample of 1000 species; ¶ Hillis & Green (1990), Schmid (1983); †† Ferrier *et al.* (1983); ‡‡ Francis (1992), for a general review see Bull (1983).

marker. There are several methods for this but the following are the best three options.

**1** *Use a known sex-linked marker.* This requires a literature review or, better still, a search of EMBL or GenBank databases. The latter is easily performed on the Internet through the National Centre for Biotechnology Information (http://www.ncbi.nlm.nih.gov/index.html). It is possible to search there either for keywords or sequences related to known sex-linked markers. Try to aim for a gene or if this is not possible look for sex-linked DNA from the organism in question or a closely related species. Once candidates are found, some laboratory-based research may be necessary before selecting a test procedure, which may be based on polymerase chain reaction (PCR) or DNA hybridization. Examples are found in sections 12.5–12.9.

**2** *Look for your own sex-linked sequence.* In humans the Y chromosome takes up 2–3% of the genome but not all this fraction is sex-specific. In many other organisms the sex-specific component is a far smaller fraction than this so markers become increasingly difficult to find. One search strategy is to compare an array of DNA fragments in males and females that are selected using the same technique. Most fragments will be shared but those unique to male are sex markers. This requires a method that will allow you to scan a fraction of the DNA in the genome but will select the same fraction in a repeatable manner. The random amplification of polymorphic DNA (RAPD) and AFLP (this is not a true acronym for Amplified Fragment Length Polymorphism as restriction enzyme cutting sites also have a central role in this technique; Vos *et al.* 1995) techniques were designed to scan the genome for polymorphic markers (Welsh & McClelland 1990; Williams *et al.* 1990; Vos *et al.* 1995) but can easily be adopted in a search for sex-specific markers (sections 12.5 and 12.6).

**3** *Find a known repetitive sequence that is sex-linked.* Both micro- and minisatellites are likely to be concentrated on the Y chromosome. Several that have been found have undergone huge size amplification. This means a Southern blot of known male and female organisms can be screened with synthesized microsatellite probes like (CT)n and (GATA)n (where the latter is particularly useful in snakes; Jones & Singh 1985) or cloned minisatellite probes such as 33.15 or 33.6. Unfortunately, hybridization may occur to sex-specific fragments as large as 20–60 kb (Longmire *et al.* 1993). Consequently the resultant test has to be performed using Southern blot hybridization (Chapter 2), which is an accurate technique but relatively slow and expensive.

## 12.4  Can a reliable test be designed?

Before searching for a sex-linked marker, first consider the test. It should be accurate, reliable, sensitive, easy, and cheap. The option that covers most of these variables is the PCR (Chapter 3) and, where possible, I shall suggest this method is used.

If we consider 'accuracy' with a PCR test in a little more depth, it can be broken down into two parts (Lessells & Mateman 1998). During the isolation of the DNA marker we have to ensure the candidate is actually sex-specific and not just a polymorphic sequence. To do this, it has to be tried on individuals of known sex. The probability of the test being right is $q^m \cdot (1-q)^f$, where m=the number of males, f=number of females and q=m/m+f. This is a simple test and if you correctly sex four males and four females there is a probability of 0.996 that you have located a sex-specific marker.

The second part of 'accuracy' is more difficult. If I design a set of PCR primers that will amplify a Y-linked sequence, there is a chance that a Y-linked polymorphism will prevent the test from working in some males. As Lessells & Mateman (1998) describe, a confidence limit for the accuracy of the test within a population can be calculated. Unfortunately, 100 presexed males are needed before you can state that 96–100% of males have the same Y linked sequence with 95% confidence. Obviously this is difficult but the sexing test should be as accurate as possible. If such a test is impossible, then there are a couple of precautions that should be taken. First, try to design a sexing test based on coding sequences using two ≈24nt primers. This is far more likely to be reliable than those based on 10nt primers in a fast evolving, noncoding sequence. This means using a gene (12.7, 12.8) rather than a RAPD (12.5) or AFLP marker (12.6). Second, if the work is being carried out on families with a single father, an indirect test should be carried out by looking for all-female broods. These could be all females but if they occur more frequently than expected, it could be a sign that the males have inherited a null or nonamplifying allele from their father.

'Reliability' is addressed by the use of controls. Negative controls ensure that no contamination was introduced at the DNA extraction phase or when setting up the test (PCR; Chapter 3). There should also be positive controls to ensure that the test does work; for example, if you design a PCR test where the production of one band means it is a male and of zero bands a female, then a failed PCR would also look like a female. A PCR reaction with a positive control will produce a band in all individuals, while the males get an extra band to indicate their sex. Where possible, the positive control reaction should operate less well than the sexing reaction. This ensures that when the control band can be seen, the bird can definitely be sexed. In some cases, designing the positive control takes a great deal of effort but it results in a more reliable test.

## 12.5 RAPD: to identify and use a sex-linked marker

Polymerase chain reaction usually employs two ≈24nt primers to target and amplify a precise segment of DNA. In contrast RAPD uses a single, 10nt primer at a low annealing temperature and this amplifies a range of miscellaneous genomic fragments (Welsh & McClelland 1990; Williams *et al.* 1990). In effect it is a genomic sampling procedure. This means that if an amplified band occurs in males but not females then this is the product of a Y-specific marker (Griffiths & Tiwari 1993; Lessells & Mateman 1998).

There are worries that the RAPD technique is erratic (Ellesworth *et al.* 1993; Bielawski *et al.* 1995) but if the PCR conditions are duplicated faithfully then the major bands will consistently be amplified in an individual (Lessells & Mateman 1998). There is a further complication because RAPD was designed to sample polymorphisms, so bands appear in some individuals and not in others (Welsh & McClelland 1990; Williams *et al.* 1990). This confusion is usually removed if samples of 3–5 males are mixed and compared with a sample of 3–5 females. The sexes do have to be identified correctly for this strategy to work. However, as Michelmore *et al.* (1991), who devised this Bulk Segregant Analysis, state: 'RAPD analysis was unexpectedly insensitive in its ability to detect a rare allele (<10%).' This means occasional sex identification mistakes will have no effect on sex-specific marker selection if the pool of samples is above 10 or more individuals. This does of course rely on good sample collection!

Randomly amplified polymorphic DNA analysis with a single primer amplifies 1–20 bands so it is unlikely that a Y-linked product will be discovered with a single primer. A usual procedure is to buy a range of 40 or so RAPD primers (from e.g. Operon Technologies, Alameda CA 94501, USA) and screen each of them. These primers are cheap, they have 10 arbitrarily chosen nucleotides and a median G+C content of 50–70%.

---

### Box 12.1  Polymerase chain reaction amplifications

I describe conditions cited by other authors or myself to carry out PCR reactions (if adjustments to the reaction conditions are required, try annealing temperature, $MgCl_2$ or primer concentration). The description starts with a reference that describes the design of the test. It is followed by a PCR reaction volume and continues with the 5' to 3' description of the primers. If wobble is required at a particular nucleotide, the alternatives are given within parentheses e.g. (A/T) and companies are able to construct primers that integrate this wobble (if two pairs of primers are being used in a reaction and one provides excessive

*continued*

---

**Box 12.1** *contd*

amplification, lower this pair's concentration). 'Std Conditions' are the reaction conditions within a 10 μL reaction which are: 6 pmol primers, 200 μM of each dNTP, 50–100 ng target DNA, 0.35 units Taq polymerase, 1.5 mmol MgCl$_2$, 50 mM KCl, 10 mM TrisHCl (pH 8.8 at 25 °C), 0.1% Triton X-100 (where the last four are supplied by Promega in 'with Mg/Taq buffer'). Any variations from this are given. The thermal conditions are °C for S and the cycling conditions lie within brackets prefixed with the cycles required. Each description ends with the type and concentration of gel required for analysis.

---

### 12.5.1 A RAPD analysis

Using:

    RAPD primers
    Agarose gel
    Taq polymerase
    TBE buffer

**1**  Extract DNA from 10 individuals of each sex using a standard procedure such as the phenol/chloroform extraction (Chapter 2). Five males and five females will be used to isolate the Y marker and the remainder to confirm the test.

**2**  For each primer make two 10 μL PCR reactions which contain 20–100 ng of genomic DNA from a mix of either five males or five females. The PCR reaction contains 5–10 pmol of primer and uses standard RAPD conditions; 95 °C for 120 s, 38×(35–37 °C for 60 s, 72 °C for 180 s (with a ramp rate of 6 s °C$^{-1}$ between 37 °C and 72 °C), 94 °C for 60 s), 37 °C for 60 s, 72 °C for 300 s (see Box 12.1). The success of a RAPD reaction is best influenced by annealing temperature and concentrations of genomic DNA, MgCl$_2$ and primer. Control the first three carefully and standardize the DNA extraction procedure to repress differences in sample concentration.

**3**  Run the products on a 1.5% TBE (89 mM Tris. Cl, 89 mM Boric acid, 2 mM EDTA (pH8)) agarose gel containing ethidium bromide (40 ng mL$^{-1}$) at 3–4 V cm$^{-1}$ (Chapter 2).

Examine and photograph the gel under ultraviolet (UV) light. When the correct primer has been found you will obtain a result that shows several bands shared between male and female lanes but one that is male specific. Confirm this result by carrying out RAPD amplifications on the DNA from the remainder of the birds. This is performed on each individual sample. An example is shown in Fig. 12.1. These are results taken from work on the great tit (*Parus major*) and show several bands that occur in each individual. Around 400 bp in size there are polymorphisms, most individuals have a band of

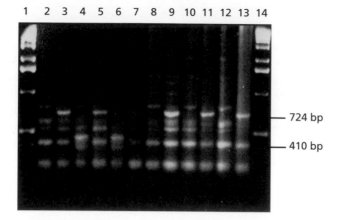

1  2  3  4  5  6  7  8  9  10  11  12  13  14

— 724 bp

— 410 bp

**Fig. 12.1** (a) Randomly amplified polymorphic DNA (RAPD) amplification of the genomic DNA of the great tit (*Parus major*) which produces a female, W chromosome marker of 724 bp (Griffiths & Tiwari 1993). The odd-numbered lanes are females, the even lanes male and the external lanes contain Gibco BRL (maryland, USA) 1 kb marker.

410 bp but birds 8 and 10 have a doublet of 350 bp and 450 bp. However, at 724 bp there is a strong band that occurs only in the females and is derived from an intron of the *CHD1-W* gene from the W chromosome (remember birds are mZZ, fWZ; Griffiths & Tiwari 1993).

The Y-linked marker should amplify sufficiently well so that it is among the top four bands in intensity. This is because weak bands will disappear if the RAPD reaction does not perform well and this would make sex identification unreliable (Smith *et al.* 1994). If the sex-linked band is of suitable strength, choose a weaker band as a positive control. Only score an organism as female if the control appears but there is no Y-linked band. This confirms that the result you achieve is reliable even if the amplification is not optimal. If the sex link marker is weak the only option is to clone and sequence this DNA fragment and design primers directly to this. This starts with the isolation of the band from the gel. Note that the UV light that illuminates the gel can be extremely destructive to the DNA so keep the exposure short or use the 'low power' function on the UV light box.

### 12.5.2  Cloning a RAPD fragment

Using:

    TAE buffer
    Dissection knife
    Flat-blade tweezers

Qiaquick kit (or similar; Qi gen, California, USA)

PCR purification

Sterile plastic wool (pet shops stock this to use as a filter in fish tanks)

Run the RAPD PCR product on a Tris-acetate (TAE: 40 mM Tris acetate, 2 mM EDTA (pH8)) rather than a Tris-borate (TBE) gel (TAE substitutes for the TBE in both the gel and the buffer) as the boric acid in TBE can inhibit subsequent reactions. It also helps if the gel is <1% and the fragment to be isolated is smaller than 1 kb.

1   With a sterile knife cut the fragment out of the gel under low-power UV and extract the gel slice with a pair of flat tweezers. The fragment can often be trimmed once removed from the gel.

2   Chop the fragment and add 50 µL of TAE, incubate at 60 °C for 1 h or overnight on the bench.

3   Punch a small hole in the bottom of a 0.5-mL Eppendorf tube with needle, plug with sterile plastic wool and place in a 1.5-mL Eppendorf tube.

4   Put the gel slice/liquid on top of the wool and spin 6000 $g$ for 5 min increasing to 8000 $g$ for 5 min.

5   Purify the product in the 1.5 mL tube (e.g. Qiaquick kit).

The rest of a sequencing protocol which involves DNA cloning and the design of primers can be found in section 12.6.1 (point 5 to the end of the AFLP section). The only difference is that when the RAPD band sequence is used to make dedicated primers, you do not have to ignore the DNA sequence at the ends of the fragments that complement the RAPD primers. These sequences actually exist in the genome and can be used as the basis of a primer. This idea has a superb acronym of Sequence Characterized Amplified Regions (SCARs; Paran & Michelmore 1993) which allows you to impress colleagues.

Bear in mind that the RAPD technique does not *have* to find a sex-linked marker (Lessells & Mateman 1998). Unlike the primers, the DNA on the Y chromosome does not have an arbitrary nucleotide sequence; for example 65% of the chicken W chromosome comprises two families of short repeat (Suka *et al.* 1993) which RAPD will not locate. The chicken W is about 1.5% of the genome and many of the remaining sequences could also occur on other chromosomes. Together these factors act against the isolation of a W-specific marker. Nevertheless this method remains one of the best ways to isolate and use sex-linked DNA for sexing.

## 12.6  AFLP: to identify and use a sex-linked marker

AFLP (see Chapter 9) is a second technique that can be used to search for sex-linked markers (Vos *et al.* 1995; Griffiths & Orr 1999). Like RAPD it was designed to locate polymorphisms for use in population analysis. It is more complicated and expensive than RAPD but more powerful, generating approximately 100 fragments as opposed to approximately 10. If the array of fragments produced by male and female AFLPs are compared, then many bands are monomorphic (M) as they are of the same size in different individuals (Fig. 12.2). Some bands are polymorphic (P) but there are a few that are sex-linked as they are confined to a single gender (W). Although the AFLP technique is described in the following section, this book provides a comprehensive chapter on the use of AFLP (Chapter 9). There is also the extensive, if noncritical, paper by Vos *et al.* (1995) for consultation.

Like RAPD the strategy of AFLP is to select and display a small proportion of the genomic DNA using PCR. The protocol can be divided into a number of key stages.

**1  Restrict** the genomic DNA with a frequent, 4-base cutter such as *Mse*I (cuts every approximately 254 bp) and a rare, 6-base cutter like *Eco*RI (cuts every approximately 1024 bp).

**2  Ligate** two different double-stranded oligonucleotide adaptors, one specific to the cut *Mse*I site, the other specific to the *Eco*RI site.

**3**  Perform **preselective PCR amplification** with M-N and E-N primers. These anneal exactly to each adaptor but have a nucleotide (N), e.g. E-T that extends from the 3′ end over the genomic DNA. Only the genomic DNA that complements this base will amplify.

**4**  Perform **selective PCR amplification** with the same basic primers but there are three 3′ nucleotides rather than one: M-NNN and E-NNN. These add to the selective effect of the previous stage, e.g. **E-T**AG, and reduce the portion of the genomic DNA that is amplified. Moreover, only the E-NNN primer is radioactively labelled, so the 90% of the fragments that have been restricted by just *Mse*I cannot be visualized (Vos *et al.* 1995).

**5  Separate** the PCR product on an acrylamide gel and prepare an autoradiograph.

The AFLP protocol is relatively complex and requires that a range of chemicals, particularly oligonucleotides, are purchased. An alternative course is to buy a kit (Griffiths & Orr 1999). The 'AFLP analysis system 1' (cat. no. 10544–13) is marketed by Gibco BRL (Maryland, USA) and comes complete with clear instructions and the essential components. Note that this kit is marketed for the recognition of polymorphisms so the protocol does need adapting, while the amount of radioactive marker they suggest can be reduced.

An example of the results of AFLP can be seen in the autoradiograph in Fig. 12.2. This features the entire AFLP profile that is produced by one primer pair using DNA from the ostrich (*Struthio camelus*). Through a comparison of male and female lanes, W-linked sex-specific bands (W) can be identified among the monomorphic (M) and polymorphic (P) alternatives. The female-specific

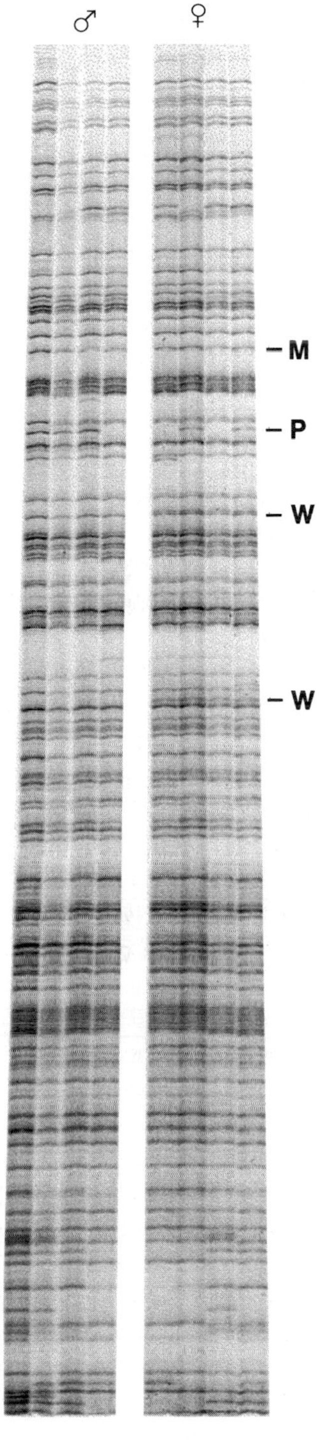

**Fig. 12.2** An AFLP profile using the *E*-AGG/*M*-CAG selective primers on four male and four female ostriches. Around 120 markers can be seen, of which two are W-linked and female-specific (W), some are polymorphic (e.g. P), although most are monomorphic (e.g. M).

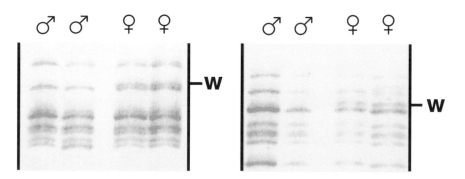

**Fig. 12.3** A close-up of an amplified fragment-length polymorphism (AFLP) profile (see Fig. 12.2) highlighting the W-linked markers (W) in two male and two female ostriches.

bands can be seen more clearly in the enlargements in Fig. 12.3. If a single pair of ostriches are used in this screening, many AFLP primers appear to produce a sex-linked marker. When these are checked using more birds, these turn out to be the result of polymorphisms. In RAPD this effect was counteracted by the use of pooled samples that contained DNA from five males or five females. If this approach is taken with AFLP, the results do not always reflect each of the samples in the pool. As a consequence, the screening of three males and three females individually is a better way to survey the primers. The numbers are limited to three of each sex to speed up the screening of the AFLP primers (Fig. 12.2). Once a putative sex-linked fragment is identified this result is confirmed with the additional samples of known-sex birds.

Sometimes the AFLP procedure fails to work. One of the most common causes is the impurity of the primers. This can be remedied by a change in the primer supplier or in having them further purified by high-pressure liquid chromatography (HPLC). An advantage of the kits is that this problem has been anticipated as the primers have already undergone HPLC purification. Another problem is that some pairs of AFLP primers fail to produce a result, while others perform admirably. This can be a result of the differential annealing of the selective AFLP primers to the sample DNA and highlights the nonrandom sequence of nucleotides in the genome; for example, most eukaryotes, particularly birds, are AT-rich hence the use of the *Mse*I restriction enzyme (TTAA). There is little that can be done about this, apart from selecting a different primer pair.

When a sex-linked fragment is discovered it can be employed in two ways. It can be used to sex animals directly as long as precautions are taken. In the paper by Vos *et al.* (1995) they suggest that AFLP produces fragments that amplify equally. In our work we have found some variation. Consequently, the sex-linked fragment has to amplify sufficiently well to be reliable. As the

AFLP bands do maintain a similar relative intensity, a positive control should be used. This is a band that is less intense than the sex-linked fragment so only when it is visible should the homogametic sex be identified from the autoradiograph.

Although using AFLP to sex animals directly seems obvious, an alternative is to use the sex-linked fragment as the basis for a normal PCR test. This has the advantage of removing the complexity and expense of the AFLP technique. It requires that the sex-linked DNA is cloned and sequenced and sex-specific primers are designed. The procedures to complete this are outlined below.

### 12.6.1 Cloning an AFLP fragment

Using:

> AFLP primers
> T4 polynucleotide kinase (PNK)
> 10× PNK buffer
> AFLP PCR buffer
> Qiaquick (or similar) PCR purification kit
> Taq polymerase
> Klenow
> dNTPs
> Chloroform
> Phenol

**1** Align the autoradiograph and the original acrylamide gel to identify the site of the sex-linked fragment.

**2** Cut out the AFLP marker on the 3MM paper, chop and soak in 100 μL TE (10 mM Tris (pH 7.6), 1 mM EDTA (pH 8)) for at least 1 h, vortex occasionally.

**3** Perform a 20 μL PCR with 2 μL of the isolated DNA using the selective primers and reactants as the original AFLP reaction. It requires 20 thermal cycles: 94 °C for 30 s, 56 °C for 30 s and 72 °C for 60 s.

**4** Purify the product (e.g. Qiaquick kit).

**5** Phosphorylate the 5′ termini by taking 20 μL of purified PCR product, 2.5 μL of 10× PNK buffer (0.5 mM Tris.Cl (pH 7.6); 0.1 M $MgCl_2$; 50 mM Dithiothereitol; 1 mM Spermidine HCl; 1 mM EDTA (pH 8.0)), 0.5 μL of 10 mM dATP and 0.5 μL of T4 polynucleotide kinase and incubate for 30 min at 37 °C. Subsequently add 0.5 μL of dNTP (25 mM each) and 0.25 units of Klenow and continue incubation for 30 min to create blunt ends. Scale this reaction if required, but precision with dATP, PNK, dNTP and Klenow is not vital.

**6** Add an equal volume of 1:1 phenol/chloroform, vortex then mix for 20 min. Centrifuge at 14000 $g$ for 5 min and retain the aqueous layer.

**7** Repeat step 6, using chloroform rather than phenol/chloroform.

**8** Add 1 vol. 4 mol $NH_4$ acetate (pH 4.8) and 2 vol. of ethanol, shake then place on ice for 5 min.

**9** Centrifuge at 14 000$g$ for 5 min, gently pour off the ethanol, allow to dry for 15 min and add 10 μL of TE.

## 12.6.2  Ligation and transformation

Using:

> *Sma*I cut pUC18 vector (Amersham Biotech, Uppsala, Sweden)
> 5× ligation buffer
> SOB ampicillin plates
> T4 DNA ligase
> JM109 competent cells ($10^7$ colony founding units μg$^{-1}$)
> SOB media

**1** Ligate 5 μL of purified DNA (10–20 ng) into 100 ng of dephosphorylated *Sma*I cut pUC18 plasmid (this aims for a 1 : 1 insert/vector ratio in terms of fragment numbers) with 2 μL 5× ligation buffer, 1 unit of T4 DNA ligase (Promega) and dH$_2$O to 10 μL and incubate at 15 °C, for 4 h or overnight.

**2** Chill 17×100 mm polypropylene tubes (Falcon 2059; Alpha, Hampshire, UK) in ice.

**3** Remove subcloning efficiency competent JM109 cells ($10^7$cfu μg$^{-1}$; Promega) from –70 °C and place on ice until they just thaw.

**4** Gently mix the cells and transfer 50 μL to each of the prechilled tubes with 5 μL of the ligation reaction and lightly mix again.

**5** In a fresh tube repeat step 4 with the control DNA supplied by Promega to ensure effective transformation.

**6** Return control and experimental tubes to ice for 10 min.

**7** Heat-shock the tubes for 45 s in a 42 °C waterbath without shaking then replace in the ice for 2 min.

**8** Add 450 μL of cold (4 °C) SOB medium (0.58 gL$^{-1}$ NaCl, 0.19 gL$^{-1}$ KCl, 2 gL$^{-1}$ MgCl$_2$, 2.5 gL$^{-1}$ MgSO$_4$, 20 gL$^{-1}$ bacto-tryptone, 5 gL$^{-1}$ bacto-yeast extract; pH adjusted to 7. Autoclave 20 min at 103.4 kPa) to each transformation and incubate for 1 h at 37 °C while shaking at approximately 225 r.p.m.

**9** Spread 100 μL of the undiluted and 100 μL 1 : 10 dilution of each transformation on separate SOB antibiotic plates (SOB agar/ampicillin plate: Autoclave SOB broth with agar (15 gL$^{-1}$). Allow to cool to 50 °C, add ampicillin to 100 μg mL$^{-1}$ and prepare plates and incubate at 37 °C overnight.

This should lead to a plate covered with colonies of transformed cells. The protocol below describes how the plasmids from around 12 colonies per plate should be purified and the inserts they carry sized.

### 12.6.3 Plasmid extraction by alkaline lysis miniprep

Using:

Solution I: 250 mM Glucose; 25 mM Tris Cl (pH 8), 10 mM EDTA (pH 8), filter 0.45 μm.

Solution II: 1% SDS, 0.2 M NaOH (make fresh)

3 M NaOAc adjusted to pH 5.2 with glacial acetic acid

9 : 1 chloroform : phenol

SOB media

100% ethanol

70% ethanol

2% agarose gel

1  Incubate 24 separate colonies in 3 mL of SOB media (see section 12.6.2) for 8 h to overnight.

2  Transfer 1.5 mL into an Eppendorf tube and spin at 14 000 $g$ for 1 min.

3  Suck off broth from pellet using an aspirator. Remove as much broth as possible.

4  Resuspend in 100 μL of Solution I by vortexing. Incubate at room temp for 5 min.

5  Prepare Solution II. This must be freshly prepared. Add 200 μL and mix with a gentle shake. Place on ice for 5 min.

6  Add 150 μL of 3 M NaOAc pH 5.2, shake and keep on ice for 5 min.

7  Spin at 14 000 $g$ for 5 min. Remove supernatant to a fresh tube.

8  Extract with 100 μL of 1 : 9 phenol : chloroform. Spin at 14 000 $g$ for 5 min.

9  Remove aqueous layer to fresh tube and add 2× vol. 100% ethanol and vortex.

10  Spin at 14 000 $g$ for 5 min and pour off alcohol.

11  Wash pellet with 300 μL of 70% ethanol. Spin at 14 000 $g$ for 3 min, pour off alcohol and dry the DNA by leaving the tube open for 15 min.

12  Resuspend in 30 μL of dH$_2$O or TE and add 1 μL of 10 mg mL$^{-1}$ RNAse.

13  Restrict 5 μL with 0.5 μL of *Eco*RI and 0.5 μL of *Pst*I in 1 μL of 10×H buffer (Promega, Michigan, USA) and 3 μL of H$_2$O for 30 min at 37 °C. Run this on a 1.5% agarose gel and visualize under UV light.

The inserts in most of the clones should be of the same size as the sex-linked AFLP band. The isolation of nearby bands may be a problem so a second screening is necessary. To do this, sequence three or four of the potential candidates using pUC/M13 forward and reverse primers (Promega, Michigan, USA). If all have similar sequences, this will ensure that you are examining the correct AFLP fragment.

To design the specific primers make sure you remove the AFLP primers, either those described in Chapter 9 or the versions marketed by BRL (E 5′-CGACTGCGTACCAATTC and M 5′-GATGAGTCCTGAGTAA), from the sequence. Then use a program such as Oligo Primers! (Rychlik & Rhoads 1989; very good but expensive), on the internet (Rozen & Skaletsky 1996, 1997; http://www.williamstone.com/primers/javascript/3) or follow thesimple rules described by Rychlik (1995) to help with the sex-specific primer design. After the primers are synthesized, they should be used in a PCR reaction at a reasonably high stringency (Chapter 3) as this will reduce the amplification of contaminants. They should produce a sex-linked band. Confirm the reliability of the primers by testing on 20 or more animals of known sex.

There is a chance that these specific primers will not work. This could be a result either of an accident due to cloning and analysing the wrong sequence, or because AFLP has amplified a sequence that is closely related to others in the genome. The latter can easily occur, as AFLP requires only a single nucleotide difference for it to identify a sex-specific sequence. If the primers do fail, try to clone the AFLP fragment again to rule out miscloning. If this produces clones with the same DNA sequence, either find a different sex-linked AFLP marker or resort to sexing the animals directly using AFLP.

If the production of the specific sex identification primers has been successful, a positive control PCR is still required. This has to be carried out in the same reaction but should produce a band that is larger and hopefully fainter than that produced by the sex-linked primers. This is necessary because the amplification of two products is a competitive reaction and a large fragment is just more difficult to make than the smaller sex-linked band. Moreover, a fainter band will always be the last to appear and so forms a better control. Under these conditions the appearance of the positive control will rule out the chance of incorrect sexing (Fig. 12.4).

To select control primers, the first option is to try any primers to that species that are already available in the laboratory. Second, undertake a literature or a DNA search (GenBank: http://www.ncbi.nlm.nih.gov/index.html) for sequence or primer data for your species. If there is still nothing available, select a highly conserved gene like histone, actin or insulin (Page & Holmes 1998) and use GenBank sequences taken from the nearest possible relatives to create primers that should be effective over a wide evolutionary range.

Once positive control primers are available, they may not work in the same reaction as the sexing primers as a result of a negative interaction. To combat this, the most recent Oligo program (Rychlik & Rhoads 1989) allows the design of complementary primers. Perhaps the best approach is to design and try a range of different control and sexing primers so the best combination is selected. If, for example, the control primers produce a band that is considerably more intense than the sex-linked primers, reduce the concentration of the control primers as this may even out the competition. Alternatively, reduce the PCR extension period as again this will favour the amplification of the smaller sex-linked product.

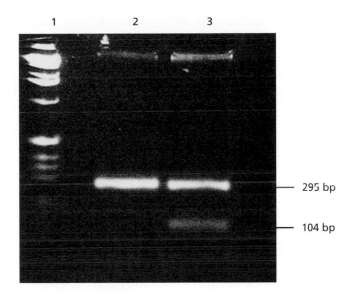

**Fig. 12.4** Sex identification in the ostrich with two different multiplex polymerase chain reactions (PCRs). Lane 1 is a 1 kb marker (BRL), lanes 2 and 3 are ostriches sexed with ScW1F 5'-GAATTCAGGACCTTGGTGAA and ScW1R 5'-ACAAGATGTTTTGGAAAGAAGAG primers (161 bp band). In lanes 2–3 the primers P8 5'-CTCCCAAGGATGAG (A/G) AA (C/T) TG and P18 5'-GAGATGGAGTCACTATCAGATCC provide the internal positive controls ≈ 295 bp (Griffiths & Orr 1999) although it is too intense.

Although the production of a specific test does appear to be a complicated procedure, it will produce a sex identification test that is faster, cheaper and easier to use than AFLP, both by you and colleagues who may work on your study animal.

## 12.7 RAPD or AFLP?

Whether to employ RAPD or AFLP to isolate and use sex-specific markers is a difficult decision. RAPD is simple, cheap, uses less radioactive or toxic chemicals and can produce results extremely rapidly (Griffiths & Tiwari 1993; Lessells & Mateman 1998), while AFLP is more complex and expensive (Vos *et al.* 1995). However, AFLP is 10 times as powerful, delivering 100 rather than the 10 markers per reaction. This is definitely an advantage if attempting to sex an organism that is known to have poorly differentiated sex chromosomes; for example, the ostrich arose at an early stage in bird evolution and the reliable avian W marker CHD1 (see section 12.9.1) fails to operate. However, AFLP has produced three alternative sex-linked markers, two of which were made into simple PCR tests (Griffiths & Orr 1999). AFLP has also been shown to be more consistent in the production of the amplified fragments despite large changes in template concentration and PCR annealing temperature (Vos *et al.* 1995). This contrasts favourably with RAPD, which is frequently criticized on this point (Ellesworth *et al.* 1993; Smith *et al.* 1994; Bielawski *et al.* 1995) despite the fact that RAPD will reliably produce a well amplified product if both temperature and the reaction conditions are reasonably consistent (Lessells & Mateman 1998).

Randomly amplified polymorphic DNA has also been used directly to sex over 700 individual birds as part of an ecological experiment and was reliable (Lessells *et al.* 1996). AFLP has not yet undergone such a direct test, although primers designed from AFLP fragments are now being used to sex ostriches on a commercial basis (University Diagnostics, personal communication). In summary, both techniques, when used with the suggested precautions, offer a reliable performance. The choice relies on the sexual genetic differences in the considered species and the laboratory experience and facilities that are available.

## 12.8 Mammalian Y-linked genes

Back in 1967, Ohno rightly suggested that the sex chromosomes were once a pair of autosomes but the genetic control of gender had split their evolutionary pathways (Ohno 1967). The X would remain important as a carrier of genes, while the Y becomes ruled by sex determination. Once the split has happened various mechanisms begin to operate to degrade the rest of the Y chromosome (Graves 1995b; Charlesworth 1996). As a result, the Y is usually smaller in size than the X and contains a mere sprinkling of genes. In this section it is our intention to locate and use one of these genes as a male-specific marker.

For our purpose, mammals have a few advantages. The first is that from the most anciently evolved form like the platypus to the most recent version, all mammals are based on the same X/Y sex determination system. In a very few species the Y has been lost or sex determination changed but generally we are dealing with one system. Nevertheless, things have not stood still, as there are several internal rearrangements of the X or the Y chromosome in the mammalian lineage. This may be part of an 'addition–attrition' mechanism as outlined by Graves (1995a,b), where rare mutations add a part of an autosome to each sex chromosome. The autosomal fragments are preserved on the X but the ex-autosomal genes are gently disposed of on the Y chromosome. This means the sex determination system has been preserved but Y linked genes are usually on the edge of 'attrition'.

The second advantage of mammals is that humans are part of this class. This has sparked a great deal of genetic work. One of the genes discovered is the sex-determining gene *SRY* (Sex Determining Region, Y) and this is the switch that makes a mammal male. There is a structurally similar gene on the X chromosome called *Sox3* that used to be the allele of *SRY* when the sex chromosomes were autosomes (Foster & Graves 1994). Another gene *ZFY* (Zinc Finger, Y) and its ex-allele *ZFX* were 'added' to the sex chromosome shortly after the marsupials split from the eutherians (Sinclair *et al.* 1988). *ZFY* appears to have avoided attrition because of its involvement with spermatogenesis. This is not true of the *AMGY* (Amelogenin, Y) gene which forms tooth enamel. It is sometimes on the Y, but if it is not, its function is taken over by *AMG* its old partner on the X chromosome.

Despite their differences these three genes *SRY*, *ZFY* and *AMGY* are conserved and have been used in standard sexing tests. Most of these techniques are based on a PCR which is fast and requires little DNA. If time and tissue are less pressing, Southern blot hybridization (Chapter 2) remains an option.

### 12.8.1 *AMG*

*AMG* has not been comprehensively tested but it is Y linked in cattle and some primates but not in rats or mice. In humans *AMGY* and *AMG* cover a short genomic region ($\approx 2700$ bp) where introns and exons have a similarity of around 90% (Nakahori *et al.* 1991). As a result of an insertion/deletion (indel) in the X, a standard set of human forensic primers produces an *AMG* band of 106 bp which contrasts nicely to the male-specific band of *AMGY* at 112 bp (Sullivan *et al.* 1993). Other larger human indels are available and can be found (Nakahori *et al.* 1991) if required. Primers are also available for cattle (Ennis & Gallagher 1994), but further work is required to extend this test to other taxa.

---

**Box 12.2** *AMG* sex identification of mammals

Sullivan *et al.* (1993) 10–20 µL; Amel-A CCCTGGGCTCTGTAAAGAATAGTG, Amel-B ATCAGAGCTTAAACTGGGAAGCTG; standard conditions; 95 °C for 120 s, 35×(60 °C for 60 s, 72 °C for 60 s, 94 °C for 60 s), 60 °C for 60 s, 72 °C for 300 s; 4% agarose or 4% 3 : 1 NuSeive/SeaPlaque for gel clarity (FMC Maine, USA). A PCR test is also marketed by Promega (TB251) which is analysed using a fluorescent scanner.

---

### 12.8.2 *ZFY*

This gene was once thought to determine sex but its autosomal location in the marsupials, along with other evidence, undermined this hypothesis. Again it is conserved and primers designed by Aasen & Medrano (1990) will amplify *ZFY* and *ZFX* in Primates, Artiodactyls, Carnivora, Perissodactyla and Rodentia. Unfortunately the indels range from 0 to 2 bp in an ≈445bp product so discrimination involves cutting with restriction enzymes. Selecting the appropriate enzyme will rely on:

1    literature searches;
2    trial cutting with a range of enzymes; or
3    sequence data. The extra restriction step does increase the chance of test failure.

---

**Box 12.3  *ZFY* sex identification of mammals**

Aasen & Medrano (1990) 10–20 μL; P1-5EZ ATAATCACATGGAGAGCCA-CAAGCT P2-3EZ GCACTTCTTTGGTATCTGAGAAAGT; standard conditions; 95 °C for 120 s, 30×(57–60 °C for 45 s, 72 °C for 60 s, 94 °C for 45), 60 °C for 60 s, 72 °C for 300 s; clean up product if required, cut with appropriate enzyme and run on approximately 3% agarose gel.

---

### 12.8.3  *SRY*

The *SRY* gene evolves in a strange manner. This is because it initiates male development, which has become a position maintained by belligerence. It is built around a conserved 237 bp Sox box which makes it a member of the *SOX* gene family. However, *SRY*, particularly outside the Sox box, evolves at a remarkably rapid rate for a gene and this has hindered the identification of *SOX3* as its ex-allele (Foster & Graves 1994). The PCR primers specific to *SRY* rarely amplify in more than a few mammalian orders because of the speed of the gene's evolution. The primers of Pomp *et al.* (1995) are particularly useful and cover examples in the Primates, Carnivora, Perissodactyla and Rodentia where the *ZFY/ZFX* primers described above can be included as positive controls. Fortunately *SRY* has been sequenced across the class so there is the opportunity to check GenBank (http://www.ncbi.nlm.nih.gov/index.html), modify the Pomp *et al.* (1995) primers and continue to use the *ZFY/ZFX* primers as a control.

---

**Box 12.4  *ZFY* sex identification of mammals**

Pomp *et al.* (1995) 10–20 μL; SRYA-5 TGAACGCATTCATGGTGTGGT SRYA-3 AATCTCTGTGCCTCCTGGAA (200 nM), P1-5EZ, P2-3EZ *ZFY/ZFX* primers (200 nM); standard conditions 2.25 mM MgCl$_2$. 95 °C for 120 s, 30×(55 °C for 60 s, 72 °C for 120 s, 94 °C for 60 s), 55 °C for 60 s, 72 °C for 300 s; 3–4% agarose.

## 12.9 Avian W-linked genes

As opposed to the mammals, the birds have a W/Z sex determination system (fWZ, mZZ). Again this system is shared throughout the class, but less work has been undertaken on the birds so the genes and mechanisms of sex determination are unknown. We do know that the sex chromosomes are well differentiated in all species apart from the ratites like ostriches, emu, etc., so sex-linked DNA is available. Various examples of such sequences have been identified, and most are repeats including mini- and microsatellites, which tend to evolve quickly. Two genes and a possible pseudogene have also been found. These evolve more slowly but none have a W-specific form in the ratites and are described below.

### 12.9.1  Chromo-helicase-DNA binding-W

The best option for sexing birds employs the two avian *CHD1* genes (Griffiths & Tiwari 1995; Ellegren 1996; Griffiths *et al.* 1996). The *CHD1* gene was first described in the mouse (Delmas *et al.* 1993) and is extremely well conserved with homologues in humans, *Drosophila* and yeast (Woodage *et al.* 1997). The example in the yeast is particularly striking as it split from the lineage leading to birds about 1 billion years ago, so its similarity emphasizes the conservation of this gene. The function of *CHD1* is not yet clear. Nevertheless its name describes the three functional motifs upon which it is based, and their individual roles are better understood (Delmas *et al.* 1993). From this and other more recent work (Woodage *et al.* 1997) it is likely that *CHD1* may modify the structure of stored DNA to alter the access of transcriptional factors and so control the production of proteins. In most other animals the genome contains a single *CHD1* gene. The birds differ, as they have two. One is situated on the W chromosome, *CHD1-W* (Griffiths & Tiwari 1995), while the other is on the Z, *CHD1-Z* (Griffiths & Korn 1997). The DNA sequence of the *CHD1-W* and *CHD1-Z* are distinct but remain remarkably similar. In terms of PCR sex identification this raises a problem as it is difficult to design a pair of primers that have sufficient scope to work on a range of species but are constrained so that they will only amplify *CHD1-W*. A different approach is to use *CHD1-Z* as a positive control. This has been made use of in three tests, all of which use introns to aid gene recognition (Griffiths *et al.* 1998; Kahn *et al.* 1998; Fridolfsson & Ellegren 1999). It is possible, because genes are built of exons, which are functional elements whose DNA sequence is highly conserved. Between the exons are introns, which have little purpose and so evolve rapidly. As a consequence a pair of primers can be made that anneal to the exons of both *CHD1-W* and *CHD1-Z* but amplify across an intron. Consequently, the PCR products from *CHD1-W* and *CHD1-Z* are of a different size so

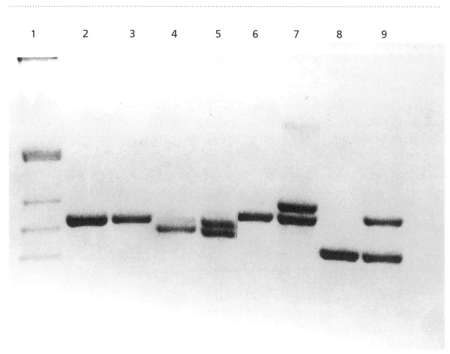

**Fig. 12.5** Sex identification in a range of birds using polymerase chain reaction (PCR) with the P2 and P8 *CHD1* gene primers. Two bands appear in females and one in males in all the species barring the ostrich. Lane 1 contains Gibco BRL 1 kb marker: 2 male (M) and 3 female (F) ostrich; 4M and 5F chicken; 6M and 7F Spix's macaw; 8M and 9F black guillemot; and 10F and 11M blue tit.

they can easily be discriminated by size on an agarose or an acrylamide gel. This provides a simple but effective sexing test.

The sex identification tests described by Griffiths *et al.* (1998) and Kahn *et al.* (1998) are based on two sets of primers that flank the same intron. The second test, given by Fridolfsson & Ellegren (1999), is based on a slightly larger intron that is closer to the 5′ end of the gene. Table 12.2 describes each test in more detail. The choice between them is initially based on the study species. Fortunately, in those taxa where one test fails, the others perform well. If there remains a choice of test, it is wise to find the primers to each and make a choice based on the quality of the results.

Once the type of test is selected there may still be problems and these can often be resolved. For instance, the *CHD1-Z* product can amplify better than the *CHD1-W* version. Consequently, in a poor PCR reaction the *CHD1-Z* band can be visible, although there has not been enough amplification to produce a *CHD1-W* band. At this point the appearance of the *CHD1-Z* product ceases to be a positive control, as sex identification has not been achieved. This is a serious fault but can often be remedied by a reduction in the annealing temperature

**Table 12.2** A comparison of the three published *CHD1* gene avian sexing tests. To indicate the position of the primers and introns the mouse *CHD1* gene is used as a reference. This can be aligned to the birds homologue and the nucleotides are numbered as they are found in the mouse: GenBank: no L10410. The number given with a primer indicates the position of the 5′ nucleotide

| Primers | Avian orders tested (*n*) | Site of intron | Characteristics | Size of product |
|---|---|---|---|---|
| P2/P8† | 11 | 3399-00 | Identical band size in Apodiformes, Strigiformes | 300–400 bp |
| 1237L/1272H‡ | 7 | 3399-00 | Identical band size in Apodiformes, Strigiformes and Falconidae | 200–300 bp |
| 2550F/2718R§ | 11 | 2660-61 | Single Z band in Paridae and some Passeridae | 400–650 bp |

† Griffiths *et al.* 1998. ‡ Kahn *et al.* 1998. § Fridolfsson & Ellegren 1999.
See http://r3-griffiths.zoology.gla.ac.uk/chd.html for the CHD1 gene alignment.

and/or the reduction of the extension time. A further problem is a result of the relative annealing efficiency of the primers. In some cases the intron may have little size difference between *CHD1-W* and *CHD1-Z* so gene identification on an agarose gel can be difficult. This can be resolved by the use of an acrylamide gel, which often provides sufficient resolution (Griffiths *et al.* 1998).

---

**Box 12.5** *CHD1* **sex identification in birds**

Griffiths *et al.* (1998) 10 μL, P2 TCTGCATCGCTAAATCCTTT, P8 CTCCCAAG-GATGAG(A/G)AA(C/T)TG; standard conditions; 94 °C for 90 s, 30×(48 °C for 45 s, 72 °C for 45 s, 94 °C for 30 s), 48 °C for 60 s, 72 °C for 300 s; 3% agarose.

Kahn *et al.* (1998) 10 μL, 1237L GAGAAACTGTGCAAAACAG, 1272H TCCAGAATATCTTCTGCTCC; standard conditions; 94 °C for 90 s, 30×(56 °C for 60 s, 72 °C for 60 s, 94 °C for 30 s), 48 °C for 60 s, 72 °C for 300 s; 3% agarose.

Fridolfsson & Ellegren (1999) 10 μL, 2550F GTTACTGATTCGTCTACGAGA, 2718R ATTGAAATGATCCAGTGCTTG; standard conditions but no triton, primers at 2 pM, MgCl$_2$ 1.75 mM non-passerines or 3 mM passerines; 94 °C for 120 s, 10 ×(60– > 50 °C for 30 s, 72 °C for 30–40 s, 94 °C for 30 s), where the annealing stage has a 1 °C touchdown from 60 °C to 51 °C, 25–35×(50 °C for 30 s, 72 °C for 30–40 s, 94 °C for 30 s), 50 °C for 30 s, 72 °C for 300 s, 2–3% agarose (this works well under the thermal conditions given for P2/P8).

---

### 12.9.2 ATPase W

This gene has only been published as a patent (Halverson & Dvorak 1994) but it is used regularly in avian literature where it is known as *pMg1*. Rather like the mammalian *AMGY* gene, its sex linkage is variable. This demonstrates that the evolution of the avian sex chromosomes gives them a structural mobility similar to that of the mammal X/Y chromosomes. In some Galliformes, Anseriformes, Ciconiiformes, Psittaciformes and Passeriformes different forms of the ATPase gene lie on the W and Z and these can be distinguished through Southern blots probed with *pMg1*. However, in other bird species with representatives from the Psittaciformes, etc., there is only a Z-linked ATPase so sex identification becomes more difficult.

### 12.9.3 EE0.6

This is a recent discovery, so EE0.6 is the designation of the clone as the gene may represent a new, unnamed type (Ogawa *et al.* 1997). It seems likely that the gene is nonfunctional so evolution is not constrained. The EE0.6 primers produce a W-specific ≈380 bp band in Galliformes, Anseriformes, Columbiformes and Passeriformes. No positive control band is produced so if this method is used a second set of primers will have to fulfil this role. In Psittaciformes and Sphenisciformes, sex identification is not clear.

---

**Box 12.6 EE0.6 sex identification of birds**

Ogawa *et al.* (1997), 10–20 µL, USP1 CTATGCCTACCAC(A/C)TTCCTATTTGC USP3 AGCTGGA(T/C)TTCAG(A/T)(C/G)CATCTTCT; standard conditions; 95 °C for 120 s, 35 ×(58–60 °C for 90 s, 72 °C for 120 s, 95 °C for 80 s), 60 °C for 60 s, 72 °C for 300 s; 2% agarose.

---

## 12.10 Summary

'What gender is that individual?' may provide a serious problem. Perhaps the sexes may be monomorphic or the sample is a leaf or feather. This chapter suggests that DNA can be a solution to sex identification. It is not a universal solution, as genetics will be difficult when sex is determined by the environment or by haplodiploidy. However, the chromosomal sex determination systems, X:Y and W:Z, offer a whole chromosome (Y or W) that contains genetic markers unique to one sex. Even in species where there are no obvious sex chromosomes, there may be enough sex-linked DNA to allow sex identification.

Various options to search for sex-linked markers are available. In some species, markers have already been isolated so the background literature should always be consulted. A new technique that may become available with time is the laser destruction of all but the Y chromosome (Ogawa *et al.* 1997). The remaining chromosome can be directly searched for male-specific sequences. At present RAPD or AFLP provide the best, simple manipulations to provide sex-specific DNA.

These isolation techniques often provide noncoding DNA. This DNA evolves quickly, so will provide a test that will only work on one or a few closely related species. If the test is designed for a single species, this limitation is irrelevant but for a wide-ranging test, a sex-linked gene is a better answer. Genes are conserved as a result of their function and can provide a technique that works on most of the species within a class. This is the situation in the birds (*CHD1* gene) and mammals (*SRY, ZFY* and *AMGY*). In most other large taxa like plants and fish, sex and sex determination is a more recent phenomenon. Different methods of gender control like environmental sex determination, X:Y W:Z, etc. operate and any sex-specific chromosomes will have evolved from different autosomes. As a result the sex-linked genes will vary so a sexing test based on a single gene is impossible.

Once a sex-linked marker is isolated, standard PCR forms the best technique for gender identification. Recent advances suggest new methods may rely on DNA annealing to identify sex-linked markers. However, such techniques are still being developed or are too costly. Despite this, PCR-based sex identification offers a fast, cheap method to provide a large amount of data. This data examines the role of each sex in evolution and, as recent research has shown, this is an exciting new field.

## References

Aasen, E. & Medrano, J. (1990) Amplification of the ZFY and ZFX genes for sex identification in humans, cattle, sheep and goats. *Bio/Technology* **8**, 1279–1281.

Bielawski, J., Noack, K. & Pumo, D. (1995) Reproducible amplification of RAPD markers from vertebrate DNA. *Biotechniques* **18**, 856–860.

Bull, J.J. (1983) *Evolution of Sex Determining Mechanisms*. Benjamin/Cummings, Menlo Park, California.

Charlesworth, B. (1991) The evolution of sex chromosomes. *Science* **251**, 1030–1033.

Charlesworth, B. (1996) The evolution of chromosomal sex determination and dosage compensation. *Current Biology* **6**, 149–162.

Christidis, L. (1990). *Animal Cytogenetics 4: Chordata 3; B, Aves*. Gebrüder Borntraeger, Berlin.

Delmas, V., Stokes, D.G. & Perry, R.P. (1993) A mammalian DNA binding protein that contains a chromodomain and an SNF2/SW12-like helicase domain. *Proceedings of the National Academy of Sciences USA* **90**, 2414–2418.

Ellegren, H. (1996) First gene on the avian W chromosome (CHD) provides a tag for universal sexing of non-ratite birds. *Proceedings of the Royal Society of London B* **263**, 1635–1644.

Ellesworth, D.L., Rittenhouse, K. & Honeycutt, R. (1993) Artifactual variation in randomly amplified polymorphic DNA banding patterns. *Biotechniques* **14**, 214–217.

Ennis, S. & Gallagher, T. (1994) A PCR-based sex determination assay in cattle based on the bovine amelogenin locus. *Animal Genetics* **25**, 425–427.

Ferrier, V. & Gasser, F., Jaylet, A. & Cayrol, C. (1983) A genetic study of various enzyme polymorphisms in *Pleurodeles watlii* (Urodele Amphibian). II Pepetidases: demonstration of sex linkage. *Biochemistry and Genetics* **21**, 535–549.

Foster, J.W. & Graves, J.A.M. (1994) An *SRY*-related sequence on the marsupial X chromosome: Implications for the evolution of the mammalian testis determining gene. *Proceedings of the National Academy of Sciences USA* **91**, 1927–1931.

Francis, R. (1992) Sexual lability in teleosts: developmental factors. *Quarterly Review of Biology* **67**, 1–18.

Fridolfsson, A. & Ellegren (1999) A simple and universal method for molecular sexing of non-ratite birds. *Journal of Avian Biology* **30**, 116–121.

Grant, S., Houben, A., Vyskot, B., Siroky, J., Pan, W.-H., Macas, J. & Saedler, H. (1994) Genetics of sex determination in flowering plants. *Developmental Genetics* **15**, 214–230.

Graves, J. (1995a) The evolution of mammalian sex chromosomes and the origin of sex determining genes. *Philosophical Transactions of the Royal Society of London B* **350**, 305–312.

Graves, J. (1995b) The origin and function of the mammalian Y chromosome and Y borne genes—an evolving understanding. *Bioessays* **17**, 311–321.

Griffiths, R. & Holland, P.W.H. (1990) A novel avian W chromosome DNA repeat sequence in the lesser black-backed gull (*Larus fuscus*). *Chromosoma* **99**, 243–250.

Griffiths, R. & Korn, R. (1997) A CHD1 gene is Z chromosome linked in the Chicken *Gallus domesticus*. *Gene* **197**, 225–229.

Griffiths, R. & Orr, K. (1999) The use of AFLPs to identify a sex-linked marker. *Molecular Ecology* **8**, 671–674.

Griffiths, R. & Tiwari, B. (1993) The isolation of molecular genetic markers for the identification of sex. *Proceedings of the National Academy of Sciences USA* **90**, 8324–8326.

Griffiths, R. & Tiwari, B. (1995) Sex of the last wild Spix's macaw. *Nature* **375**, 454.

Griffiths, R., Daan, S. & Dijkstra, C. (1996) Sex identification in birds using two CHD genes. *Proceedings of the Royal Society of London B* **263**, 1249–1254.

Griffiths, R., Double, M., Orr, K. & Dawson, R. (1998) A simple DNA test to sex most birds. *Molecular Ecology* **7**, 1071–1076.

Halverson, J. & Dvorak, J. (1994) Avian sex identification probes. Pat. Application no. PCT/US92/08284.

Hillis, D. & Green, D. (1990) Evolutionary changes of heterogametic sex in the phylogenetic history of amphibians. *Journal of Evolutionary Biology* **3**, 49–64.

Janzen, F. & Paukstis, G. (1991) Environmental sex determination in reptiles: ecology, evolution and experimental design. *Quarterly Review of Biology* **66**, 149–179.

Johnston, C.M., Barnett, M. & Sharpe, P.T. (1995) The molecular biology of temperature-dependent sex determination. *Philosophical Transactions of the Royal Society of London B* **350**, 297–304.

Jones, K. & Singh, L. (1985) Snakes and the evolution of sex chromosomes. *Trends in Genetics* **1**, 55–61.

Kahn, N., StJohn, J. & Quinn, T. (1998) Chromosome-specific intron size differences in the avian CHD gene provides a simple and efficient method for sex identification. *Auk* **115**, 1074–1078.

Lessells, C. & Mateman, A. (1998) Sexing birds using random amplified polymorphic DNA (RAPD) markers. *Molecular Ecology* **7**, 187–195.

Lessells, C., Mateman, A. & Visser, J. (1996) Great tit hatchling sex ratios. *Journal of Avian Biology* **27**, 135–142.

Longmire, J.L., Ambrose, R.E., Brown, N.C. *et al.* (1991) Use of sex-linked minisatellite fragments to investigate genetic differentiation and migration of North American populations of the peregrine falcon *Falco peregrinus*. In: *DNA Fingerprinting: Approaches and Applications* (eds.)

Longmire, J.L., Maltbie, M., Pavelka, R.W. *et al.* (1993) Gender identification in birds using microsatellite DNA fingerprint analysis. *Auk* **110**, 378–381.

Michelmore, R.W., Paran, I. & Kesseli, R.V. (1991) Identification of markers linked to disease-resistant genes by bulked segregant analysis. *Proceedings of the National Academy of Sciences USA* **88**, 9828–9832.

Nakahori, Y., Takenaka, O. & Nakagome, Y. (1991) A human X-Y homologous region encodes 'amelogenin'. *Genomics* **9**, 264–269.

Ogawa, A., Solovei, I., Hutchinson, N., Saitoh, Y., Ikeda, J., Macgregor, H. & Mizuno, S. (1997) Molecular characterization and cytological mapping of a non-repetitive DNA sequence region from the W chromosome of chicken and its use as a universal probe for sexing Carinatae birds. *Chromosome Research* **5**, 93–101.

Ohno, S. (1967) *Sex Chromosomes and Sex-Linked Genes*. Springer Verlag, Berlin.

Page, R. & Holmes, E. (1998) *Molecular Evolution: a Phylogenetic Approach*. Blackwell Science Ltd, Oxford.

Paran, I. & Michelmore, R.W. (1993) Development of reliable PCR-based markers linked to downy mildew resistance genes in lettuce. *Theoretical and Applied Genetics* **85**, 985–993.

Pomp, D., Good, B., Geisert, R., Corbin, C. & Conley, A. (1995) Sex identification in mammals with polymerase chain reaction and its use to examine sex effects on diameter of day-10 or day-11 pig embryos. *Journal of Animal Science* **73**, 1408–1425.

Rozen, S. & Skaletsky, H. (1996, 1997) *Primer3*. Code available at http://www–genome.wi.mit.edu/genome_software/other/primer3.html

Rychlik, W. (1995) Selection of primers for polymerase chain-reaction. *Molecular Biotechnology* **3**, 129–134.

Rychlik, W. & Rhoads, R.E. (1989) A computer program for choosing optimal oligonucleotides for filter hybridization, sequencing and *in vitro* amplification of DNA. *Nucleic Acids Research* **17**, 8543–8551.

Schmid, M. (1983) Evolution of sex chromosomes and heterogametic systems in Amphibia. *Differentiation* **23** (Suppl.), S13–S22.

Sinclair, A.H., Foster, J.W., Spencer, J.A., Page, D.C., Palmer, M., Goodfellow, P.N. & Graves, J.A.M. (1988) Sequences homologous to ZFY, a candidate human sex determining gene, are autosomal in marsupials. *Nature* **336**, 780–783.

Smith, J., Scott-Craig, J., Leadbetter, J., Bush, G., Roberts, D. & Fulbright, D. (1994) Characterization of random amplified polymorphic DNA (RAPD) products from *Xanthomonas campestris* and some comments on the use of RAPD products in phylogenetic analysis. *Molecular Phylogenetics and Evolution* **3**, 135–145.

Smith, K.D., Young, K.E., Talbot, C.C. & Schmeckpeper, B.J. (1987) Repeated DNA of the human Y chromosome. *Development Suppl* **101**, 77–92.

Suka, N., Shinohara, Y., Saitoh, Y., Ohtomo, K., Harata, M., Sphigelman, E. & Mizuno, S. (1993) W Heterochromatin of chicken its unusual DNA components, late replication and chromatin structure. *Genetica* **88**, 93–105.

Sullivan, K., Mannucci, A., Kimpton, C. & Gill, P. (1993) A rapid and quantitative DNA test: fluorescence-based PCR analysis of X-Y homologous gene amelogenin. *Biotechniques* **15**, 636–642.

Vos, P., Hogers, R., Bleaker, M. *et al.* (1995) AFLP: a new technique for DNA fingerprinting. *Nucleic Acids Research* **23**, 4407–4414.

Welsh, J. & McClelland, M. (1990) Fingerprinting genomes using PCR with arbitrary primers. *Nucleic Acids Research* **18**, 7213–7218.

Williams, J.G.K., Kubelik, A.R., Livak, K.J., Rafalski, J.A. & Tingey, S.V. (1990) DNA polymorphisms amplified by arbitrary primers are useful as genetic markers. *Nucleic Acids Research* **18**, 6531–6535.

Woodage, T., Basrai, M., Baxevanis, A., Hieter, P. & Collins, F. (1997) Characterization of the CHD family of proteins. *Proceedings of the National Academy of Sciences USA* **94**, 11472–11477.

# Index